Shock Wave and High Pressure Phenomena

More information about this series at http://www.springer.com/series/1774

Shock Wave and High Pressure Phenomena

Seán Prunty

Introduction to Simple Shock Waves in Air

With Numerical Solutions Using Artificial Viscosity

Second Edition

 Springer

Seán Prunty
Ballincollig, Cork, Ireland

ISSN 2197-9529   ISSN 2197-9537 (electronic)
Shock Wave and High Pressure Phenomena
ISBN 978-3-030-63608-1   ISBN 978-3-030-63606-7 (eBook)
https://doi.org/10.1007/978-3-030-63606-7

This Springer imprint is published by the registered company Springer Nature Switzerland AG
The registered company address is: Gewerbestrasse 11, 6330 Cham, Switzerland

Preface to the Second Edition

Date: January 2021

This second edition incorporates a number of changes with the inclusion of new material in each of the chapters, while a substantial quantity of additional material has been included in Chap. 4. A section dealing with typical sound wave parameters is provided in Chap. 1 and new sections dealing with weak shock waves and the thickness of the shock wave region have been included in Chap. 3. Additional material on amplitude effects in wave propagation, short duration impulsive piston motion and shock decay, as well as some numerical results for shock wave interactions have been included in Chap. 4. The section in Chap. 5 dealing with Taylor's estimation of the yield of the first atomic explosion has been extended and a small amount of additional material has been included in Chap. 6. Two appendices have been added; the first deals with the piston withdrawal problem with more emphasis on numerical calculations and computations, and the second deals with the results of some numerical calculations for the variation in pressure, particle velocity and density for the shock tube over an extended time interval. Since the first edition of the text appeared, I have corrected several errors that I discovered, and I trust that I have identified all errors and that no new ones have crept in with the addition of the new material. The provision of this new material as outlined above is anticipated to enhance students' understanding of shock wave propagation and to increase student interest in the subject area. I am most grateful to Dr. Sam Harrison, editor at Springer (Physics), and his production team for all their help in getting this new edition of the book published.

Ballincollig, Cork, Ireland   Seán Prunty

Preface to the First Edition

This book provides an elementary introduction to some one-dimensional fluid flow problems involving shock waves in air. The differential equations of fluid flow are approximated by finite difference equations, and these in turn are numerically integrated in a stepwise manner. Artificial viscosity is introduced into the numerical calculations in order to deal with shocks. The presentation is restricted to the *finite difference* approach to solve the coupled differential equations of fluid flow as distinct from *finite volume* or *finite element* methods. It presents the results arising from the numerical solution using Mathcad programming, and, as I had Mathcad installed on my computer, it was natural for me to use it in order to obtain solutions to the examples presented here. Both plane and spherical shock waves are discussed with particular emphasis on very strong explosive shocks in air.

I am not an expert in fluid dynamics, and I only took an interest in this area within the past 3 years in an effort to solve some specific problems in compressible fluid flow involving shock waves in air. The very strong shocks produced by explosions became a particular interest after reading the book by Bruce Cameron Reed, entitled *The Physics of the Manhattan Project*. The book provides an excellent account of the basic physics in relation to the enormous amount of energy released in nuclear reactions. My primary interest was not specifically in the area of critical mass calculations and their ramifications but, instead, on very large quantity of energy released and the propagation of its effects on the surrounding atmosphere. The learning process in coming to terms with the subject of gas dynamics was an interesting adventure for one who had no exposure to the subject at undergraduate physics level. In fact, my lack of experience in the area is no different from other physics graduates since the pressure on physics departments to teach other subjects has meant that the important subject of gas dynamics has received very little attention in the physics curriculum for many decades. It is important to emphasize that this book is not an introduction to gas dynamics or to computational fluid dynamics (CFD), and the method of solution to the problems presented here is a personal one and makes no attempt to emulate or make reference to the numerical techniques employed in modern-day CFD. Instead, it presents the results arising from the numerical solution to several simple examples of compressible fluid flow involving shock waves that were of interest to the author. Accordingly, the reader is strongly advised to consult the standard texts on fluid or gas dynamics for a comprehensive account of fluid motion as well the rapidly growing area of CFD by the experts in the area. Several textbooks and articles that I found useful can be found in the references at the end of each chapter.

The book is suitable for both graduate and advanced undergraduate students in applied mathematics, engineering and physics who are taking courses in fluid dynamics and who require an introduction to a specific numerical technique for dealing with shock waves. I decided very early on that the program listings would not be included for the very good reason that there are good and bad programming techniques and I would probably fall into the latter category. Nonetheless, the *finite difference* representation of the differential equations is presented, and it is hoped that this will be a starting point to encourage interested students to obtain solutions to similar problems in compressible flow using their preferred programming language. It is hoped that the material presented here will renew interest in gas dynamics and shock waves in the undergraduate physics curriculum.

The book is structured in such a manner as to allow the reader to review some basic material in relation to the equations of fluid flow. In this respect, a brief review of the one-dimensional form of the equations is presented in Chap. 1 together with the propagation of small amplitude disturbances. Chapter 1 also includes some basic thermodynamic relationships that are relevant to gas dynamics. Waves of finite amplitude and the formation of shock waves are discussed in Chap. 2. This chapter also includes a basic introduction to the method of characteristics and to Riemann invariants. Some important relationships arising from the conservation of mass, momentum and energy across the shock front are presented in Chap. 3, and these relationships are used in the subsequent chapters to ascertain the accuracy of the numerical results obtained. Chapter 4 outlines the numerical procedure employed to solve some examples of plane shock waves using artificial viscosity: The differential equations in Lagrangian form are derived, and the corresponding difference equations are presented. Stability issues in relation to these equations are briefly discussed as well as the choice of grid interval to be used in the numerical procedure. Several simple examples of plane shocks arising from piston motion are also presented and discussed. The remaining chapters deal with spherical shock waves: Chap. 5 has an almost independent character and deals exclusively with the strong-shock, point-source solution and its ramifications. Chapter 6 describes the numerical procedure used when dealing with spherical shock waves, and the appropriate differential equations in Lagrangian form are derived, and the difference form of these equations incorporating artificial viscosity is presented. The equations are numerically integrated to predict the pressure, density and particle velocity as functions of position outside the strong-shock regime by using the strong-shock, point-source solution as initial conditions. Finally, the shock waves generated following the sudden expansion of a high-pressure, high-temperature sphere of air into the surrounding atmosphere are presented and discussed.

It is assumed that the reader is familiar with differential and integral calculus and has a solid understanding of the physical principles underpinning the basic equations of fluid dynamics. Some knowledge of numerical methods would be an advantage, particularly, to do with the stability issues in relation to the difference equations involved.

I wish to acknowledge the help received from Diarmuid O'Riordáin in relation to a computer file storage issue with embedded graphics.

Ballincollig, Cork, Ireland

Seán Prunty

Contents

Chapter 1
Brief Outline of the Equations of Fluid Flow

1.1 Introduction

Those who are familiar with compressible fluid flow are aware that the equations of
motion are nonlinear and, as such, it is very difficult to obtain analytical solutions.
As a consequence, numerical methods are generally employed and the differential
equations are approximated by finite difference equations and these in turn are
solved in a stepwise manner. In many examples involving compressible fluid flow
shocks appear and their presence is a complicating factor since they are characterized
by very steep gradients in the variables describing the flow, such as, in the velocity,
density, pressure and temperature. In fact, the gradients become infinitely steep when
the effects of viscosity and thermal conduction are neglected: this introduces discontinuities in the solutions and, as a result, it requires the application of boundary
conditions connecting the values across the shock front but the implementation of
this technique can be quite complex. However, the need for any boundary conditions
can be avoided by using a method proposed in 1950 by Von Neumann and
Richtmyer [1] where an artificially large viscosity is introduced into the numerical
calculations: the present text utilizes this technique. Instead of obtaining a discontinuous solution at the shock front, the shock acquires a thickness comparable to the
spacing of the grid points used in the numerical procedure so that the shock appears
as a near-discontinuity and across which velocity, pressure etc. vary rapidly but
continuously.

A brief review of the fundamental equations of fluid dynamics [2–6] is provided
in this chapter so that the reader can have to-hand the appropriate governing
equations. The one-dimensional form of the equations is presented as they apply
to non-viscous[1] flow, so that any physical effects involving friction and thermal
conduction are neglected. A treatment involving the derivation of the full

---

[1]Although the term "non-viscous" implies the absence of viscosity or friction, it also implies "non-
conducting" as well.

© The Author(s), under exclusive license to Springer Nature Switzerland AG 2021   1
S. Prunty, *Introduction to Simple Shock Waves in Air*, Shock Wave and High
Pressure Phenomena, https://doi.org/10.1007/978-3-030-63606-7_1

3-dimensional fluid dynamic equations can be found in the excellent text by Anderson [7] where viscous forces and thermal conduction are included. However, as this present text deals with fluid motion where shocks occur, the neglect of friction and thermal conduction in the neighbourhood of the shock must be included and, accordingly, an *artificial viscosity* is introduced, as we shall see in due course, into the set of difference equations that approximate the differential equations for the flow.

1.2   Eulerian and Lagrangian Form of the Equations

The equations governing a fluid or gas in motion are mathematical expressions of the laws of conservation of mass, momentum and energy. In the case of an ideal gas, for example, these equations are supplemented by thermodynamic equations; one called the *equation of state* and the other called the *caloric equation of state*. The equations describing the motion of the fluid can be written in one of two coordinate systems [6]; one called the Eulerian system and the other called the Lagrangian system (each named after the mathematicians, Euler and Lagrange). In the Lagrangian system we follow the path taken by individual particles of fluid and determine the velocity, pressure, density etc. as a function of the path taken. Accordingly, the Lagrangian description of motion is connected with a definite particle or mass element of the fluid; each particle is assigned a symbol x_0, for example, indicating its position at some initial time which is usually taken at time $t = 0$, while at a later time t the position of the particle is $x(x_0, t)$ and, clearly, $x(x_0, 0) = x_0$. The motion of the particle satisfies Newton's second law of motion, namely; $m(d^2x/dt^2) = F$ where m is the mass of this particle of fluid lying between x_0 and $x_0 + dx_0$ and F is the force acting on the particle. Suppose, for example, we wish to determine the flow of air over a fixed surface, we could imagine a weightless soap bubble released into the air. The path taken by this bubble as it moves over the surface provides a Lagrangian description of the flow. To indicate the time-rate of change following the fluid particle the *material derivative*, D/Dt rather than the usual d/dt has become standard notation in fluid dynamics. However, in the present text we will adopt a slightly different notation as we follow specific particles of fluid and we will be returning to this aspect in Chaps. 4 and 6.

In the Eulerian description, on the other hand, one is not interested in the motion of individual particles; instead, one is interested in, say, the velocity at points in space. This is analogous to setting up a fixed and very fine grid throughout space and noting, for example, the velocity, pressure or density etc. at each grid point, so in the case of the Eulerian system we are interested in the properties of the fluid, such as, velocity, pressure, density etc. as they pass fixed points in space.

Another good example, the source of which cannot be recalled, that distinguishes the Lagrangian and Eulerian systems involves the acceleration of a log on a steadily flowing river which has a section of rapids on it. By concentrating on the log one observes that it accelerates as it enters the rapids which is the Lagrangian acceleration (following the particle). However, an observer on the bank of the river who concentrates on the velocity at fixed points in the flow will not see any acceleration as a succession of logs that pass the same point will do so at the same velocity since the river as a whole is not accelerating.

In mathematical terms, Du/Dt is the Lagrangian acceleration, while $\partial u/\partial t$ or more specifically, $(\partial u/\partial t)_x$ is the Eulerian acceleration, where the subscript indicates that the acceleration is measured at a particular point x. However, $u \equiv u(x, t)$, so that

$$\frac{du}{dt} \equiv \frac{Du}{Dt} = \frac{\partial u}{\partial t} + u\frac{\partial u}{\partial x}.$$

In relation to the log entering the rapids we have, $\partial u/\partial t = 0$, so that $Du/Dt = u$ $(\partial u/\partial x)$; implying that the river exhibits a spatial variation in velocity, thereby accounting for the Lagrangian acceleration.

1.3 Some Elements of Thermodynamics

A concise review of some elements of thermodynamics is provided in this section. Only those aspects of thermodynamics that are relevant to gas dynamics are considered and, in particular, to air which is assumed in this text to behave as an ideal gas.

1.3.1 Ideal Gas Equation

The equation of state for an ideal gas is given by the equation

$$pV = \frac{m}{M}\mathfrak{R}T, \tag{1.1}$$

where p is the pressure, V is the volume of gas of mass m, M is the molecular weight, $\mathfrak{R}$ is the universal gas constant ($\mathfrak{R} = 8.31 JK^{-1}mole^{-1}$) and T is the temperature. In this text we will assume that air behaves as an ideal gas. One mole of air has an approximate molecular weight (based on its composition which contains largely nitrogen and oxygen) of $28.97 \times 10^{-3}kg$, so that $\mathfrak{R}/M = 287 Jkg^{-1}K^{-1}$. Accordingly, Eq. (1.1) for air can be written as

$$pV = mRT, \tag{1.2}$$

where $R = 287 JK^{-1}kg^{-1}$, or we can also write it as

$$p = \rho RT \quad \text{or} \quad pv = RT \tag{1.3}$$

where ρ is the density and v is the specific volume (that is, the inverse of the density). For example, the density of air is $\rho = 1.29 kgm^{-3}$ at a pressure of one atmosphere (equal to $1.013 \times 10^5 Nm^{-2}$) and at a temperature of 273 K according to Eq. (1.3).

1.3.2   The First Law of Thermodynamics

The first law of thermodynamics states that the energy of an isolated system is conserved; however, the energy may be transformed from one form to another but the energy cannot be created nor destroyed. The mathematical form of the first law of thermodynamics for an infinitesimal change is given by the equation,

$$dQ = dE + dW, \tag{1.4}$$

where dQ is the quantity of heat exchanged between the system and its surroundings, dE^2 is the change in the internal energy of the system and dW is the work done. For a system of constant mass m that exerts a uniform pressure p on its surroundings, the first law can be written as

$$dQ = mde + pdV,$$

where de is the change in the specific internal energy, that is, the change in the internal energy per unit mass and dV is the change in volume. Dividing across by m, gives,

$$dq = de + pdv, \tag{1.5}$$

where dq is the quantity of heat transferred per unit mass and dv is the change in the specific volume.

1.3.3   Heat Capacity

If a system undergoes a temperature increase of dT following a transfer of heat dq, the heat capacity of the system is defined as

---

[2]The symbol E is used to represent the internal energy of a system rather than the more commonly used symbol U, as this symbol is used in this text to represent the velocity of air motion and the velocity of shock waves.

$$c = \frac{dq}{dT}.$$

If the process takes place at constant volume, then the heat capacity at constant volume is

$$c_V = \left(\frac{dq}{dT}\right)_V \tag{1.6}$$

and if the process takes place at constant pressure, the heat capacity is

$$c_P = \left(\frac{dq}{dT}\right)_P. \tag{1.7}$$

If the specific internal energy e is expressed in terms of v and T, then

$$de = \left(\frac{\partial e}{\partial v}\right)_T dv + \left(\frac{\partial e}{\partial T}\right)_v dT \tag{1.8}$$

and the mathematical statement of the first law can be written as

$$dq = \left(\frac{\partial e}{\partial T}\right)_v dT + \left[p + \left(\frac{\partial e}{\partial v}\right)_T\right] dv.$$

For a constant volume process, we have

$$\frac{dq}{dT} = \left(\frac{\partial e}{\partial T}\right)_v,$$

hence,

$$c_V = \left(\frac{\partial e}{\partial T}\right)_v \tag{1.9}$$

Defining the specific enthalpy,

$$h = e + pv, \tag{1.10}$$

then

$$dh = de + pdv + vdp$$
$$= dq + vdp,$$

hence,

$$dq = dh - v\,dp. \tag{1.11}$$

If h is considered a function of p and T, then

$$dh = \left(\frac{\partial h}{\partial p}\right)_T dp + \left(\frac{\partial h}{\partial T}\right)_p dT, \tag{1.12}$$

hence,

$$dq = \left(\frac{\partial h}{\partial T}\right)_p dT + \left[\left(\frac{\partial h}{\partial p}\right)_T - v\right] dp, \tag{1.13}$$

so that the specific heat at constant pressure is

$$c_P = \left(\frac{\partial h}{\partial T}\right)_p. \tag{1.14}$$

Also, for an ideal gas the specific internal energy is a function of the temperature only, hence, $E = f(T)$ and, therefore,

$$c_V = \left(\frac{\partial e}{\partial T}\right)_v = \frac{de}{dT},$$

hence,

$$e = c_V T \tag{1.15}$$

and similarly,

$$h = c_P T. \tag{1.16}$$

Substituting these relationships in the equation, $h = e + pv$, we have

$$c_P T = c_V T + RT,$$

so that

$$c_P = c_V + R. \tag{1.17}$$

1.3.4   Isothermal Expansion or Compression of an Ideal Gas

Work in thermodynamics is concerned only with the changes that take place between a system and its surroundings. An infinitesimal amount of work dW is said to be done by a system when the system undergoes a change in volume dV under the action of the pressure p that the system exerts on its surroundings, hence,

$$dW = pdV \tag{1.18}$$

and if the volume changes from an initial value V_i to a final value V_f the work done by the system is given by

$$W = \int_{V_i}^{V_f} pdV. \tag{1.19}$$

This latter equation cannot be integrated until the pressure is known as a function of V. This means that dW is not an exact differential; it depends on the path, unlike, for example, the internal energy function, E, which is an exact differential and only depends on the initial and final states and therefore independent of the path taken in going from, say, E_i to E_f.

In the case of an *isothermal* expansion or compression of an ideal gas whose equation of state is given by

$$pV = mRT.$$

The work done can be calculated by substituting for the pressure, p, hence,

$$W = \int_{V_i}^{V_f} \frac{mRT}{V} dV. \tag{1.20}$$

Since the process of expansion or compression is isothermal (the temperature T remaining constant during the process), then by performing the integration we obtain

$$W = mRT \ln\left(\frac{V_f}{V_i}\right), \tag{1.21}$$

where ln in the latter equation denotes the natural logarithm. If $V_f > V_i$ the system does work on its surroundings and, on the other hand, if $V_f < V_i$, the surroundings does work on the system, that is, the system is compressed.

1.3.5 Reversible Adiabatic Process for an Ideal Gas

Let us now consider a change in a system that undergoes an *adiabatic* process, that is, changes take place to the system without the addition or removal of heat, hence, $dq = 0$. During the process the system departs very little from thermodynamic equilibrium, so that the change taking place in going from some initial state to some final state goes through a sequence of equilibrium states. The mathematical statement of the first law of thermodynamics in infinitesimal form is

$$dq = de + pdv$$
$$= c_V dT + pdv. \tag{1.22}$$

Using the ideal gas law, we have

$$dT = \frac{1}{R}(pdv + vdp)$$

and substituting this latter relationship in Eq. (1.22), gives

$$dq = \frac{c_V}{R}(pdv + vdp) + pdv$$

$$= \frac{c_V}{R}vdp + \frac{c_P}{R}pdv.$$

If the process is adiabatic, $dq = 0$, hence,

$$\frac{dp}{p} + \gamma\frac{dv}{v} = 0, \tag{1.23}$$

where $\gamma = c_P/c_V$ is the ratio of the specific heats. Integrating Eq. (1.23), gives

$$pv^\gamma = \text{ constant} \tag{1.24}$$

or in terms of the density, we have

$$\frac{p}{\rho^\gamma} = \text{ constant.} \tag{1.25}$$

By using the equation for an ideal gas, one can show that the following relationships apply for an adiabatic process;

$$\frac{T}{p^{\frac{\gamma-1}{\gamma}}} = \text{ constant and } Tv^{\gamma-1} = \text{ constant.} \tag{1.26}$$

1.3.6 Work Done by an Ideal Gas During an Adiabatic Expansion

Having established the relationships between the thermodynamic coordinates during an adiabatic process, let us now determine the work done during an adiabatic expansion. Writing again the mathematical expression for the first law of thermodynamics as

$$dq = c_V dT + dW. \tag{1.27}$$

Since the process is adiabatic, $dq = 0$, hence,

$$dW = -c_V dT \tag{1.28}$$

and integrating gives

$$W = c_V (T_i - T_f), \tag{1.29}$$

where T_i and T_f are the initial and final temperatures, respectively. Since the process in going from T_i to T_f passes through equilibrium states, we can use the ideal gas equation and write the latter equation as

$$W = \frac{c_V}{R} \left(p_i v_i - p_f v_f \right) \tag{1.30}$$

and using the fact that $R = c_P - c_V$ in conjunction with $\gamma = c_P / c_V$ Eq. (1.30) becomes

$$W = \frac{p_i v_i - p_f v_f}{\gamma - 1}. \tag{1.31}$$

Alternatively, by using the relationship between p and v for an adiabatic process, one can show that the work done by an ideal gas during an adiabatic expansion is

$$W = \frac{p_i v_i}{\gamma - 1} \left[1 - \left(\frac{p_f}{p_i} \right)^{\frac{\gamma - 1}{\gamma}} \right]. \tag{1.32}$$

1.3.7 Alternate Form of the Equations for Specific Internal Energy and Enthalpy

Having defined γ as the ratio of the specific heats we can write

$$e = c_V T = \frac{c_V}{R} p\upsilon = \frac{p\upsilon}{\gamma - 1}, \tag{1.33}$$

or in terms of the density ρ as

$$e = \frac{p}{(\gamma - 1)\rho}. \tag{1.34}$$

Similarly,

$$h = \frac{\gamma p}{(\gamma - 1)\rho}. \tag{1.35}$$

1.3.8 Ratio of the Specific Heats for Air

Let us now consider the values for c_V and c_P or, more specifically, the value of γ for air. In relation to the Kinetic Theory of Gases, the *principle of equipartition of energy* states that the energy associated with each degree of freedom of an atom or molecule is $(1/2)k_B T$, where k_B is Boltzmann's constant ($k_B = 1.38 \times 10^{-23} JK^{-1}$). Each atom or molecule has three *translational* degrees of freedom; namely, in the x, y and z-directions, giving $(3/2)k_B T$ for its internal energy.

One mole, corresponding to the molecular weight M, contains N_A atoms or molecules ($N_A = 6.02 \times 10^{23}$), so that the specific internal energy of one mole is $(3/2)N_A k_B T = (3/2)\Re T$, where $\Re = N_A k_B$ is the universal gas constant.

Air comprises largely N_2 and O_2 molecules and each molecule contributes two *rotational* degrees of freedom in addition to the translational degrees, hence, the specific internal energy for air amounts to $(5/2)RT$ and the specific enthalpy amounts to $(7/2)RT$, hence,

$$e = \frac{5}{2}RT \;\text{ and }\; h = \frac{7}{2}RT \tag{1.36}$$

and therefore $\gamma = 7/5 = 1.4$ for air.

1.3.9   The Second Law of Thermodynamics

The second law of thermodynamics was formulated following many attempts undertaken for the efficient conversion of heat into work and, as such, its early development was very much focused on engineering applications related to heat engine efficiency. The second law can be stated in many different ways. The Kelvin-Planck statement in the context of heat engines can be expressed in the following manner; *"No process is possible whose sole effect is the absorption of heat from a temperature reservoir and the conversion of this heat completely into work"*. The second law, like the first, is expressed in negative terms like "it is not possible" and, consequently, the second law places limits on what can be achieved.

When the isothermal expansion of an ideal gas was considered in Sect. 1.3.4 it was evident that there was no change in the internal energy E of the system. In this context, one can regard the system as a cylinder fitted with a piston and in contact with an external reservoir at a constant temperature T as shown in Fig. 1.1.

Since $\Delta E = 0$ for the isothermal process, this implies that $Q = W$ so that all the heat transfer from the single reservoir has been converted into work and it would appear that the second law has been violated in this process. Certainly, all the heat has been converted into work but this is not the *"the sole effect"* for the process as specified in the Kelvin-Planck statement of the law: instead, the piston has moved from some initial position to some final position during the expansion and, accordingly, the absorption of heat and the conversion of this heat completely into work is not the *"sole effect"*. It would be necessary for the piston to return to its initial position following the process of heat transfer and this explains why heat engines work in a cyclic manner.

A thermodynamic property of a system exists, called the *entropy*, S, which was introduced into thermodynamics by Clausius, and an infinitesimal change in this property is given by

$$dS = \frac{dQ_{rev}}{T},\qquad(1.37)$$

Fig. 1.1 Isothermal expansion of an ideal gas in contact with a constant temperature reservoir

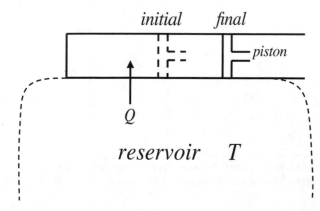

where the subscript *rev* on *dQ* implies that the heat transfer must occur *reversibly*.[3] In the case of a finite change from some initial (i) state to a final (f) state, the entropy change of the system is given by

$$S_f - S_i = \int_i^f \frac{dQ_{rev}}{T} \tag{1.38}$$

and the integral must be evaluated along *any* reversible path connecting the initial and final states. Entropy, like internal energy, is a property of the system; hence, the change $S_f - S_i$ is independent of the path connecting the two states. For a reversible adiabatic process, $dQ_{rev} = 0$, so that $dS = 0$, hence, the entropy remains constant and the process is called *isentropic*. An adiabatic process is one in which no heat exchange occurs between the system and its surroundings, so, in general, an adiabatic process is not necessarily isentropic; the process must be reversible for it to be isentropic.

If a system undergoes an *irreversible* process between an initial and a final state as illustrated in Fig. 1.2, then the change in entropy ΔS is given by

$$\Delta S = S_f - S_i,$$

where it is assumed that these initial and final states are equilibrium states. One can calculate this change in entropy by replacing the irreversible path (broken line in Fig. 1.2) by *any* reversible path connecting these initial and final states. The change in entropy can be obtained by evaluating the integral,

Fig. 1.2 The reversible path *iAf* replaces the irreversible path to calculate the entropy change for the irreversible process shown (see text)

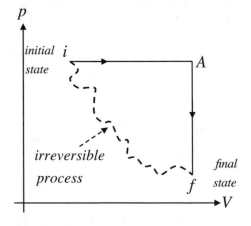

$$\int\limits_{iAf} \frac{dQ}{T}$$

where, for example, the integration is carried out for the reversible path iAf that is shown in Fig. 1.2: the path, $i \rightarrow A$ is taken as a constant pressure process and the path $A \rightarrow f$ is taken as a constant volume process.

In calculating any entropy changes one must also include the entropy change of the surroundings that interact with the system. Hence, the total entropy change is specified as the entropy change of the *universe* which is equal to the entropy change of the system plus the entropy change of the surrounding environment and is given by

$$\Delta S_{Total} = \Delta S_{system} + \Delta S_{surroundings}. \qquad (1.39)$$

When the second law for a closed system is expressed in terms of entropy change, we have

$$\Delta S_{Total} \geq 0, \qquad (1.40)$$

where the equality sign applies to a *reversible* process and the inequality sign applies to an *irreversible* process. Therefore, reversible processes produce no change in the entropy of the universe while all irreversible processes result in an increase in the entropy of the universe.

1.4   Conservation Equations in Plane Geometry

Let us now provide a brief review of the one-dimensional equations of fluid flow for plane geometry. These are the conservation equations for mass, momentum and energy.

1.4.1   Equation of Mass Conservation: The Continuity Equation

The continuity equation can be obtained by considering the mass of fluid entering and leaving a small element of volume Adx lying between x and $x + dx$ as shown in Fig. 1.3. It is assumed that the motion is one-dimensional so that velocity, density and pressure are constant over the cross-sectional area A. Mass conservation implies that the difference between the mass flow rate into the volume element and the mass flow rate out of the volume element is equal to the rate of accumulation of mass within the volume element.

Fig. 1.3 Fluid entering and
leaving a small volume
element is shown

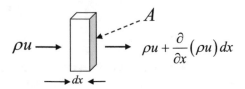

The mass of fluid entering from the left-hand side in a time interval dt is $A\rho u dt$ and the mass leaving is $A\left[\rho u + \frac{\partial}{\partial x}(\rho u)dx\right] dt$. Hence, the net increase of mass per unit time in this volume element is

$$-A\left[\frac{\partial}{\partial x}(\rho u)dx\right]$$

and this must be equal to the rate of increase of mass within this element, that is, $A dx$ $(\partial \rho/\partial t)$, hence the one-dimensional continuity equation in Eulerian form is

$$\frac{\partial \rho}{\partial t} + \frac{\partial}{\partial x}(\rho u) = 0. \tag{1.41}$$

By writing this latter equation as,

$$\frac{\partial \rho}{\partial t} + u\frac{\partial \rho}{\partial x} + \rho\frac{\partial u}{\partial x} = 0,$$

we have the following Lagrangian form of the continuity equation,

$$\frac{D\rho}{Dt} = -\rho\frac{\partial u}{\partial x}, \tag{1.42}$$

while the 3-dimensional form of the latter equation is

$$\frac{D\rho}{Dt} = -\rho\vec{\nabla} \cdot \vec{V}. \tag{1.43}$$

with

$$\vec{V} = (u, v, w)$$

where u, v and w are the velocity components in the x, y and z-directions, respectively, and the gradient operator is

$$\vec{\nabla} = \left(\frac{\partial}{\partial x}, \frac{\partial}{\partial y}, \frac{\partial}{\partial z}\right).$$

In terms of its components Eq. (1.43) becomes,

$$\frac{\partial \rho}{\partial t} + u\frac{\partial \rho}{\partial x} + v\frac{\partial \rho}{\partial y} + w\frac{\partial \rho}{\partial z} = -\rho\left(\frac{\partial u}{\partial x} + \frac{\partial v}{\partial y} + \frac{\partial w}{\partial z}\right)$$

or

$$\frac{\partial \rho}{\partial t} + \frac{\partial(\rho u)}{\partial x} + \frac{\partial(\rho v)}{\partial y} + \frac{\partial(\rho w)}{\partial z} = 0.$$

By writing the derivative in Eq. (1.42) in terms of the specific volume $v = 1/\rho$ rather than the density ρ we have,

$$\rho\frac{Dv}{Dt} = \frac{\partial u}{\partial x}. \tag{1.44}$$

1.4.2   Equation of Motion: The Momentum Equation

The law of the conservation of momentum implies that the rate of change in momentum of the fluid in a volume element between x and $x + dx$ is equal to the rate at which momentum flows into the volume element at x minus the rate at which momentum flows out at $x + dx$ plus the net force acting on the volume element. Expressing this mathematically, we have [8],

$$\frac{\partial}{\partial t}(\rho u)Adx = \rho(x, t)u^2(x, t)A - \rho(x + dx, t)u^2(x + dx, t)A + p(x, t)A$$
$$- p(x + dx, t)A$$

hence,

$$\frac{\partial}{\partial t}(\rho u)Adx = -\frac{\partial}{\partial x}\left[\rho(x, t)u^2(x, t)\right]Adx - \frac{\partial p(x, t)}{\partial x}Adx,$$

so that

$$\frac{\partial}{\partial t}(\rho u) + \frac{\partial}{\partial x}(\rho u^2) = -\frac{\partial p}{\partial x}, \tag{1.45}$$

which is the momentum equation. By using the continuity equation it is easy to show that the latter equation can be written in final form as

$$\frac{\partial u}{\partial t} + u \frac{\partial u}{\partial x} = -\frac{1}{\rho} \frac{\partial p}{\partial x}, \tag{1.46}$$

which is the one-dimensional form of Euler's equation of motion. The Lagrangian form of this one-dimensional equation is

$$\rho \frac{Du}{Dt} = -\frac{\partial p}{\partial x}, \tag{1.47}$$

and for the sake of completeness, the 3-dimensional form of Eq. (1.47) can be obtained by noting that u, in general, is a function of x, y and z, hence,

$$u = u(x, y, z, t)$$

and, therefore,

$$du = \frac{\partial u}{\partial t} dt + \frac{\partial u}{\partial x} dx + \frac{\partial u}{\partial y} dy + \frac{\partial u}{\partial z} dz,$$

accordingly,

$$\frac{Du}{Dt} \equiv \frac{du}{dt} = \frac{\partial u}{\partial t} + u \frac{\partial u}{\partial x} + v \frac{\partial u}{\partial y} + w \frac{\partial u}{\partial z},$$

which can be written in vector notation as,

$$\frac{Du}{Dt} = \frac{\partial u}{\partial t} + \vec{V} \cdot \vec{\nabla} u,$$

where $\vec{V}$ is given by

$$\vec{V} = u\hat{x} + v\hat{y} + w\hat{z}$$

and (u, v, w) are the velocity components in the x, y and z-directions, respectively, and $(\hat{x}, \hat{y}, \hat{z})$ are unit vectors in these directions, hence, Eq. (1.47) becomes,

$$\rho \frac{Du}{Dt} = \rho \frac{\partial u}{\partial t} + \rho \vec{V} \cdot \vec{\nabla} u = -\frac{\partial p}{\partial x} \tag{1.48a}$$

Similarly, with $v = v(x, y, z, t)$ and $w = w(x, y, z, t)$, we have the following additional equations for the other components of the velocity,

$$\rho \frac{Dv}{Dt} = \rho \frac{\partial v}{\partial t} + \rho \vec{V} \cdot \vec{\nabla} v = -\frac{\partial p}{\partial y} \tag{1.48b}$$

$$\rho \frac{Dw}{Dt} = \rho \frac{\partial w}{\partial t} + \rho \vec{V} \cdot \vec{\nabla} w = -\frac{\partial p}{\partial z}; \tag{1.48c}$$

leading to the following momentum equations in the x, y and z-directions, respectively.

$$\rho \left(\frac{\partial u}{\partial t} + u \frac{\partial u}{\partial x} + v \frac{\partial u}{\partial y} + w \frac{\partial u}{\partial z} \right) = -\frac{\partial p}{\partial x} \tag{1.49a}$$

$$\rho \left(\frac{\partial v}{\partial t} + u \frac{\partial v}{\partial x} + v \frac{\partial v}{\partial y} + w \frac{\partial v}{\partial z} \right) = -\frac{\partial p}{\partial y} \tag{1.49b}$$

$$\rho \left(\frac{\partial w}{\partial t} + u \frac{\partial w}{\partial x} + v \frac{\partial w}{\partial y} + w \frac{\partial w}{\partial z} \right) = -\frac{\partial p}{\partial z} \tag{1.49c}$$

1.4.3   Energy Balance Equation

Let us now consider how the energy within the small volume element Adx changes with time, this energy is comprised of two parts; kinetic energy due to gas motion and internal energy due to molecular motion. The rate of change of this energy with time is

$$\frac{\partial}{\partial t} \left[\rho \left(\frac{u^2}{2} + e \right) \right] Adx$$

where e is the internal energy per unit mass (for an ideal gas $e = p/(\gamma - 1)\rho$). This rate is equal to the rate of flow of energy into this volume element at x minus the rate of flow of energy out of the volume element at $x + dx$, plus the rate at which pressure forces does work at x minus the rate at which pressure forces does work at $x + dx$. Expressing this mathematically, we have [8],

$$\frac{\partial}{\partial t} \left[\rho \left(\frac{u^2}{2} + e \right) \right] Adx = \left[\frac{1}{2} \rho(x, t) u^2(x, t) + \rho(x, t) e(x, t) \right] u(x, t) A -$$

$$\left[\frac{1}{2}\rho(x+dx,t)u^2(x+dx,t) + \rho(x+dx,t)e(x+dx,t)\right]u(x+dx,t)A +$$

$$p(x,t)u(x,t)A - p(x+dx,t)u(x+dx,t)A.$$

By carrying out a Taylor expansion to first order in dx, namely, for example, $\rho(x+dx,t) = \rho(x,t) + \frac{\partial \rho}{\partial x}dx$ with similar expansions for $u(x+dx,t)$ and $e(x+dx,t)$, it is straightforward to show (after neglecting terms of the order of dx^2) that

$$\frac{\partial}{\partial t}\left[\rho\left(\frac{u^2}{2}+e\right)\right]Adx = -Adx\left[\frac{3}{2}\rho u^2\frac{\partial u}{\partial x} + \frac{u^3}{2}\frac{\partial \rho}{\partial x} + \rho u\frac{\partial e}{\partial x} + \rho e\frac{\partial u}{\partial x} + ue\frac{\partial \rho}{\partial x}\right]$$

$$- Adx\frac{\partial(\rho u)}{\partial x}$$

$$= -Adx\left[\frac{\partial}{\partial x}\left(\frac{\rho u^3}{2}\right) + \frac{\partial}{\partial x}(u\rho e)\right] - Adx\frac{\partial(\rho u)}{\partial x}$$

$$= -Adx\frac{\partial}{\partial x}\left[\rho u\left(\frac{u^2}{2}+e\right)\right] - Adx\frac{\partial(\rho u)}{\partial x}.$$

Hence,

$$\frac{\partial}{\partial t}\left[\rho\left(\frac{u^2}{2}+e\right)\right] + \frac{\partial}{\partial x}\left[\rho u\left(\frac{u^2}{2}+e\right) + pu\right] = 0, \tag{1.50}$$

which is the one-dimensional *energy balance equation* in Eulerian form. The three equations; continuity, motion and energy are supplemented by two other equations; an *equation of state* of the form, $p = p(\rho, T)$ and a *caloric equation* of the form, $e = e(\rho, T)$. For an ideal gas we have already seen that these equations are $p = \rho RT$ and $e = c_V T = p/\rho(\gamma - 1)$, respectively. These five equations are sufficient to determine all five quantities, u, p, ρ, T and e. It is important to note that external forces such as gravity as well viscous forces have been neglected in the momentum equation; similarly, heat transport arising from temperature gradients as well as viscous forces have also been neglected in the energy balance equation. In real fluids, however, these quantities are never quite zero but in the case of idealized flow the neglect of these quantities forms a substantial and useful part of fluid dynamics.

Writing Eq. (1.50) as

$$\frac{\partial}{\partial t}[\rho\varpi] + \frac{\partial}{\partial x}[\rho u\varpi + pu] = 0$$

where $\varpi = \frac{u^2}{2} + e$. Hence,

$$\frac{\partial}{\partial t}[\rho\varpi] + \frac{\partial}{\partial x}[u(\rho\varpi + p)] = 0$$

and carrying out the differentiation we have

$$\frac{\partial}{\partial t}(\rho\varpi) + u\frac{\partial}{\partial x}(\rho\varpi + p) + (\rho\varpi + p)\frac{\partial u}{\partial x} = 0$$

and expanding we obtain,

$$\frac{\partial}{\partial t}(\rho\varpi) + u\frac{\partial}{\partial x}(\rho\varpi) + u\frac{\partial p}{\partial x} + \rho\varpi\frac{\partial u}{\partial x} + p\frac{\partial u}{\partial x} = 0.$$

Hence,

$$\rho\frac{\partial\varpi}{\partial t} + \varpi\frac{\partial\rho}{\partial t} + u\rho\frac{\partial\varpi}{\partial x} + u\varpi\frac{\partial\rho}{\partial x} + u\frac{\partial p}{\partial x} + \rho\varpi\frac{\partial u}{\partial x} + p\frac{\partial u}{\partial x} = 0$$

and collecting terms we have

$$\rho\left(\frac{\partial\varpi}{\partial t} + u\frac{\partial\varpi}{\partial x}\right) + \varpi\left(\frac{\partial\rho}{\partial t} + u\frac{\partial\rho}{\partial x}\right) + u\frac{\partial p}{\partial x} + \rho\varpi\frac{\partial u}{\partial x} + p\frac{\partial u}{\partial x} = 0,$$

which can be written as

$$\rho\left(\frac{\partial\varpi}{\partial t} + u\frac{\partial\varpi}{\partial x}\right) + \varpi\left[\frac{\partial\rho}{\partial t} + u\frac{\partial\rho}{\partial x} + \rho\frac{\partial u}{\partial x}\right] + u\frac{\partial p}{\partial x} + p\frac{\partial u}{\partial x} = 0.$$

The quantity in square brackets in the latter equation goes to zero as we recognise it as the terms appearing in the continuity equation, hence,

$$\rho\left(\frac{\partial\varpi}{\partial t} + u\frac{\partial\varpi}{\partial x}\right) + u\frac{\partial p}{\partial x} + p\frac{\partial u}{\partial x} = 0 \tag{1.51}$$

Now using; $\varpi = \frac{u^2}{2} + e$, we have

$$\frac{\partial\varpi}{\partial t} = u\frac{\partial u}{\partial t} + \frac{\partial e}{\partial t} \quad \text{and} \quad \frac{\partial\varpi}{\partial x} = u\frac{\partial u}{\partial x} + \frac{\partial e}{\partial x}$$

and substituting these relationships in Eq. (1.51) we have

$$\rho\left(\frac{\partial e}{\partial t} + u\frac{\partial e}{\partial x}\right) + u\left[\rho\left(\frac{\partial u}{\partial t} + u\frac{\partial u}{\partial x}\right) + \frac{\partial p}{\partial x}\right] + p\frac{\partial u}{\partial x} = 0.$$

Similarly, the quantity in square brackets in this latter equation goes to zero according to the momentum equation; hence,

$$\rho\left(\frac{\partial e}{\partial t} + u\frac{\partial e}{\partial x}\right) + p\frac{\partial u}{\partial x} = 0.$$

By substituting for $\partial u/\partial x$ using the continuity equation this latter equation can be written in the following form;

$$\rho\left(\frac{\partial e}{\partial t} + u\frac{\partial e}{\partial x}\right) - \frac{p}{\rho}\left(\frac{\partial \rho}{\partial t} + u\frac{\partial \rho}{\partial x}\right) = 0 \qquad (1.52)$$

or writing it in Lagrangian form as

$$\rho\frac{De}{Dt} - \frac{p}{\rho}\frac{D\rho}{Dt} = 0. \qquad (1.53)$$

As $\rho = 1/v$, where v is the specific volume (that is, the volume per unit mass of material), then

$$\frac{D\rho}{Dt} = -\rho^2\frac{Dv}{Dt},$$

so that the energy balance equation in Lagrangian form becomes,

$$\frac{De}{Dt} = -p\frac{Dv}{Dt}. \qquad (1.54)$$

If the *caloric equation* for an ideal gas in the form,

$$e = \frac{pv}{\gamma - 1}$$

is substituted in Eq. (1.54) it is easy to verify that

$$\frac{D(pv^\gamma)}{Dt} = 0,$$

so that

$$\left(\frac{\partial}{\partial t} + u\frac{\partial}{\partial x}\right)(pv^\gamma) = 0.$$

It is important to appreciate that this result has been obtained by assuming that the fluid element is non-conducting, devoid of viscosity and obeys the ideal gas equation with constant heat capacities. The latter equation implies that the product pv^γ remains constant as we follow the fluid element in its motion. If, for example, the fluid element has pressure p_0 and specific volume v_0 at some instant in time, then its pressure p and specific volume v at later times are related according to the equation,

$$pv^\gamma = p_0 v_0^\gamma.$$

Consequently, the quantity pv^γ remains constant in the flow and, as such, it assumes the status of a state variable which, we will see in the subsequent discussion, is related to another state variable called the specific entropy s, where we show in Sect. 1.6 that

$$s = c_V \ln\left(pv^\gamma\right) + \quad \text{constant.}$$

1.5   Constancy of the Entropy with Time for a Fluid Element

Provided the fluid motion experiences no abrupt changes in any of the quantities, u, p or ρ etc., the conservation of energy implies the constancy of entropy [9, 10]. Accordingly, if the fluid element is in thermodynamic equilibrium during the motion the entropy of a fluid element will remain constant with time. In order to see this we need to consider reversible changes taking place in thermodynamic systems. For a reversible thermodynamic change we have [11],

$$TdS = dE + pdV$$

where dS is the change in entropy, dE is the change in internal energy and pdV represents the work done in the process. Writing the latter equation as

$$dE = TdS - pdV$$

and dividing both sides of this latter equation by the mass m of a fluid element we have

$$de = Tds - pdv$$

where e is the internal energy per unit mass, ds is the entropy change per unit mass and v is the specific volume. As $v = 1/\rho$, we have $dv = -(1/\rho^2)d\rho$, so that

$$de = Tds + \frac{p}{\rho^2}d\rho. \tag{1.55}$$

Expressing the internal energy e as a function of s and ρ, we can write

$$e = e(s,\rho),$$

so that

$$de = \left(\frac{\partial e}{\partial s}\right)_\rho ds + \left(\frac{\partial e}{\partial \rho}\right)_s d\rho. \tag{1.56}$$

Hence, by comparing Eqs. (1.55) and (1.56) we have,

$$T = \left(\frac{\partial e}{\partial s}\right)_\rho \quad \text{and} \quad \frac{p}{\rho^2} = \left(\frac{\partial e}{\partial \rho}\right)_s \tag{1.57}$$

From Eq. (1.56) we can write the following two equations;

$$\frac{\partial e}{\partial t} = \left(\frac{\partial e}{\partial s}\right)_\rho \frac{\partial s}{\partial t} + \left(\frac{\partial e}{\partial \rho}\right)_s \frac{\partial \rho}{\partial t}$$

and

$$\frac{\partial e}{\partial x} = \left(\frac{\partial e}{\partial s}\right)_\rho \frac{\partial s}{\partial x} + \left(\frac{\partial e}{\partial \rho}\right)_s \frac{\partial \rho}{\partial x},$$

hence

$$\frac{\partial e}{\partial t} = T\frac{\partial s}{\partial t} + \frac{p}{\rho^2}\frac{\partial \rho}{\partial t}$$

and

$$\frac{\partial e}{\partial x} = T\frac{\partial s}{\partial x} + \frac{p}{\rho^2}\frac{\partial \rho}{\partial x},$$

after using Eq. (1.57). Substituting these latter two equations in Eq. (1.52) we obtain,

$$\rho\left(T\frac{\partial s}{\partial t} + \frac{p}{\rho^2}\frac{\partial \rho}{\partial t} + uT\frac{\partial s}{\partial x} + \frac{up}{\rho^2}\frac{\partial \rho}{\partial x}\right) - \frac{p}{\rho}\left(\frac{\partial \rho}{\partial t} + u\frac{\partial \rho}{\partial x}\right) = 0,$$

so that

$$\rho T \left(\frac{\partial s}{\partial t} + u \frac{\partial s}{\partial x} \right) = 0$$

or

$$\frac{Ds}{Dt} = 0. \tag{1.58}$$

This equation implies that the entropy of the fluid element or particle does not change with time along the element's path of motion and, as a result, the flow is called *isentropic*. However, if discontinuities, such as, shock waves occur in the flow, even in the case of an ideal fluid, there will be an increase in entropy across the discontinuity and the equation $Ds/Dt = 0$ no longer applies.

1.6   Entropy Change for an Ideal Gas

If the fluid is an ideal gas, we can calculate the entropy change (per unit mass) as follows;

$$Tds = de + pdv \ \text{ and } \ pv = RT$$

Hence,

$$ds = \frac{c_V dT}{T} + \frac{pdv}{T},$$

where $de = c_V dT$, therefore,

$$ds = c_V \left(\frac{pdv + vdp}{pv} \right) + \frac{Rdv}{v}$$

$$= (c_V + R)\frac{dv}{v} + c_V \frac{dp}{p},$$

where $c_P = c_V + R$, hence,

$$ds = c_P \frac{dv}{v} + c_V \frac{dp}{p},$$

therefore,

$$\frac{ds}{c_V} = \gamma \frac{dv}{v} + \frac{dp}{p},$$

where $\gamma = c_P/c_V$. Integrating the latter equation we have

$$\int_{s_0}^{s} \frac{ds}{c_V} = \gamma \int_{v_0}^{v} \frac{dv}{v} + \int_{p_0}^{p} \frac{dp}{p},$$

hence,

$$\frac{s - s_0}{c_V} = \gamma \ln \frac{v}{v_0} + \ln \frac{p}{p_0} = \ln \frac{pv^\gamma}{p_0 v_0^\gamma},$$

so that

$$\frac{pv^\gamma}{p_0 v_0^\gamma} = e^{\frac{s-s_0}{c_V}}.$$

Accordingly, constancy of the entropy as expressed by Eq. (1.58) implies that the fluid element lies on the same adiabatic curve when it commenced its motion, hence,

$$\left(\frac{\partial}{\partial t} + u \frac{\partial}{\partial x} \right)(pv^\gamma) = 0, \tag{1.59}$$

or expressed in terms of the density ρ as,

$$\left(\frac{\partial}{\partial t} + u \frac{\partial}{\partial x} \right)(p\rho^{-\gamma}) = 0. \tag{1.60}$$

1.7   Spherical Geometry

Let us now turn our attention to the conservation equations in spherical geometry as we will be requiring these equations later on when spherical motion is discussed. Initially, we will present the general form of these equations in spherical coordinates before considering the case where the motion is confined to take place in the radial direction and the equations reduce to their one-dimensional form.

Initially, we will spend some time developing various *vector relationships* that will be required for writing the conservation equations in spherical geometry.

Relationships Between the Unit Vectors

In a Cartesian coordinate system the position vector $\vec{r}$ is given by the equation,

$$\vec{r} = x\hat{x} + y\hat{y} + z\hat{z},$$

where $\hat{x}$, $\hat{y}$ and $\hat{z}$ are unit vectors in the x, y and z directions, respectively. In terms of the spherical coordinate system as shown in Fig. 1.4, the latter equation becomes,

$$\vec{r} = (rSin\theta Cos\phi)\hat{x} + (rSin\theta Sin\phi)\hat{y} + (rCos\theta)\hat{z},$$

since $x = rSin\theta Cos\phi$, $y = rSin\theta Sin\phi$ and $z = rCos\theta$ from Fig. 1.4. Tangent vectors in the r, θ, ϕ directions are given by $\partial\vec{r}/\partial r$, $\partial\vec{r}/\partial\theta$ and $\partial\vec{r}/\partial\phi$, respectively, and unit vectors $(\hat{r}, \hat{\theta}, \hat{\phi})$ in these directions as shown in Fig. 1.4 are given by

$$\hat{r} = \frac{\partial\vec{r}/\partial r}{|\partial\vec{r}/\partial r|}, \quad \hat{\theta} = \frac{\partial\vec{r}/\partial\theta}{|\partial\vec{r}/\partial\theta|} \quad \text{and} \quad \hat{\phi} = \frac{\partial\vec{r}/\partial\phi}{|\partial\vec{r}/\partial\phi|}.$$

By carrying out the differentiation of the vector $\vec{r}$ as prescribed by these latter equations, we obtain the following set of relationships for the unit vectors in the spherical coordinate system in terms of the unit vectors in the Cartesian coordinate system,

Fig. 1.4 Spherical (r, θ, ϕ) coordinate system is shown with unit vectors in the $\hat{r}, \hat{\theta}$ and $\hat{\phi}$ directions

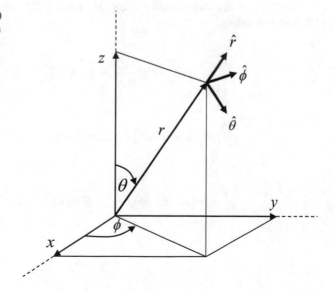

$$\widehat{r} = (Sin\theta Cos\phi)\widehat{x} + (Sin\theta Sin\phi)\widehat{y} + (Cos\theta)\widehat{z}$$

$$\widehat{\theta} = (Cos\theta Cos\phi)\widehat{x} + (Cos\theta Sin\phi)\widehat{y} - (Sin\theta)\widehat{z}$$

$$\widehat{\phi} = -(Sin\phi)\widehat{x} + (Cos\phi)\widehat{y}$$

or in matrix form the latter three equations can be written as

$$\begin{pmatrix} \widehat{r} \\ \widehat{\theta} \\ \widehat{\phi} \end{pmatrix} = \begin{pmatrix} Sin\theta Cos\phi & Sin\theta Sin\phi & Cos\theta \\ Cos\theta Cos\phi & Cos\theta Sin\phi & -Sin\theta \\ -Sin\phi & Cos\phi & 0 \end{pmatrix} \begin{pmatrix} \widehat{x} \\ \widehat{y} \\ \widehat{z} \end{pmatrix}.$$

By forming the following dot products, namely, $\widehat{r} \cdot \widehat{\theta}, \widehat{r} \cdot \widehat{\phi}$ and $\widehat{\phi} \cdot \widehat{\theta}$, it is easy to show that these unit vectors are orthogonal, that is, $\widehat{r} \cdot \widehat{\theta} = 0, \widehat{r} \cdot \widehat{\phi} = 0$ and $\widehat{\phi} \cdot \widehat{\theta} = 0$. Inverting the above matrix we find that

$$\begin{pmatrix} \widehat{x} \\ \widehat{y} \\ \widehat{z} \end{pmatrix} = \begin{pmatrix} Sin\theta Cos\phi & Cos\theta Cos\phi & -Sin\phi \\ Sin\theta Sin\phi & Cos\theta Sin\phi & Cos\phi \\ Cos\theta & -Sin\theta & 0 \end{pmatrix} \begin{pmatrix} \widehat{r} \\ \widehat{\theta} \\ \widehat{\phi} \end{pmatrix}.$$

Derivatives of the Unit Vectors

Later on we will be requiring various derivatives when dealing with the momentum and continuity equations; accordingly, we obtain the following results from the previous equations;

$$\frac{\partial \widehat{r}}{\partial r} = 0, \quad \frac{\partial \widehat{\theta}}{\partial r} = 0 \text{ and } \frac{\partial \widehat{\phi}}{\partial r} = 0$$

$$\frac{\partial \widehat{r}}{\partial \theta} = (Cos\theta Cos\phi)\widehat{x} + (Cos\theta Sin\phi)\widehat{y} - (Sin\theta)\widehat{z} = \widehat{\theta}$$

$$\frac{\partial \widehat{\theta}}{\partial \theta} = -(Sin\theta Cos\phi)\widehat{x} - (Sin\theta Sin\phi)\widehat{y} - (Cos\theta)\widehat{z} = -\widehat{r}$$

$$\frac{\partial \widehat{\phi}}{\partial \theta} = 0$$

$$\frac{\partial \widehat{r}}{\partial \phi} = -(Sin\theta Sin\phi)\widehat{x} + (Sin\theta Cos\phi)\widehat{y} = \widehat{\phi} Sin\theta$$

$$\frac{\partial \widehat{\theta}}{\partial \phi} = -(Cos\theta Sin\phi)\widehat{x} + (Cos\theta Cos\phi)\widehat{y} = \widehat{\phi} Cos\theta$$

$$\frac{\partial \widehat{\phi}}{\partial \phi} = -(Cos\phi)\widehat{x} - (Sin\phi)\widehat{y} = -(Sin\theta)\widehat{r} - (Cos\theta)\widehat{\theta}$$

Incremental Vector Path in Spherical Coordinates

An expression for a small increment in the vector path length $d\vec{r}$ in terms of the spherical coordinates is given by

$$d\vec{r} = d(r\widehat{r}) = \widehat{r}dr + rd\widehat{r} = \widehat{r}dr + r\left(\frac{\partial \widehat{r}}{\partial r}dr + \frac{\partial \widehat{r}}{\partial \theta}d\theta + \frac{\partial \widehat{r}}{\partial \phi}d\phi\right)$$

and by using the previous relationships for the derivatives of these unit vectors we can write the previous equation in the form,

$$d\vec{r} = \widehat{r}dr + \widehat{\theta}rd\theta + \widehat{\phi}rSin\theta d\phi.$$

The Vector Differential Operator del, Written as $\vec{\nabla}$, in Spherical Coordinates

The differential operator $\vec{\nabla}$ in a Cartesian coordinate system is

$$\vec{\nabla} = \widehat{x}\frac{\partial}{\partial x} + \widehat{y}\frac{\partial}{\partial y} + \widehat{z}\frac{\partial}{\partial z},$$

and we need to express this operator in terms of the unit vectors $\widehat{r}, \widehat{\theta}, \widehat{\phi}$ in the spherical coordinate system. In order to so this, let us consider a function Ψ that depends on r, θ, ϕ, that is, $\Psi = \Psi(r, \theta, \phi)$, then the derivative $d\Psi$ is

$$d\Psi = \frac{\partial \Psi}{\partial r}dr + \frac{\partial \Psi}{\partial \theta}d\theta + \frac{\partial \Psi}{\partial \phi}d\phi$$

and this can also be written as

$$d\Psi = \vec{\nabla}\Psi \cdot d\vec{r}$$

$$= \left(\vec{\nabla}\Psi\right)_r dr + \left(\vec{\nabla}\Psi\right)_\theta r d\theta + \left(\vec{\nabla}\Psi\right)_\phi rSin\theta d\phi,$$

after substituting for $d\vec{r}$, hence,

$$\left(\vec{\nabla}\Psi\right)_r = \frac{\partial\Psi}{\partial r}, \quad \left(\vec{\nabla}\Psi\right)_\theta = \frac{1}{r}\frac{\partial\Psi}{\partial\theta} \quad \text{and} \quad \left(\vec{\nabla}\Psi\right)_\phi = \frac{1}{rSin\theta}\frac{\partial\Psi}{\partial\phi}$$

and, therefore, the vector differential operator in spherical coordinates becomes,

$$\vec{\nabla} = \hat{r}\frac{\partial}{\partial r} + \hat{\theta}\frac{1}{r}\frac{\partial}{\partial\theta} + \hat{\phi}\frac{1}{rSin\theta}\frac{\partial}{\partial\phi}. \tag{1.61}$$

Divergence of a Vector $\vec{A}$ in Spherical Coordinates

Performing the following dot product of $\vec{\nabla}$ with the vector $\vec{A}$ we have

$$\vec{\nabla}\cdot\vec{A} = \left(\hat{r}\frac{\partial}{\partial r} + \hat{\theta}\frac{1}{r}\frac{\partial}{\partial\theta} + \hat{\phi}\frac{1}{rSin\theta}\frac{\partial}{\partial\phi}\right)\cdot\left(A_r\hat{r} + A_\theta\hat{\theta} + A_\phi\hat{\phi}\right)$$

$$= \hat{r}\cdot\frac{\partial}{\partial r}(A_r\hat{r}) + \hat{r}\cdot\frac{\partial}{\partial r}\left(A_\theta\hat{\theta}\right) + \hat{r}\cdot\frac{\partial}{\partial r}\left(A_\phi\hat{\phi}\right)$$

$$+ \hat{\theta}\cdot\frac{1}{r}\frac{\partial}{\partial\theta}(A_r\hat{r}) + \hat{\theta}\cdot\frac{1}{r}\frac{\partial}{\partial\theta}\left(A_\theta\hat{\theta}\right) + \hat{\theta}\cdot\frac{1}{r}\frac{\partial}{\partial\theta}\left(A_\phi\hat{\phi}\right)$$

$$+ \hat{\phi}\cdot\frac{1}{rSin\theta}\frac{\partial}{\partial\phi}(A_r\hat{r}) + \hat{\phi}\cdot\frac{1}{rSin\theta}\frac{\partial}{\partial\phi}\left(A_\theta\hat{\theta}\right) + \hat{\phi}\cdot\frac{1}{rSin\theta}\frac{\partial}{\partial\phi}\left(A_\phi\hat{\phi}\right)$$

Carrying out the differentiation of the various terms and by using the previously developed derivatives of the unit vectors, we find that the latter equation reduces to,

$$\vec{\nabla}\cdot\vec{A} = \frac{\partial A_r}{\partial r} + \frac{A_r}{r} + \frac{1}{r}\frac{\partial A_\theta}{\partial\theta} + \frac{A_r}{r} + \frac{A_\theta Cos\theta}{rSin\theta} + \frac{1}{rSin\theta}\frac{\partial A_\phi}{\partial\phi},$$

and this is generally written in the following standard form,

$$\vec{\nabla}\cdot\vec{A} = \frac{1}{r^2}\frac{\partial}{\partial r}\left(r^2 A_r\right) + \frac{1}{rSin\theta}\frac{\partial}{\partial\theta}(A_\theta Sin\theta) + \frac{1}{rSin\theta}\frac{\partial A_\phi}{\partial\phi}. \tag{1.62}$$

This completes our brief look at the various relationships that are required for the subsequent discussion of the conservation equations in spherical geometry.

1.7.1   Continuity Equation

From Eq. (1.43) we have

$$\frac{\partial \rho}{\partial t} + u\frac{\partial \rho}{\partial x} + v\frac{\partial \rho}{\partial y} + w\frac{\partial \rho}{\partial z} + \rho\vec{\nabla}\cdot\vec{V} = 0$$

which can be written as

$$\frac{\partial \rho}{\partial t} + \vec{V}\cdot\vec{\nabla}\rho + \rho\vec{\nabla}\cdot\vec{V} = 0$$

and, therefore, this latter equation becomes

$$\frac{\partial \rho}{\partial t} + \vec{\nabla}\cdot\left(\rho\vec{V}\right) = 0.$$

This is the vector form of the continuity equation. Using Eq. (1.62) in this latter equation we obtain the following continuity equation in spherical geometry;

$$\frac{\partial \rho}{\partial t} + \frac{1}{r^2}\frac{\partial(\rho r^2 u_r)}{\partial r} + \frac{1}{rSin\theta}\frac{\partial(\rho u_\theta Sin\theta)}{\partial \theta} + \frac{1}{rSin\theta}\frac{\partial\left(\rho u_\phi\right)}{\partial \phi} = 0,$$

with $\vec{V} = u_r\hat{r} + u_\theta\hat{\theta} + u_\phi\hat{\phi}$, where u_r, u_θ and u_ϕ are the components of the velocity vector in the $\hat{r}$, $\hat{\theta}$ and $\hat{\phi}$ directions, respectively. If the motion takes place purely in the radial direction ($u_\theta = 0$ and $u_\phi = 0$), then the latter equation becomes

$$\frac{\partial \rho}{\partial t} + \frac{1}{r^2}\frac{\partial(\rho r^2 u_r)}{\partial r} = 0,$$

and by carrying out the differentiation we obtain the following continuity equation,

$$\frac{\partial \rho}{\partial t} + u\frac{\partial \rho}{\partial r} + \rho\left(\frac{\partial u}{\partial r} + \frac{2u}{r}\right) = 0, \tag{1.63}$$

where we have set $u = u_r$ as the material velocity is assumed to be wholly in the radial direction.

1.7.2   Equation of Motion

Recalling Eq. (1.48) and writing the components of the equation of motion in the form,

$$\frac{\partial u}{\partial t} + \left(\vec{V} \cdot \vec{\nabla}\right)u = -\frac{1}{\rho}\frac{\partial p}{\partial x}$$

$$\frac{\partial v}{\partial t} + \left(\vec{V} \cdot \vec{\nabla}\right)v = -\frac{1}{\rho}\frac{\partial p}{\partial y}$$

$$\frac{\partial w}{\partial t} + \left(\vec{V} \cdot \vec{\nabla}\right)w = -\frac{1}{\rho}\frac{\partial p}{\partial z}.$$

Multiplying the first of these equations above by $\hat{x}$, the second equation by $\hat{y}$ and the third equation by $\hat{z}$, yielding,

$$\frac{\partial u\hat{x}}{\partial t} + \left(\vec{V} \cdot \vec{\nabla}\right)u\hat{x} = -\frac{1}{\rho}\hat{x}\frac{\partial p}{\partial x},$$

$$\frac{\partial v\hat{y}}{\partial t} + \left(\vec{V} \cdot \vec{\nabla}\right)v\hat{y} = -\frac{1}{\rho}\hat{y}\frac{\partial p}{\partial y}$$

$$\frac{\partial w\hat{z}}{\partial t} + \left(\vec{V} \cdot \vec{\nabla}\right)w\hat{z} = -\frac{1}{\rho}\hat{z}\frac{\partial p}{\partial z},$$

by adding all three components of the above equation we can write it as a single equation in the form,

$$\frac{\partial}{\partial t}\left(u\hat{x} + v\hat{y} + w\hat{z}\right) + \left(\vec{V} \cdot \vec{\nabla}\right)\left(u\hat{x} + v\hat{y} + w\hat{z}\right) = -\frac{1}{\rho}\left(\hat{x}\frac{\partial}{\partial x} + \hat{y}\frac{\partial}{\partial y} + \hat{z}\frac{\partial}{\partial z}\right)p,$$

which gives the following momentum equation for inviscid flow in vector notation,

$$\frac{\partial \vec{V}}{\partial t} + \left(\vec{V} \cdot \vec{\nabla}\right)\vec{V} = -\frac{1}{\rho}\vec{\nabla}p, \tag{1.64}$$

where $\vec{V} = u\hat{x} + v\hat{y} + w\hat{z}$ and $\vec{\nabla} = \hat{x}(\partial/\partial x) + \hat{y}(\partial/\partial y) + \hat{z}(\partial/\partial z)$.

We have already noted that the velocity vector $\vec{V}$ in terms of the orthogonal spherical velocity components, u_r, u_θ, u_ϕ, is given by

$$\vec{V} = u_r\hat{r} + u_\theta\hat{\theta} + u_\phi\hat{\phi}$$

and we have shown that the vector differential operator $\vec{\nabla}$ for this spherical coordinate system is

$$\vec{\nabla} = \hat{r}\frac{\partial}{\partial r} + \hat{\theta}\frac{1}{r}\frac{\partial}{\partial \theta} + \hat{\phi}\frac{1}{rSin\theta}\frac{\partial}{\partial \phi},$$

hence $\left(\vec{V} \cdot \vec{\nabla}\right)$ in Eq. (1.64) becomes,

$$\left(\vec{V} \cdot \vec{\nabla}\right) = u_r\frac{\partial}{\partial r} + \frac{u_\theta}{r}\frac{\partial}{\partial \theta} + \frac{u_\phi}{rSin\theta}\frac{\partial}{\partial \phi}$$

and forming the product $\left(\vec{V} \cdot \vec{\nabla}\right)\vec{V}$ we have

$$\left(\vec{V} \cdot \vec{\nabla}\right)\vec{V} = \left(u_r\frac{\partial}{\partial r} + \frac{u_\theta}{r}\frac{\partial}{\partial \theta} + \frac{u_\phi}{rSin\theta}\frac{\partial}{\partial \phi}\right)\left(u_r\hat{r} + u_\theta\hat{\theta} + u_\phi\hat{\phi}\right)$$

$$= u_r\frac{\partial}{\partial r}(u_r\hat{r}) + u_r\frac{\partial}{\partial r}\left(u_\theta\hat{\theta}\right) + \frac{\partial}{\partial r}\left(u_\phi\hat{\phi}\right) + \frac{u_\theta}{r}\frac{\partial}{\partial \theta}(u_r\hat{r}) + \frac{u_\theta}{r}\frac{\partial}{\partial \theta}\left(u_\theta\hat{\theta}\right)$$

$$+ \frac{u_\theta}{r}\frac{\partial}{\partial \theta}\left(u_\phi\hat{\phi}\right) + \frac{u_\phi}{rSin\theta}\frac{\partial}{\partial \phi}(u_r\hat{r}) + \frac{u_\phi}{rSin\theta}\frac{\partial}{\partial \phi}\left(u_\theta\hat{\theta}\right) + \frac{u_\phi}{rSin\theta}\frac{\partial}{\partial \phi}\left(u_\phi\hat{\phi}\right).$$

Carrying out the differentiation in this latter equation and using the previously established results for the derivatives of the unit vectors, we can write the *r-component*, the *θ-component* and the *φ-component* of Eq.(1.64), in the following form,

$$\frac{\partial u_r}{\partial t} + u_r\frac{\partial u_r}{\partial r} + \frac{u_\theta}{r}\frac{\partial u_r}{\partial \theta} + \frac{u_\phi}{rSin\theta}\frac{\partial u_r}{\partial \phi} - \frac{\left(u_\theta^2 + u_\phi^2\right)}{r} = -\frac{1}{\rho}\frac{\partial p}{\partial r}$$

$$\frac{\partial u_\theta}{\partial t} + u_r\frac{\partial u_\theta}{\partial r} + \frac{u_\theta}{r}\frac{\partial u_\theta}{\partial \theta} + \frac{u_\phi}{rSin\theta}\frac{\partial u_\theta}{\partial \phi} + \frac{u_r u_\theta}{r} - \frac{u_\phi^2}{r}Cot\theta = -\frac{1}{\rho r}\frac{\partial p}{\partial \theta}$$

$$\frac{\partial u_\phi}{\partial t} + u_r\frac{\partial u_\phi}{\partial r} + \frac{u_\theta}{r}\frac{\partial u_\phi}{\partial \theta} + \frac{u_\phi}{rSin\theta}\frac{\partial u_\phi}{\partial \phi} + \frac{u_r u_\phi}{r} + \frac{u_\theta u_\phi}{r}Cot\theta = -\frac{1}{\rho rSin\theta}\frac{\partial p}{\partial \phi}.$$

If the motion takes place purely in the radial direction we obtain the following momentum equation,

$$\frac{\partial u}{\partial t} + u \frac{\partial u}{\partial r} = -\frac{1}{\rho} \frac{\partial p}{\partial r}, \tag{1.65}$$

where we have set $u = u_r$ as the material velocity is assumed to be wholly in the radial direction.

1.7.3   Equation of Energy Conservation

We have seen that the energy balance equation in plane geometry where the motion takes place purely in the x-direction is

$$\frac{\partial}{\partial t} \left[\rho \left(\frac{u^2}{2} + e \right) \right] + \frac{\partial}{\partial x} \left[\rho u \left(\frac{u^2}{2} + e \right) + pu \right] = 0.$$

Generalizing this equation for 3-dimensional motion, the following equation in vector form is obtained,

$$\frac{\partial}{\partial t} \left[\rho \left(e + \frac{V^2}{2} \right) \right] + \vec{\nabla} \cdot \left[\rho \left(e + \frac{V^2}{2} \right) \vec{V} + p\vec{V} \right] = 0,$$

where $\vec{V}$ is given by

$$\vec{V} = u\hat{x} + v\hat{y} + w\hat{z},$$

and (u, v, w) are the velocity components in the x, y and z-directions, respectively, and $(\hat{x}, \hat{y}, \hat{z})$ are unit vectors in these directions. In spherical geometry, where the motion takes place purely in the radial direction with velocity u, the 3-dimensional form of the energy balance equation reduces to the following 1-dimensional form;

$$\frac{\partial}{\partial t} \left[\rho \left(\frac{1}{2} u^2 + e \right) \right] + \frac{1}{r^2} \frac{\partial}{\partial r} \left[r^2 \left\{ \rho u \left(\frac{1}{2} u^2 + e \right) + pu \right\} \right] = 0 \tag{1.66a}$$

after noting that

$$\vec{\nabla} \cdot u\hat{r} = \frac{1}{r^2} \frac{\partial}{\partial r} \left(r^2 u \right),$$

where $\hat{r}$ is a unit vector in the radial direction. We can also write this energy balance equation as

$$\frac{\partial}{\partial t}\left(\frac{1}{2}\rho u^2 + \frac{p}{\gamma - 1}\right) + \frac{1}{r^2}\frac{\partial}{\partial r}\left[r^2\left(\frac{1}{2}\rho u^2 + \frac{\gamma p}{\gamma - 1}\right)u\right] = 0 \qquad (1.66b)$$

by using the relation, $e = p/(\gamma - 1)\rho$, for an ideal gas. However, assuming the flow is *isentropic* we can use the equation, $Ds/Dt = 0$, in place of the energy equation and write the energy conservation equation as,

$$\left(\frac{\partial}{\partial t} + u\frac{\partial}{\partial r}\right)(p\rho^{-\gamma}) = 0. \qquad (1.67)$$

When spherical shock waves are discussed in Chap. 5 we will be returning to Eqs. (1.63), (1.65), (1.6.6) and (1.67).

1.8 Small Amplitude Disturbances: Sound Waves

The speed at which small amplitude disturbances are propagated to other parts of the fluid is called the *acoustic speed* or the *speed of sound*, c_0 and it is a fundamental parameter in compressible fluid dynamics [12–14]. The disturbance produced by ordinary sound waves is so small that any changes experienced by fluid particles are slow enough that any gradients generated in the flow parameters of pressure, density, temperature etc. are very small; as a result, the flow can be regarded as *isentropic*.
The continuity equation is

$$\frac{\partial \rho}{\partial t} + \rho\frac{\partial u}{\partial x} + u\frac{\partial \rho}{\partial x} = 0 \qquad (1.68)$$

and the momentum equation is

$$\frac{\partial u}{\partial t} + u\frac{\partial u}{\partial x} + \frac{1}{\rho}\frac{\partial p}{\partial x} = 0. \qquad (1.69)$$

For the pressure p we can express it as a function of other thermodynamic variables, such as, the density ρ and the entropy s, that is, $p = p(\rho, s)$, so that

$$dp = \left(\frac{\partial p}{\partial \rho}\right)_s d\rho + \left(\frac{\partial p}{\partial s}\right)_\rho ds.$$

For *isentropic* flow $ds = 0$, hence,

$$dp = \left(\frac{\partial p}{\partial \rho}\right)_s d\rho$$

and accordingly,

$$\frac{\partial p}{\partial x} = \left(\frac{\partial p}{\partial \rho}\right)_s \frac{\partial \rho}{\partial x} = c^2 \frac{\partial \rho}{\partial x}, \tag{1.70}$$

where c^2 is defined according to the equation,

$$c^2 = \left(\frac{\partial p}{\partial \rho}\right)_s. \tag{1.71}$$

Substituting Eq. (1.70) in Eq. (1.69), yields,

$$\frac{\partial u}{\partial t} + u\frac{\partial u}{\partial x} + \frac{c^2}{\rho}\frac{\partial \rho}{\partial x} = 0. \tag{1.72}$$

The two coupled Eqs. (1.68) and (1.72), are difficult to solve due to the presence of the nonlinear term, $u(\partial u/\partial x)$. However, let us assume that we have small perturbations about ambient values such that

$$u(x, t) = u_0 + \Delta u(x, t)$$

and

$$\rho(x, t) = \rho_0 + \Delta \rho(x, t),$$

where it is assumed that $\Delta \rho(x, t) \ll \rho_0$, so that $\Delta \rho(x, t)$ is very small in comparison to the ambient density ρ_0 and clearly, $u_0 = 0$ as the particle velocity is zero under ambient conditions. Similarly, by expanding c^2 we have,

$$c^2 = c_0^2 + \left(\frac{\partial c^2}{\partial \rho}\right)_{\rho_0} \Delta \rho + \cdots$$

where c_0 is equal to c under ambient conditions, accordingly,

$$c_0^2 = \left(\frac{\partial p}{\partial \rho}\right)_{s,\rho_0}.$$

Since isentropic conditions apply we have,

$$\frac{p}{\rho^\gamma} = \text{constant},$$

hence,

$$\left(\frac{\partial p}{\partial \rho}\right)_{s,\rho_0} = \frac{\gamma p_0}{\rho_0}$$

so that

$$c_0 = \left(\frac{\gamma p_0}{\rho_0}\right)^{1/2}. \tag{1.73}$$

Substituting these values of $u(x, t)$, $\rho(x, t)$ and c^2 in Eqs. (1.68) and (1.72) and neglecting terms of the order of Δu^2 and product terms like $\Delta u \Delta \rho$ etc., we have the resulting acoustic equations,

$$\frac{1}{\rho_0}\frac{\partial \Delta \rho}{\partial t} + \frac{\partial \Delta u}{\partial x} = 0 \tag{1.74}$$

and

$$\frac{\partial \Delta u}{\partial t} + \frac{c_0^2}{\rho_0}\frac{\partial \Delta \rho}{\partial x} = 0 \tag{1.75}$$

where only quantities to first order have been retained. Differentiating each of the equations above we have

$$\frac{1}{\rho_0}\frac{\partial^2 \Delta \rho}{\partial x \partial t} + \frac{\partial^2 \Delta u}{\partial x^2} = 0 \text{ and } \frac{\partial^2 \Delta u}{\partial t^2} + \frac{c_0^2}{\rho_0}\frac{\partial^2 \Delta \rho}{\partial t \partial x} = 0,$$

so that

$$\frac{\partial^2 \Delta u}{\partial x^2} = \frac{1}{c_0^2}\frac{\partial^2 \Delta u}{\partial t^2}. \tag{1.76}$$

Similarly, eliminating Δu from Eqs. (1.74) and (1.75) we have

$$\frac{\partial^2 \Delta \rho}{\partial x^2} = \frac{1}{c_0^2}\frac{\partial^2 \Delta \rho}{\partial t^2}, \tag{1.77}$$

and by using the isentropic conditions it is straightforward to show that similar equations are obtained for the pressure and temperature perturbations.

We recognise these latter equations as the one-dimensional *wave equations* where c_0 has the dimensions of speed which we identify as the speed of sound or the speed of propagation of small amplitude disturbances and where the disturbance propagates without change in shape. In air at standard temperature and pressure (STP) we have $p_0 = 1.01 \times 10^5 Nm^{-2}$, $\rho_0 = 1.21 kgm^{-3}$ and $\gamma = 1.4$ we find that $c_0 \approx 340ms^{-1}$. Using the equation of state for an ideal gas, namely, $p = \rho RT$, we have $c_0 = \sqrt{\gamma RT}$, which states that the speed of sound in an ideal gas is proportional to the square-root of the absolute temperature. The solution of the *wave equation* for Δu above can be written in the general form [15];

$$\Delta u(x,t) = f(x - c_0 t) + g(x + c_0 t), \qquad (1.78)$$

where f and g are arbitrary functions. The solution above represents the sum of a right-travelling wave f and a left-travelling wave g; each wave travels at constant speed c_0 with unchanging form. Let us consider the wave propagating in the positive x-direction so that we can set $g = 0$, hence, $\Delta u(x,t) = f(x - c_0 t)$ and differentiating this with respect to x we obtain,

$$\frac{\partial \Delta u}{\partial x} = f',$$

and differentiating with respect to t we have,

$$\frac{\partial \Delta u}{\partial t} = -c_0 f',$$

and combining these latter two equations gives,

$$\frac{\partial \Delta u}{\partial x} = -\frac{1}{c_0} \frac{\partial \Delta u}{\partial t}.$$

Substituting this latter equation in Eq. (1.74) we obtain

$$\frac{1}{\rho_0} \frac{\partial \Delta \rho}{\partial t} - \frac{1}{c_0} \frac{\partial \Delta u}{\partial t} = \frac{\partial}{\partial t} \left(\frac{\Delta \rho}{\rho_0} - \frac{\Delta u}{c_0} \right) = 0,$$

and integrating yields,

$$\Delta u = c_0 \frac{\Delta \rho}{\rho_0}, \qquad (1.79)$$

where the constant of integration is set to zero as $\Delta u = 0$ when $\Delta \rho = 0$. Similarly, by considering the wave propagating in the negative x-direction we obtain,

$$\Delta u = -c_0 \frac{\Delta \rho}{\rho_0}. \qquad (1.80)$$

Combining Eqs. (1.79) and (1.80) gives us the following general relationship between the density perturbation $\Delta\rho$ and velocity perturbation Δu,

$$\Delta u = \pm c_0 \frac{\Delta\rho}{\rho_0}, \tag{1.81}$$

where the positive sign refers to the wave travelling in the positive x-direction and the negative sign refers to the wave travelling in the negative x-direction. If the density perturbation $\Delta\rho$ rises above the ambient value ρ_0 we have, $\Delta u > 0$, hence, the particle velocity is in the same direction as the wave motion. On the other hand, if the density perturbation falls below the ambient density the particle velocity moves in a direction opposite to the direction of wave motion.

By using the fact that the flow is isentropic we have

$$\Delta p = c_0^2 \Delta\rho \text{ or } \Delta p/p_0 = \gamma(\Delta\rho/\rho_0),$$

hence, from Eq. (1.81) we obtain,

$$\Delta u = \pm \frac{\Delta p}{c_0\rho_0} = \pm \frac{c_0\Delta p}{\gamma p_0}. \tag{1.82}$$

Similarly, using the equation of state for a perfect gas, namely, $p = \rho RT$, in conjunction with the isentropic condition we have

$$\Delta T/T_0 = (\gamma - 1)(\Delta\rho/\rho_0),$$

so that

$$\Delta u = \pm c_0 \frac{\Delta T}{(\gamma - 1)T_0}. \tag{1.83}$$

We can see from these relationships that all perturbations in the physical parameters are functions of a single argument, $x \pm c_0 t$, for small amplitude disturbances and each can be expressed as a function of the other.

Once the amplitude of the disturbance increases, however, the simple wave theory above no longer apply and it is necessary to solve the coupled nonlinear Eqs. (1.68) and (1.69). When this is implemented it is found that different parts of the wave profile travel at different speeds and the wave profile distorts as it propagates. Temkin [16] has presented an analysis of the distortion of acoustic waves by considering the propagation of a source of plane waves produced by a piston oscillating at a frequency ω. By using the nonlinear equations, namely, Eqs. (1.68) and (1.69) and, assuming the motion to be isentropic, Temkin adopted a perturbation procedure in which the dependent variables are approximated by the following expansions,

$$p = p_0 + \varepsilon p_1 + \varepsilon^2 p_2 + \ldots$$

$$\rho = \rho_0 + \varepsilon \rho_1 + \varepsilon^2 \rho_2 + \ldots$$

$$u = \varepsilon u_1 + \varepsilon^2 u_2 + \ldots$$

where ε is taken to be much smaller than unity. When the above approximations are substituted in the nonlinear equations it is found that the first approximation, p_1, gives rise to a monochromatic wave at frequency ω. It is found, however, that the second approximation, p_2, is also a monochromatic wave, but at twice the frequency of the first approximation due to nonlinear effects which is known as second harmonic distortion. Temkin showed that the amplitude of the second harmonic increased linearly with distance, indicating that the profile of the wave becomes increasingly distorted as it moves away from the source. The inclusion of more terms in the approximation is possible but too cumbersome as pointed out by Temkin but goes on to state that, with sufficiently large initial amplitude, it is experimentally found that the initial monochromatic wave develops a discontinuous profile. We will be returning to wave profile distortion in the next chapter when we consider the propagation of waves of finite amplitude.

1.9 Typical Sound Wave Parameters

Let us now consider the magnitude of some typical parameters involved in the propagation of sound waves. We can regard a sound wave as a slight pressure perturbation of magnitude Δp that is detectible by the human ear. In the case of harmonic sound waves we can write the perturbation in the parameters as

$$f = f_m Sin(\omega t - kx)$$

where f can denote the perturbation in pressure, density, amplitude etc. and ω is the radian frequency, $k = \omega/c_0$ and $\omega = 2\pi \nu$, where ν is the frequency in cycles per second, c_0 is the sound speed and f_m is the maximum amplitude of the perturbation. Sound waves audible to the human ear range in frequency from 20 to 20,000 Hz.

Atoms or molecules are set in motion as a result of a sound wave. For unit volume of material, the kinetic energy is

$$E = \frac{1}{2}\rho_0 (\Delta V)^2, \tag{1.84}$$

where ρ_0 is the density and ΔV is the perturbation in the particle velocity while the sound intensity I is given by

$$I = c_0 E. \tag{1.85}$$

If the displacement of the atoms or molecules due to the harmonic sound wave is given by

$$a = a_m Sin(\omega t - kz),$$

then the particle velocity is

$$\Delta V = \frac{da}{dt} = \omega a_m Cos(\omega t - kz),$$

which gives a maximum velocity of $\Delta V_m = \omega a_m$. Substituting these results in the equation for the intensity, yields,

$$I = \frac{1}{2}\rho_0 c_0 (\Delta V_m)^2. \tag{1.86}$$

When this latter equation is compared to its electrical equivalent, namely, $(1/2)$ Ri^2, where i denotes the current and R the resistance or impedance, we can identify the quantity, $\rho_0 c_0$, as an impedance, which is called the acoustic impedance and, consequently, the pressure perturbation (analogous to the voltage) is given by

$$\Delta p = \rho_0 c_0 \Delta V_m. \tag{1.87}$$

Alternatively, this latter equation can be obtained by using Newton's second law of motion as shown below. In order to see this, let us consider a sound wave that is generated by applying a small force of magnitude ΔF to a piston in a tube of cross-sectional area A containing, for example, air as shown in Fig. 1.5. The force will impart a compression wave that moves to the right-hand side, and in a time interval of dt the forward front of the wave will moves a short distance, $dx = c_0 dt$, where c_0 is the speed of sound. The piston imparts a small particle velocity of magnitude ΔV as shown in Fig. 1.5 and the air molecules are set in motion. In this small-time interval of dt we can assume that all the molecules in the dotted region moves at this particle velocity. Hence, the mass of material set in motion is given by, $dm = \rho_0 A dx$, where ρ_0 is the density of air. The momentum of this mass motion is given by,

$$dP_{mom.} = dm \times \Delta V = (\rho_0 A c_0 dt)\Delta V,$$

hence, from Newton's second law of motion, we have,

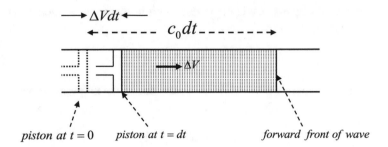

piston at $t = 0$ piston at $t = dt$ forward front of wave

Fig. 1.5 Piston motion generating a small compression wave that moves to the right-hand side at the speed of sound (see text)

$$\frac{dP_{mom.}}{dt} = \Delta F = \rho_0 A c_0 \Delta V$$

and the pressure perturbation, $\Delta p = \Delta F/A$, becomes,

$$\Delta p = \rho_0 c_0 \Delta V,$$

which is in agreement with Eq. (1.87) above.

By using Eq. (1.87) the sound intensity can also be written as

$$I = \frac{1}{2} \frac{(\Delta p)^2}{\rho_0 c_0}. \tag{1.88}$$

1.9.1   Typical Sound Intensity in Normal Conversation

The typical sound energy in a normal conversation is about $10\mu W$. A person listening about 1 m away will receive an intensity given by

$$I = \frac{10\mu W}{4\pi(1)^2} = 0.8\mu W m^{-2},$$

so that the excess pressure Δp is given by

$$\Delta p = \sqrt{2\rho_0 c_0 I} = \sqrt{2 \times 1.25 \times 340 \times 0.8 \times 10^{-6}} = 0.026 N m^{-2},$$

which is a tiny fraction of the atmospheric pressure. The velocity amplitude ΔV_m is about 0.006 cm/s and the typical molecular displacement amplitude is approximately

$0.02\mu m$. The weakest sound that can be heard by the human ear is dependent on the frequency; at about 400 Hz the threshold of audibility is approximately equal to $10^{-6}\mu Wm^{-2}$, hence, the displacement amplitude is given by

$$a_m = \sqrt{\frac{2I}{\omega^2 \rho_0 c_0}} = \sqrt{\frac{2 \times 1 \times 10^{-12}}{(2\pi \times 400)^2 \times 1.25 \times 340}}$$

$$= 2.73 \times 10^{-11} m,$$

which is about 0.25 Angstrom units!

1.9.2 Loud Sounds

A sound wave can be considered loud when the intensity approaches about $1Wm^{-2}$ and sound waves can become quite painful when the intensity reaches about $10Wm^{-2}$. The excess pressure in this latter extreme case is above $90Nm^{-2}$, which is about 0.1% of the atmospheric pressure and the maximum amplitude a_m for molecular displacement at a frequency of 400 Hz is approximately 0.08 mm. The velocity amplitude is about $0.22ms^{-1}$ or approximately 0.06% of speed of sound.

References

1. J. Von Neumann, R.D. Richtmyer, A method for the numerical calculation of hydrodynamic shocks. J. Appl. Phys. **21**, 232–237 (1950)
2. M. A. Saad, *Compressible Fluid Flow*, (Prentice-Hall, Inc., Englewood Cliffs, New Jersey 1985)
3. G. Falkovich, Fluid Mechanics: A Short Course for Physicists, (Cambridge University Press, New York, 2011)
4. V. Babu, *Fundamentals of Gas Dynamics*, 2nd edn. (John Wiley & Sons Ltd., Chichester, United Kingdom, 2015)
5. O. Regev, O.M. Umurhan, P.A. Yecko, *Modern Fluid Dynamics for Physics and Astrophysics* (Springer, New York, 2016)
6. J.H.S. Lee, *The Gas Dynamics of Explosions* (Cambridge University Press, New York, 2016) Chapter 1
7. J.D. Anderson Jr., *Computational Fluid Dynamics: The Basics with Applications* (McGraw-Hill, New York, 1995) Chapter 2
8. J.D. Logan, *An Introduction to Nonlinear Partial Differential Equations* (John Wiley & Sons, New York, 1984) Chapter 5
9. J. von Neumann, *Theory of Shock Waves, Collected Works*, vol 6 (Pergamon Press, New York, 1976), p. 178
10. H. A. Bethe et al LA-165 Report (Los Alamos Scientific Laboratory of the University of California, Los Alamos, New Mexico, October 1944), Section 1

11. M.W. Zemansky, *Heat and Thermodynamics*, 5th edn. (McGraw-Hill, New York, 1968) Chapter 11
12. D. Tabor, *Gases, Liquids and Solids* (Penguin Books, Middlesex, England, 1969), pp. 62–64
13. L.D. Landau, *E. M. Lifshitz* (Fluid Mechanics, Pergamon Press, London, 1966) Chapter 8
14. N. Curle, H.J. Davies, *Modern Fluid Dynamics*, vol 2 (Von Nostrand Reinhold Company, London, 1971) Chapter 2
15. J.D. Anderson, *Modern Compressible Flow with Historical Perspective*, 3rd edn. (McGraw-Hill, New York, 2003) Chapter 7
16. S. Temkin, *Elements of Acoustics*, (John Wiley, New York, 1981), Section 3.7

Chapter 2
Waves of Finite Amplitude

2.1 Introduction

In our brief discussion of sound waves in Chap. 1 we assumed that the wave amplitude is sufficiently small that the resulting equations are linear. As a result, it was relatively easy to solve the equations which led to travelling waves whose distribution of density, pressure, velocity etc. moved with constant velocity c_0 and the profile of the wave did not change with time. This, however, is not the case when the wave has appreciable amplitude and we will see, in due course, that the wave profile changes its shape as it propagates.

2.2 Finite Amplitude Waves

It is necessary to return to the more exact equations of motion when the wave has appreciable amplitude: as we have already seen, these are the continuity and momentum equations

$$\frac{\partial \rho}{\partial t} + \rho \frac{\partial u}{\partial x} + u \frac{\partial \rho}{\partial x} = 0 \tag{2.1}$$

and

$$\frac{\partial u}{\partial t} + u \frac{\partial u}{\partial x} + \frac{1}{\rho} \frac{\partial p}{\partial x} = 0, \tag{2.2}$$

respectively. For isentropic flow we have the further equation, $p\rho^{-\gamma} = $ constant, hence,

© The Author(s), under exclusive license to Springer Nature Switzerland AG 2021
S. Prunty, *Introduction to Simple Shock Waves in Air*, Shock Wave and High
Pressure Phenomena, https://doi.org/10.1007/978-3-030-63606-7_2

$$\frac{\partial p}{\partial x} = \frac{\gamma p}{\rho}\frac{\partial \rho}{\partial x} = c^2 \frac{\partial \rho}{\partial x},$$

and with this substitution Eq. (2.2) becomes,

$$\frac{\partial u}{\partial t} + u\frac{\partial u}{\partial x} + \frac{c^2}{\rho}\frac{\partial \rho}{\partial x} = 0. \tag{2.3}$$

By using Eqs. (2.1) and (2.3) Band [1] has shown, after eliminating the density ρ, that the following equation is obtained,

$$\frac{\partial^2 u}{\partial t^2} + 2u\frac{\partial^2 u}{\partial t \partial x} + 2\frac{\partial u}{\partial x}\left(\frac{\partial u}{\partial t} + u\frac{\partial u}{\partial x}\right) - (c^2 - u^2)\frac{\partial^2 u}{\partial x^2} = 0, \tag{2.4}$$

which is a nonlinear equation for u as a function of x and t which one must proceed to solve by successive approximations. It reduces to the simple wave equation if u is small enough so that product of more than one factor containing u can be neglected. Band substituted the zeroth order approximation;

$$u = u_0 \exp(ikx - i\omega t),$$

which is just the solution of the simple wave equation, namely,

$$\frac{\partial^2 u}{\partial t^2} - c^2 \frac{\partial^2 u}{\partial x^2} = 0$$

into Eq. (2.4), where $c = \omega/k$, and found that

$$u = u_0 \exp[i\omega(x/c' - t)]$$

was a better approximation than the zeroth order approximation in which the speed of propagation of a disturbance behaved according to the relation,

$$c' = (c^2 + u^2)^{1/2} + 2u.$$

Consequently, $c' > c$ when u is positive so that the crest of the wave travels faster than the troughs and eventually the wave distorts into a saw-tooth like wave form. This analysis of wave distortion by Band [1] is similar to the analysis of acoustic wave distortion by Temkin [2] referred to previously. The important point to note is that a wave of finite amplitude distorts as it propagates: different parts of a wave profile propagate at different speeds so that the speed of propagation depends on the amplitude [3].

We saw in the case of sound waves that all perturbations in the physical parameters, such as, pressure, density, fluid velocity etc. are functions of a single argument, $x \pm c_0 t$ and, as such, each can be expressed as a function of any other. Using this fact in relation to waves of finite amplitude and following the analysis of Landau and Lifshitz [4] let us assume that the particle velocity can be expressed as a function of density [5–7], that is, $u = u(\rho)$, and writing again the continuity and momentum equations;

$$\frac{\partial \rho}{\partial t} + \frac{\partial (\rho u)}{\partial x} = 0 \tag{2.5}$$

and

$$\frac{\partial u}{\partial t} + u \frac{\partial u}{\partial x} + \frac{1}{\rho} \frac{\partial p}{\partial x} = 0. \tag{2.6}$$

Since the particle velocity is a function of the density the continuity equation becomes;

$$\frac{\partial \rho}{\partial t} + \frac{d}{d\rho} (\rho u) \frac{\partial \rho}{\partial x} = 0 \tag{2.7}$$

and the momentum can be written as,

$$\frac{\partial u}{\partial t} + u \frac{\partial u}{\partial x} + \frac{1}{\rho} \frac{\partial p}{\partial u} \frac{\partial u}{\partial x} = 0. \tag{2.8}$$

Since $u = u(\rho)$ and according to the *isentropic* relation, p/ρ^γ=constant, hence, $p = p(\rho)$. Then p can be written explicitly as a function of u also, so that Eqs. (2.7) and (2.8) become

$$\frac{\partial \rho}{\partial t} + \left(u + \rho \frac{du}{d\rho} \right) \frac{\partial \rho}{\partial x} = 0 \tag{2.9}$$

and

$$\frac{\partial u}{\partial t} + \left(u + \frac{1}{\rho} \frac{dp}{du} \right) \frac{\partial u}{\partial x} = 0. \tag{2.10}$$

However, $\rho = \rho(x, t)$ and $u = u(x, t)$, so that

$$\frac{d\rho}{dt} = \frac{\partial \rho}{\partial t} + \frac{dx}{dt} \frac{\partial \rho}{\partial x},$$

hence,

$$\frac{\partial \rho}{\partial t} + \left(\frac{dx}{dt}\right)_\rho \frac{\partial \rho}{\partial x} = 0$$

Comparing this latter equation with Eq. (2.9), it is clear that

$$\left(\frac{dx}{dt}\right)_\rho = u + \rho \frac{du}{d\rho}. \tag{2.11}$$

Similarly, $u = u(x, t)$, so that

$$\frac{\partial u}{\partial t} + \left(\frac{dx}{dt}\right)_u \frac{\partial u}{\partial x} = 0$$

and, similarly, comparing this latter equation with Eq. (2.10) we have

$$\left(\frac{dx}{dt}\right)_u = u + \frac{1}{\rho}\frac{dp}{du}. \tag{2.12}$$

However, if u is constant so is ρ since $u = u(\rho)$, then $(dx/dt)_\rho = (dx/dt)_u$ and this implies from Eqs. (2.11) and (2.12) that

$$u + \rho \frac{du}{d\rho} = u + \frac{1}{\rho}\frac{dp}{du}$$

or

$$\rho \frac{du}{d\rho} = \frac{1}{\rho}\frac{dp}{du}.$$

The latter equation can be written as

$$\rho \frac{du}{d\rho} = \frac{1}{\rho}\frac{dp}{d\rho}\frac{d\rho}{du}. \tag{2.13}$$

However, $p/\rho^\gamma =$ constant and therefore, $dp/d\rho = \gamma p/\rho$ which we identify as the square of the *local* sound velocity; that is, $c^2 = \gamma p/\rho$, and with this substitution in Eq. (2.13), we have,

$$\rho \frac{du}{d\rho} = \frac{c^2}{\rho}\frac{d\rho}{du},$$

which yields

$$\rho^2 (du)^2 = c^2 (d\rho)^2,$$

hence,

$$du = \pm c \frac{d\rho}{\rho}, \tag{2.14}$$

which should be compared with Eq. (1.81) for ordinary sound waves in Chap. 1. However, Eq. (2.12) gives

$$\left(\frac{dx}{dt} \right)_u = u + \frac{1}{\rho} \frac{dp}{du},$$

which can be written as

$$\left(\frac{dx}{dt} \right)_u = u + \frac{1}{\rho} \frac{dp}{d\rho} \frac{d\rho}{du}.$$

With $dp/d\rho = c^2$ and using Eq. (2.14), the latter equation becomes

$$\left(\frac{dx}{dt} \right)_u = u \pm c(u), \tag{2.15}$$

where we have explicitly indicated that c is a function of u. This latter equation implies that the speed of propagation of a disturbance is dependent on the *local* particle velocity u and the *local* speed of sound $c(u)$. Consequently, an observer moving at the *local* particle velocity will observe that the speed of propagation of a disturbance travels at the *local* speed of sound. The plus and minus signs in the latter equation represent wave propagation in the positive and negative x-directions, respectively. Returning again to Eq. (2.14) we note that

$$c^2 = \frac{\gamma p}{\rho} = \frac{\gamma}{\rho} k \rho^\gamma = \gamma k \rho^{\gamma-1},$$

where k is a constant and taking logs of both sides we have

$$2 \log c = \log \gamma k + (\gamma - 1) \log \rho,$$

hence,

$$2 \frac{dc}{c} = (\gamma - 1) \frac{d\rho}{\rho}$$

so that

$$\frac{d\rho}{\rho} = \frac{2}{\gamma - 1} \frac{dc}{c}. \tag{2.16}$$

Substituting this in Eq. (2.14) and integrating gives,

$$u = \pm \frac{2}{\gamma - 1} \int dc = \pm \frac{2}{\gamma - 1} (c - c_0), \tag{2.17}$$

where the constant of integration has been chosen to be $c = c_0$ when $u = 0$. Hence, the *local* velocity of sound in terms of the *local* particle velocity is

$$c(u) = c_0 \pm \frac{(\gamma - 1)}{2} u, \tag{2.18}$$

and substituting this in Eq. (2.15) yields,

$$\left(\frac{dx}{dt}\right)_u = u \pm \left(c_0 \pm \frac{\gamma - 1}{2} u\right)$$

$$= \pm c_0 + \left(\frac{\gamma + 1}{2}\right) u, \tag{2.19}$$

which now gives the speed of propagation of the disturbance in terms of the particle velocity and the ambient or undisturbed sound speed, c_0. Let us assume that propagation takes place in the positive x-direction, then from Eq. (2.19) we can write the speed of propagation of the disturbance as,

$$c' = c_0 + \left(\frac{\gamma + 1}{2}\right) u.$$

Using Eq. (2.17) for u in the latter equation yields,

$$c' = c_0 + \left(\frac{\gamma + 1}{2}\right) \frac{2}{\gamma - 1} (c - c_0). \tag{2.20}$$

However, we note that,

$$c^2 = \frac{\gamma p}{\rho} \quad \text{and} \quad c_0^2 = \frac{\gamma p_0}{\rho_0}$$

and

$$p = k\rho^{\gamma} \text{ and } p_0 = k\rho_0^{\gamma},$$

so that

$$c = c_0 \left(\frac{p}{p_0}\right)^{\frac{\gamma-1}{2\gamma}} \text{ and } c = c_0 \left(\frac{\rho}{\rho_0}\right)^{\frac{\gamma-1}{2}}. \tag{2.21}$$

By substituting each of these latter equations in turn in Eq. (2.20) we obtain either of the following equations for the propagation speed of a disturbance in the case of an ideal gas;

$$c' = c_0 \left[1 + \left(\frac{\gamma+1}{\gamma-1}\right)\left\{\left(\frac{\rho}{\rho_0}\right)^{\frac{\gamma-1}{2}} - 1\right\}\right] \tag{2.22a}$$

and

$$c' = c_0 \left[1 + \left(\frac{\gamma+1}{\gamma-1}\right)\left\{\left(\frac{p}{p_0}\right)^{\frac{\gamma-1}{2\gamma}} - 1\right\}\right]. \tag{2.22b}$$

In regions of compression where $p > p_0$ or $\rho > \rho_0$, the propagation speed of a disturbance is greater than c_0, and in regions of rarefaction where $p < p_0$ or $\rho < \rho_0$, the propagation speed is lower than c_0. Hence, those parts of the disturbance where the pressure or density is higher move faster than those parts where the pressure or density is lower and visa versa. This means that waves of finite amplitude distort as they propagate; compression waves steepen and rarefaction waves flatten [8, 9]. Initially, when we considered the propagation of ordinary sound waves we found that the wave was monochromatic so that every part of the wave was propagated at speed c_0 and the waveform retained its shape or profile with the progression of time. This, as we see, is not the case when finite amplitude effects are taken into account, as different parts of a wave profile propagate at different speeds and, as a result, the profile of the wave changes its shape as it propagates which can lead to the formation of shock waves. In regions of compression where significant steepening occurs the velocity and temperature gradients become so large that the effects of viscosity and heat conduction which have so far been neglected would need to be taken into account.

2.3   Change in Wave Profile

In order to understand the change in wave profile let us consider a wave propagating to the right whose velocity profile u as a function of position x is shown in Fig. 2.1 at, say, $t = 0$.

Fig. 2.1 Snapshot of a
wave profile

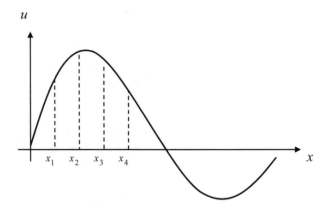

Since the velocity of the disturbance is dependent on u according to Eq. (2.19);

$$\frac{dx}{dt} = c_0 + \left(\frac{\gamma + 1}{2}\right)u,$$

then different parts of the wave travel at different speeds. Let us now consider two neighbouring point in the positive slope of the profile, namely, x_1 and x_2 where the velocities are u_1 and u_2, respectively. After a time t the point x_1 has moved to

$$x_1' = x_1 + \left[c_0 + \left(\frac{\gamma + 1}{2}\right)u_1\right]t$$

and the point x_2 has moved to

$$x_2' = x_2 + \left[c_0 + \left(\frac{\gamma + 1}{2}\right)u_2\right]t.$$

The initial separation between x_1 and x_2 is $x_2 - x_1$ while the new separation is

$$x_2' - x_1' = (x_2 - x_1) + \left(\frac{\gamma + 1}{2}\right)(u_2 - u_1)t.$$

Since $u_2 > u_1$, then $x_2' - x_1'$ will increase with time and the disturbance will spread. On the other hand, consider the points x_3 and x_4 on the negative slope as shown in Fig. 2.1. After a time t the point x_3 has moved to

$$x_3' = x_3 + \left[c_0 + \left(\frac{\gamma + 1}{2}\right)u_3\right]t$$

and x_4 has moved to

Fig. 2.2 Change in the
shape of the wave profile
(see text)

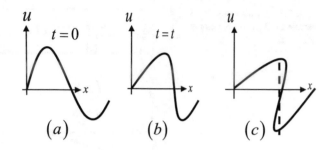

$$x_4' = x_4 + \left[c_0 + \left(\frac{\gamma + 1}{2}\right)u_4\right]t,$$

so that the new separation between x_3 and x_4 is

$$x_4' - x_3' = (x_4 - x_3) + \left(\frac{\gamma + 1}{2}\right)(u_4 - u_3)t.$$

However, in this case $u_4 < u_3$, hence, $(x_4' - x_3') < (x_4 - x_3)$ and the negative sloping edge becomes steeper: Fig. 2.2(*b*) summarises the change to the shape of the profile after a time *t*.

Eventually, the regions of compression will start to overtake the regions of rarefaction so that the wave profile tends to the form shown in Fig. 2.2(*c*). This would imply that the velocity, density or pressure would have three values at some point *x* which is physically impossible; realistically, a discontinuity is formed and the wave profile takes the form of the broken line as shown in Fig. 2.2(*c*). Prior to the onset of the discontinuity, however, the velocity, pressure and density gradients become so large that viscosity and heat conduction (which have been neglected in our previous equations) come into play to counteract the gradients. A balance is achieved between the competing effects of viscosity and heat conduction, on the one hand, and the tendency of the compressive parts to overtake the expansive parts, on the other hand. When this balance is achieved, the compressive portion of the wave profile propagates without further distortion resulting in the formation of a shock front or shock wave. In reality, the shock front is not a mathematical discontinuity but a very thin transition region of the order of a few molecular mean-free paths [10, 11] and across which there is an almost discontinuous jump in the mechanical and thermodynamic variables.

2.4 Formation of a Normal Shock Wave

Let us now investigate a little further how the disturbance speed depends on amplitude by considering the propagation of a series of disturbances in air due to the motion of a piston in a tube. Suppose the piston undergoes a uniform acceleration

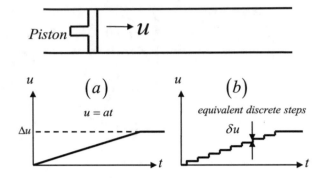

Fig. 2.3 A uniformly accelerated piston is shown (see text)

to a finite velocity Δu as illustrated in Fig. 2.3(a), and let us further imagine that Δu consists of a succession of much smaller velocity increments, each of magnitude δu, as shown in Fig. 2.3(b).

Each of these velocity increments generate a small compression wave that travels down the tube at the speed of sound. The first velocity increment produces a sound wave that travels at, say, c_1, while the disturbance generated by the second increment in velocity will travel in the wake of the first disturbance. However, the air left after the passage of the first disturbance has been compressed adiabatically, so that the temperature of the air has been increased. Since sound speed depends on temperature $(\propto \sqrt{T})$, this implies that the disturbance generated by the second velocity increment will travel at, say, speed c_2 (which is greater than c_1) with respect to the air into which the disturbance is moving. Moreover, the second disturbance travels in the air that was set in motion with velocity δu as a result of the first disturbance, so that the absolute velocity of the second disturbance relative to a fixed observer is $c_2 + \delta u$.

The continued incremental motion of the piston generates further disturbances, each travelling at a speed greater than its predecessors, so that the later disturbances will eventually catch up with those that have already travelled further down the tube. The forward fronts of each individual disturbance (denoted by small arrows) are shown in Fig. 2.4 at two successive times ($t_2 > t_1$) together the associated compression waves. With the progression of time the compression front begins to steepen as the later disturbances begin to overtake the earlier ones as illustrated in Fig. 2.5, which eventually leads to the formation of a shock wave at $t = t_5$ as shown in Fig. 2.5. Here, we show the shock as a mathematical discontinuity but, as previously stated, it is a thin transition region whose width corresponds to a few molecular mean-free paths.

2.5 Time and Place of Formation of Discontinuity

Let us now return to our discussion in relation to the change in wave profile. We can determine the time and place of formation of the discontinuity by noting that a snapshot of the profile of u versus x (see Fig. 2.2) eventually develops a vertical slope, this implies that $\partial u / \partial x$ becomes infinite or that $\partial x / \partial u = 0$ at some instance in

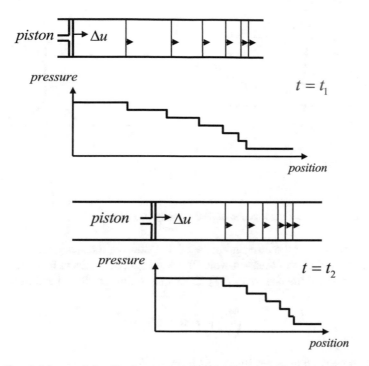

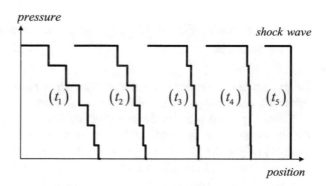

Fig. 2.4 Forward fronts of the disturbances and associated compression waves at two different times (see text)

Fig. 2.5 The compression front becomes steeper with time with the eventual formation of a shock wave

time t. In addition, the wave profile lies on both sides of the vertical tangent, so that it must also be a point of inflexion, hence, $\partial^2 x/\partial u^2 = 0$. Hence, the time and place of formation of the shock wave is determined by the solution of the following equations;

$$\left(\frac{\partial x}{\partial u}\right)_t = 0 \text{ and } \left(\frac{\partial^2 x}{\partial u^2}\right)_t = 0. \tag{2.23}$$

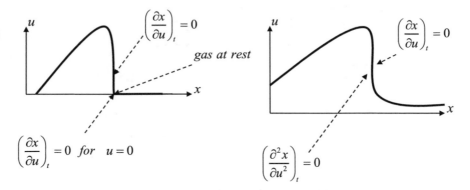

Fig. 2.6 Conditions for the formation of discontinuity (see text)

The condition for the time and place of formation of the shock wave as given by Eq. (2.23) requires modification when the wave adjoins a gas at rest and the shock wave is formed at the boundary [11]: in this case, the condition becomes:

$$\left(\frac{\partial x}{\partial u}\right)_t = 0 \ \text{for} \ u = 0 \tag{2.24}$$

both of these conditions are illustrated in Fig. 2.6.

2.5.1   Example: Piston Moving with Uniform Accelerated Velocity

Let us consider the motion of the disturbance when a piston moves into a tube with a velocity u according to the equation; $u = at$ as shown in Fig. 2.7.

Following Landau and Lifshitz's analysis [11] a compression wave is formed which propagates to the right. At the surface of the piston the gas velocity has the same velocity as the piston, then $u = at$ and integrating this we have $x = (1/2)at^2$ for the position of the piston at time t. The velocity of a disturbance in the wave profile is given by Eq. (2.19) and in terms of its integrated form we have

$$x = \left[c_0 + \left(\frac{\gamma + 1}{2}\right)u\right]t + x_0(u)$$

where $x_0(u)$ is an arbitrary function of the velocity. Substituting the above relationships, namely, $t = u/a$ and $x = (1/2)(u^2/a)$ in this latter equation we have

Fig. 2.7 Piston moving with uniform accelerated velocity in a tube

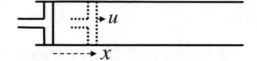

$$\frac{1}{2}\frac{u^2}{a} = \left[c_0 + \left(\frac{\gamma+1}{2}\right)u\right]\frac{u}{a} + x_0(u)$$

which determines the constant of integration $x_0(u)$ as

$$x_0(u) = \frac{1}{2}\frac{u^2}{a} - \left[c_0 + \left(\frac{\gamma+1}{2}\right)u\right]\frac{u}{a}.$$

Substituting this back in the equation for x gives

$$x(u,t) = \frac{1}{2}\frac{u^2}{a} - \frac{c_0}{a}u - \left(\frac{\gamma+1}{2}\right)\frac{u^2}{a} + \left[c_0 + \left(\frac{\gamma+1}{2}\right)u\right]t,$$

and simplifying yields,

$$x(u,t) = -\frac{c_0}{a}u - \frac{\gamma}{2}\frac{u^2}{a} + \left[c_0 + \left(\frac{\gamma+1}{2}\right)u\right]t. \tag{2.25}$$

Differentiating the latter equation we have

$$\left(\frac{\partial x}{\partial u}\right)_t = -\frac{c_0}{a} - \frac{\gamma u}{a} + \left(\frac{\gamma+1}{2}\right)t.$$

The time t_{shock} of formation of the shock wave is given by,

$$\left(\frac{\partial x}{\partial u}\right)_t = 0 \quad \text{for} \quad u = 0,$$

hence, t_{shock} is given by the equation;

$$-\frac{c_0}{a} + \left(\frac{\gamma+1}{2}\right)t = 0.$$

so that,

$$t_{shock} = \frac{2c_0}{(\gamma+1)a}, \tag{2.26}$$

and the place of formation of the shock is at the forward front of the wave, namely, at

$$x_{shock} = \frac{2c_0^2}{(\gamma + 1)a}.$$ (2.27)

Returning to Eq. (2.25) and writing it in the form,

$$2ax = -2c_0 u - \gamma u^2 + 2at\left[c_0 + \left(\frac{\gamma + 1}{2}\right)u\right],$$

which can be regarded as a quadratic equation in u with t as a parameter, so that,

$$\gamma u^2 + 2\left[c_0 - \left(\frac{\gamma + 1}{2}\right)at\right]u - 2a(c_0 t - x) = 0.$$

Hence, the following expression for u is obtained,

$$u = \frac{-\left[c_0 - \left(\frac{\gamma+1}{2}\right)at\right] + \left[\left\{c_0 - \left(\frac{\gamma+1}{2}\right)at\right\}^2 + 2\gamma a(c_0 t - x)\right]^{1/2}}{\gamma},$$ (2.28)

which is only a valid solution up to the time that the shock wave is formed. We can see from the latter equation that $u = 0$ when $x = c_0 t$. In addition,

$$u = \frac{\left[2\gamma a\left(c_0 \frac{2c_0}{(\gamma+1)a} - x\right)\right]^{1/2}}{\gamma}$$

when $t = \frac{2c_0}{(\gamma+1)a}$, which is the time that the shock wave forms. The position of the shock front is

$$x_{shock} = \frac{2c_0^2}{(\gamma + 1)a},$$

which is at the forward front of the wave, that is, at $x = c_0 t$, where $t = 2c_0/(\gamma + 1)a$ according to Eq. (2.26). In Fig. 2.8 we show plots of the particle velocity u as a function of position x according to Eq. (2.28) for three different times t. For the plots the following parameters were assumed; $\gamma = 1.4$, $c_0 = \sqrt{1.4}$ and $a = 0.01$. Using these parameters in Eqs. (2.26) and (2.27), one can see that the shock wave will form at $t = 98.6$ (arb. units), and the place of formation is at $x = 116.7$(arb. units) which corresponds to the forward front of the wave. It can be observed that the gradient $(\partial u/\partial x)$ becomes infinite at $x = 116.7$. Figure 2.8 can be compared with similarly generated plots obtained later on in Chap. 4 where the equations are numerically integrated with *artificial viscosity* included.

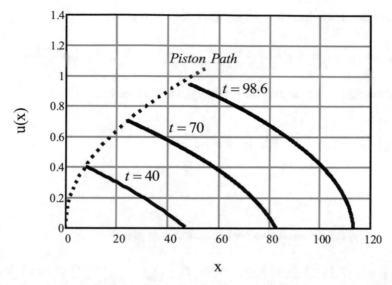

Fig. 2.8 The particle velocity u according to Eq. (2.28) as a function of position x is shown plotted with t as a parameter. The following quantities were assumed; $c_0 = \sqrt{1.4}$, $\gamma = 1.4$ and $a = 0.01$ (see text)

2.5.2   *Example: Piston Moving with a Velocity* **u = at^n, n> 0**

Let us now consider a piston moving into a tube with a velocity given by $u = at^n$ [11] and the displacement of the piston is $x = at^{n+1}/(n+1)$. This implies that

$$t = \left(\frac{u}{a}\right)^{1/n} \quad \text{and} \quad x = \frac{a}{n+1}\left(\frac{u}{a}\right)^{(n+1)/n}.$$

The fluid velocity at the piston has the same velocity as the piston itself, so by substituting these latter equations in the equation;

$$x(u, t) = \left[c_0 + \left(\frac{\gamma+1}{2}\right)u\right]t + x_0(u),$$

we can determine the constant of integration $x_0(u)$. Hence,

$$\frac{a}{n+1}\left(\frac{u}{a}\right)^{\frac{n+1}{n}} = \left[c_0 + \left(\frac{\gamma+1}{2}\right)u\right]\left(\frac{u}{a}\right)^{\frac{1}{n}} + x_0(u),$$

so that

$$x_0(u) = \frac{a}{n+1}\left(\frac{u}{a}\right)^{\frac{n+1}{n}} - \frac{c_0}{a^{1/n}}u^{\frac{1}{n}} - \frac{1}{2}\frac{(\gamma+1)}{a^{1/n}}u^{\frac{n+1}{n}}.$$

Substituting this back into the equation for $x(u, t)$ gives

$$x(u, t) = c_0 t + \frac{1}{2}(\gamma + 1)ut + \frac{1}{n+1}\frac{1}{a^{1/n}}u^{(n+1)/n} - \frac{c_0}{a^{1/n}}u^{1/n} - \frac{1}{2}\frac{(\gamma + 1)}{a^{1/n}}u^{(n+1)/n}.$$

Differentiating and setting the result equal to zero we have

$$\left(\frac{\partial x}{\partial u}\right)_t = \frac{1}{2}(\gamma + 1)t + \left(\frac{n+1}{n}\right)\frac{1}{n+1}\frac{1}{a^{1/n}}u^{1/n} - \frac{1}{n}\frac{c_0}{a^{1/n}}u^{\frac{1}{n}-1} - \frac{1}{2}\frac{(\gamma + 1)}{a^{1/n}}$$

$$\times \left(\frac{n+1}{n}\right)u^{1/n}$$

$$= 0.$$

Similarly, the second derivative equated to zero gives

$$\left(\frac{\partial^2 x}{\partial u^2}\right)_t = \left(\frac{n+1}{n}\right)\frac{1}{n+1}\frac{1}{a^{1/n}}\frac{1}{n}u^{\frac{1}{n}-1} - \frac{1}{n}\frac{c_0}{a^{1/n}}\left(\frac{1}{n}-1\right)u^{\frac{1}{n}-2} - \frac{1}{2}\frac{(\gamma + 1)}{a^{1/n}}\left(\frac{n+1}{n}\right)\frac{1}{n}u^{\frac{1}{n}-1} = 0.$$

By simplifying the latter equation we obtain

$$\frac{1}{n^2} - \frac{c_0}{n}\left(\frac{1}{n}-1\right)\frac{1}{u} - \frac{1}{2}(\gamma + 1)\left(\frac{n+1}{n^2}\right) = 0$$

and by multiplying across by n^2 we obtain

$$\frac{c_0}{u}(n - 1) = \frac{1}{2}(\gamma + 1)(n + 1) - 1, \tag{2.29}$$

so that

$$\frac{1}{u} = \frac{\left[\gamma\left(\frac{n+1}{n-1}\right) + 1\right]}{2c_0}. \tag{2.30}$$

The equation for $(\partial x/\partial u)_t = 0$ gives

$$\frac{1}{2}(\gamma + 1)t = \frac{1}{n}\frac{u^{1/n}}{a^{1/n}}\left[\frac{c_0}{u} + \frac{1}{2}(\gamma + 1)(n + 1) - 1\right]$$

and by using Eq. (2.29) this latter equation becomes

$$\frac{1}{2}(\gamma + 1)t = \frac{u^{1/n}}{a^{1/n}}\frac{c_0}{u} = \frac{c_0}{a^{1/n}}\frac{1}{u^{\frac{n-1}{n}}}.$$

However, using Eq. (2.30) in this latter equation it is straightforward to show that the time taken for the formation of the shock is given by

$$t_{shock} = \left(\frac{2c_0}{a}\right)^{\frac{1}{n}} \frac{1}{\gamma + 1} \left[\gamma\left(\frac{n+1}{n-1}\right) + 1\right]^{\frac{n-1}{n}}. \qquad (2.31)$$

The place of formation of the shock is obtained by substituting this in the equation for $x(t, u)$, yielding,

$$x_{shock} = 2c_0 \left(\frac{2c_0}{a}\right)^{\frac{1}{n}} \left[\frac{\gamma}{\gamma+1} + \frac{n-1}{n+1}\right] \frac{1}{(n-1)^{\frac{n-1}{n}}[\gamma - 1 + n(\gamma + 1)]^{\frac{1}{n}}}. \qquad (2.32)$$

In this case the shock wave is not formed at the forward front of the wave but at some intermediate point.

2.6 Another Form of the Equations: Riemann Invariants

Bernhard Riemann in 1860 developed a very powerful method for solving the one-dimensional isentropic fluid flow equations. From a physical point of view Riemann's method implies that an observer moving at the *local* particle velocity will find that the acoustic theory applies *locally* [12]. More generally, however, Riemann's method of solution is based on the *method of characteristics* which is a standard mathematical technique for solving partial differential equations. The method uses *characteristic curves* or simply *characteristics* along which the partial differential equation is transformed into a set of ordinary differential equations. Once the solution of the ordinary differential equations is obtained along the *characteristics* it can be transformed back to provide a solution of the partial differential equation. The technique has wide ranging applications not only in pure mathematics for solving partial differential equations but in such areas as traffic flow as well as fluid flow and a good starting point for the interested reader is the text by Whitham [13]. Only a very brief outline of Riemann's method is presented here and much greater details can be found in the many references cited at the end of the chapter.

The continuity equation is

$$\frac{\partial \rho}{\partial t} + \rho \frac{\partial u}{\partial x} + u \frac{\partial \rho}{\partial x} = 0,$$

and in the case of a perfect gas we have $c^2 = \gamma p/\rho$ and $p = k\rho^\gamma$ where k is a constant, hence, $c^2 = \gamma k \rho^{\gamma - 1}$. Taking logs of both sides of the latter equation yields

$$2 \log c = \log \gamma k + (\gamma - 1) \log \rho$$

therefore,

$$2\frac{\Delta c}{c} = (\gamma - 1)\frac{\Delta \rho}{\rho},$$

hence,

$$\frac{1}{\rho}\frac{\partial \rho}{\partial x} = \frac{2}{\gamma - 1}\frac{1}{c}\frac{\partial c}{\partial x}, \qquad (2.33)$$

similarly,

$$\frac{1}{\rho}\frac{\partial \rho}{\partial t} = \frac{2}{\gamma - 1}\frac{1}{c}\frac{\partial c}{\partial t}.$$

Substituting these two latter relationships in the continuity equation, we have

$$\frac{2}{\gamma - 1}\frac{\partial c}{\partial t} + c\frac{\partial u}{\partial x} + \frac{2u}{\gamma - 1}\frac{\partial c}{\partial x} = 0,$$

which is the continuity equation written in a different form. Let us now consider the momentum equation

$$\frac{\partial u}{\partial t} + u\frac{\partial u}{\partial x} + \frac{1}{\rho}\frac{\partial p}{\partial x} = 0,$$

and using the same substitutions as used for the continuity equation, namely, $c^2 = \gamma p/\rho$ and $p = k\rho^\gamma$, it is easy to verify that

$$\frac{\partial p}{\partial x} = c^2\frac{\partial \rho}{\partial x},$$

hence,

$$\frac{1}{\rho}\frac{\partial p}{\partial x} = \frac{c^2}{\rho}\frac{\partial \rho}{\partial x} = \frac{2c}{\gamma - 1}\frac{\partial c}{\partial x}$$

after using Eq. (2.33). With these substitutions the momentum equation becomes

$$\frac{\partial u}{\partial t} + u\frac{\partial u}{\partial x} + \frac{2c}{\gamma - 1}\frac{\partial c}{\partial x} = 0.$$

By adding the continuity equation to the momentum equation we obtain (after grouping various terms)

$$\frac{\partial}{\partial t}\left(u + \frac{2c}{\gamma - 1}\right) + (u + c)\frac{\partial}{\partial x}\left(u + \frac{2c}{\gamma - 1}\right) = 0, \qquad (2.34)$$

similarly, by subtracting the equations we obtain

$$\frac{\partial}{\partial t}\left(u - \frac{2c}{\gamma - 1}\right) + (u - c)\frac{\partial}{\partial x}\left(u - \frac{2c}{\gamma - 1}\right) = 0. \qquad (2.35)$$

Noting that u and c are functions of x and t, it follows that $u + 2c/(\gamma - 1)$ and $u - 2c/(\gamma - 1)$ are also functions of x and t, hence, Eqs. (2.34) and (2.35) can be written as

$$\frac{d}{dt}\left[u + \frac{2c}{\gamma - 1}\right] = \frac{\partial}{\partial t}\left[u + \frac{2c}{\gamma - 1}\right] + \frac{dx}{dt}\frac{\partial}{\partial x}\left[u + \frac{2c}{\gamma - 1}\right] = 0 \qquad (2.36)$$

and

$$\frac{d}{dt}\left[u - \frac{2c}{\gamma - 1}\right] = \frac{\partial}{\partial t}\left[u - \frac{2c}{\gamma - 1}\right] + \frac{dx}{dt}\frac{\partial}{\partial x}\left[u - \frac{2c}{\gamma - 1}\right] = 0. \qquad (2.37)$$

When the equations are written in this form, one can see that $u + 2c/(\gamma - 1)$ is constant on the curve defined by $dx/dt = u + c$ and $u - 2c/(\gamma - 1)$ is constant on the curve defined by $dx/dt = u - c$. That is, these conditions apply when we change our frame of reference so that one observes changes taking place in the fluid when our frame of reference moves with the *local* velocity of sound with respect to the moving fluid. These constants are called *Riemann invariants* [14–22]. and the curves on which the Riemann invariants are constant are called *characteristics*. We will be returning to this aspect of fluid flow after considering how the characteristic equations facilitate the solution of some simple partial differential equations.

2.6.1 Solution of some First-Order Partial Differential Equations

In this section we will solve examples of some first-order partial differential equations in order to illustrate the *method of characteristics*. Much greater detail can be found in texts [23–25] dealing specifically with the subject of partial differential equations. Initially, some linear equations will be considered and, thereafter, we will proceed to consider a *nonlinear* example that models nonlinear phenomena in gas dynamics where the solution breaks down after a finite time interval which leads to the formation of shock waves.

(a) Example 1

As an introduction to the method of characteristics, let us set out to solve the following simple first-order partial differential equation;

$$\frac{\partial f}{\partial t} + x\frac{\partial f}{\partial x} = 0 \tag{2.38}$$

with the initial condition;

$$f(x,0) = 2x.$$

From Eq. (2.38) we note that f is a function of x and t, hence, we write,

$$f = f(x,t)$$

and, therefore,

$$df = \frac{\partial f}{\partial t}dt + \frac{\partial f}{\partial x}dx,$$

accordingly,

$$\frac{df}{dt} = \frac{\partial f}{\partial t} + \frac{dx}{dt}\frac{\partial f}{\partial x}. \tag{2.39}$$

By comparing Eqs. (2.38) and (2.39) we have $df/dt = 0$ on $dx/dt = x$, hence,

$$f = \text{constant}$$

on the curve defined by the equation,

$$\frac{dx}{dt} = x. \tag{2.40}$$

Eq. (2.40) defines a family of curves in the x, t plane, called the *characteristics* of Eq. (2.38) and $f(x, t)$ is a constant along each curve (with a different constant in each case as shown below). Integrating Eq. (2.40) gives

$$\ln x = t + c,$$

where c is a constant of integration, hence,

$$t = \ln x - c. \tag{2.41}$$

Now, $f(x, t)$ is constant on this curve and the curve cuts the x-axis ($t = 0$) and "picks up" the value

$$f(x, 0) = 2x.$$

Suppose the characteristic cuts the x-axis at $x = 1$, then $c = 0$, $f(1, 0) = 2$ and the characteristic curve is $t = \ln x$. Consequently, $f = 2$ all along the characteristic $t = \ln x$ and this is illustrated in Fig. 2.9. Other sample characteristic curves are also shown that cut the x-axis at $x = 2, 3, 4$ with each one picking up a different value of f corresponding to the initial condition; $f(x, 0) = 2x$.

As each characteristic is given by $c = \ln x - t$, the function $f(x, t)$ can only have the form

$$f(x, t) = F(\ln x - t).$$

Using the initial condition, this latter equation gives

$$f(x, 0) = F(\ln x) = 2x$$

or

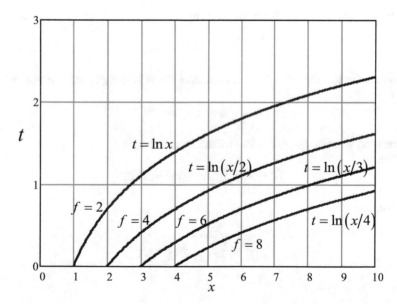

Fig. 2.9 Several characteristics for Eq. (2.38) are shown with constant values of f along each characteristic (see text)

$$F(x) = 2e^x,$$

hence, the solution of the partial differential equation is

$$f(x,t) = F(\ln x - t) = 2e^{\ln x - t} = 2xe^{-t}. \tag{2.42}$$

(b) Example 2

Let us now proceed to illustrate again the *method of characteristics* by taking a different partial differential equation but, in this case we will make a change of variables. Let us consider the equation

$$\frac{\partial f}{\partial t} + 3\frac{\partial f}{\partial x} + f = t; \quad t > 0 \tag{2.43}$$

with initial condition:

$$f(x,0) = Sin(x).$$

The characteristic equation is

$$dx/dt = 3$$

and the characteristics are the lines given by

$$x = 3t + \text{ constant}.$$

So in order to solve Eq. (2.43) let us make the following change of variables

$$\tau = t \text{ and } \zeta = x - 3t.$$

Then the function $f(x, t)$ can be written, in general, as

$$f(x,t) = f(x(\zeta,\tau), t(\tau)) = F(\zeta,\tau) \tag{2.44}$$

therefore,

$$\frac{\partial f}{\partial t} = \frac{\partial F}{\partial \zeta}\frac{\partial \zeta}{\partial t} + \frac{\partial F}{\partial \tau}\frac{\partial \tau}{\partial t}$$

and

$$\frac{\partial f}{\partial x} = \frac{\partial F}{\partial \zeta}\frac{\partial \zeta}{\partial x} + \frac{\partial F}{\partial \tau}\frac{\partial \tau}{\partial x}.$$

Hence,

$$\frac{\partial f}{\partial t} = -3\frac{\partial F}{\partial \zeta} + \frac{\partial F}{\partial \tau} \quad \text{and} \quad \frac{\partial f}{\partial x} = \frac{\partial F}{\partial \zeta}.$$

Substituting these relationships in Eq. (2.43) gives

$$-3\frac{\partial F}{\partial \zeta} + \frac{\partial F}{\partial \tau} + 3\frac{\partial F}{\partial \zeta} + F = \tau,$$

hence,

$$\frac{\partial F}{\partial \tau} + F = \tau.$$

This latter equation can be written as

$$\frac{\partial}{\partial \tau}(e^\tau F) = \tau e^\tau$$

and integrating we obtain

$$e^\tau F(\zeta, \tau) = \int \tau e^\tau d\tau + G(\zeta)$$

$$= \tau e^\tau - e^\tau + G(\zeta),$$

where the integration is carried out by parts and $G(\zeta)$ is the constant of integration which is an arbitrary function of ζ since we are integrating with respect to τ and F is just a function of ζ and τ. Hence,

$$F(\zeta, \tau) = (\tau - 1) + G(\zeta)e^{-\tau}$$

and writing this latter equation in terms of x and t gives

$$f(x, t) = (t - 1) + G(x - 3t)e^{-t}. \tag{2.45}$$

Using the initial condition, $f(x, 0) = Sin(x)$, in Eq. (2.45) gives,

$$f(x, 0) = -1 + G(x) = Sin(x),$$

therefore,

$$G(x - 3t) = 1 + Sin(x - 3t).$$

Substituting this expression for $G(x - 3t)$ in Eq. (2.45) gives

$$f(x,t) = t - 1 + e^{-t}[1 + Sin(x - 3t)] \tag{2.46}$$

as the solution of Eq. (2.43). One can see that it satisfies the initial condition and it is easy to verify that it satisfies the original partial differential equation by direct substitution.

2.6.2   Nonlinear Equation

In this section a brief introduction to nonlinear partial differential equations is presented. We will write the equations in terms of a variable u rather than a general variable f in recognition of the fact that some familiar physical property of the medium, such as, pressure, particle velocity or density, is being investigated.

Before embarking on nonlinear equations let us consider the simple first-order linear wave equation,

$$\frac{\partial u}{\partial t} + c_0 \frac{\partial u}{\partial x} = 0 \ \text{ with } \ u(x,0) = F(x), \tag{2.47}$$

and c_0 is a constant. This equation describes a wave moving in one direction with wave speed c_0. The characteristic equation is

$$\frac{dx}{dt} = c_0$$

and one can see that u=constant along characteristics defined by the latter equation. The solution of the characteristic equation is

$$x = c_0 t + x_0,$$

where x_0 is a constant of integration or the intercept on the x-axis at $t = 0$ as shown in Fig. 2.10 and all characteristics have the same slope.

For the characteristic that cuts the x-axis at $x = x_0$, we have from the initial condition,

$$u(x_0, 0) = F(x_0)$$

and this implies that the linear wave equation has the solution,

$$u(x,t) = F(x - c_0 t)$$

Fig. 2.10 Characteristics in the xt-plane for the linear wave Eq. (2.47) are shown (see text)

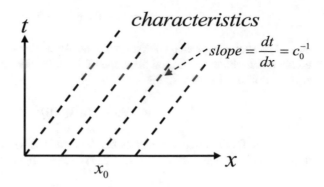

Fig. 2.11 Unchanging waveform for linear wave propagation at $t = 0$ and at $t = t'$

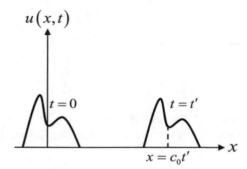

which represents a wave propagating at constant speed c_0 without any change in shape as illustrated in Fig. 2.11.

Let us now consider the following nonlinear partial differential equation,

$$\frac{\partial u}{\partial t} + c(u)\frac{\partial u}{\partial x} = 0. \tag{2.48}$$

Equation (2.48) has a wave velocity $c(u)$ that is not a constant but rather depends on the amplitude u of the disturbance itself. This implies that disturbances of larger amplitude move faster than their lower counterparts and eventually overtake them to produce distortion of the wave profile as it propagates; this has implications for the characteristics. We will investigate this aspect in due course and, unlike the examples just previously discussed, we will see that the characteristic lines intersect.

One can see from Eq. (2.48) that u is constant on the characteristic,

$$\frac{dx}{dt} = c(u)$$

and it follows that the slope $c(u)$ is also constant on the characteristic; hence, the characteristic is a straight line in the xt-plane. If we take the following initial conditions,

$$u = u(x, 0) \text{ at } t = 0,$$

and let us suppose the characteristic cuts the x-axis at $x = x_0$. Then $u(x, t) = u(x_0, 0)$ all along this characteristic and the slope of the characteristic is $c(u(x_0, 0))$, so that the equation of the characteristic is

$$x = x_0 + c(u(x_0, 0))t.$$

A whole family of characteristics can be obtained by considering x_0 as a continuous variable. Taking derivatives of $u(x, t) = u(x_0, 0)$, we have

$$\frac{\partial u}{\partial x} = \frac{\partial u}{\partial x_0} \frac{\partial x_0}{\partial x} = u'(x_0, 0) \frac{\partial x_0}{\partial x} \tag{2.49}$$

and

$$\frac{\partial u}{\partial t} = u'(x_0, 0) \frac{\partial x_0}{\partial t}, \tag{2.50}$$

where u' denotes differentiation with respect to x_0. However, along the characteristic curve we have; $x = x_0 + c(u(x_0, 0))t$, hence, differentiating with respect to x gives,

$$1 = \frac{\partial x_0}{\partial x} + c'(u(x_0, 0))t \frac{\partial x_0}{\partial x},$$

yielding,

$$\frac{\partial x_0}{\partial x} = \frac{1}{1 + c'(u(x_0, 0))t}, \tag{2.51}$$

where $c'(u(x_0, 0)) = \frac{d}{dx_0} c(u(x_0, 0))$. Similarly, differentiating $x = x_0 + c(u(x_0, 0))t$ with respect to t gives

$$\frac{\partial x_0}{\partial t} = -\frac{c(u(x_0, 0))}{1 + c'(u(x_0, 0))t} \tag{2.52}$$

Substituting Eqs. (2.51) and (2.52) in Eqs. (2.49) and (2.50) gives

$$\frac{\partial u}{\partial x} = \frac{u'(x_0, 0)}{1 + c'(u(x_0, 0))t} \tag{2.53}$$

and

$$\frac{\partial u}{\partial t} = -\frac{u'(x_0,0)c(u(x_0,0))}{1+c'(u(x_0,0))t}.$$ (2.54)

Hence,

$$\frac{\partial u}{\partial t} + c(u)\frac{\partial u}{\partial x} = -\frac{u'(x_0,0)c(u(x_0,0))}{1+c'(u(x_0,0))t} + c(u)\frac{u'(x_0,0)}{1+c'(u(x_0,0))t} = 0;$$

which illustrates that the initial condition, namely, $c(u) = c(u(x_0,0))$, satisfies the original equation.

2.6.3   An Example of Nonlinear Distortion

Let us now proceed to investigate the nonlinear distortion referred to previously by considering the following equation,

$$\frac{\partial u}{\partial t} + u\frac{\partial u}{\partial x} = 0$$ (2.55)

with the initial condition given by

$$u(x,0) = e^{-x^2}.$$

Equation (2.55) looks similar to

$$\frac{\partial u}{\partial t} + u\frac{\partial u}{\partial x} = -\frac{1}{\rho}\frac{\partial p}{\partial x},$$

which is Eq. (1.46) of Chap. 1 with the pressure variation removed. It, nonetheless, retains the nonlinear term, $u(\partial u/\partial x)$, and accounts for the nonlinear distortion that features in its solution. Eq. (2.55) is known as the inviscid form of Burger's equation.

For Eq. (2.55) we have; $du/dt = 0$ on the characteristic given by $dx/dt = u$. Hence, $u(x,t)$ is constant on the characteristic so that $dx/dt = u$ is a straight line in the xt-plane. When the characteristic cuts the x-axis ($t = 0$) it picks up the value $u(x,0)$, as a result, the equation for the characteristic line that cuts the x-axis at x_0 is

$$x = u(x_0,0)t + x_0$$

$$= e^{-x_0^2}t + x_0,$$ (2.56)

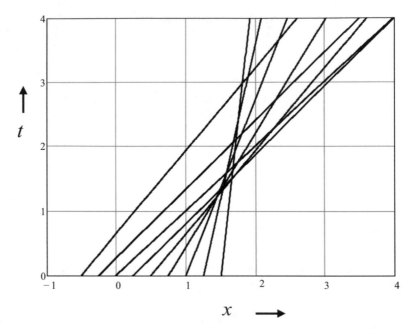

Fig. 2.12 Characteristics for Eq. (2.55) with initial condition given by $u(x,0) = e^{-x^2}$ (see text)

Therefore,

$$u(x,t) = e^{-x_0^2} \quad \text{on the line} \quad t = e^{x_0^2}(x - x_0).$$

Some sample characteristic lines are shown in Fig. 2.12 starting at $x_0 = -0.5$ and ending at $x_0 = 1.5$ with increments of 0.25. The important feature to note here is that the characteristic lines intersect. However, the slope of each characteristic line is equal to the value of u on that line and, therefore, a line with a different slope must have a different value of u. The fact that the lines intersect at some (x,t) implies that the solution is required to have two different values. As quantities like $u(x,t)$ are supposed to represent physical quantities like pressure, particle velocity or density and can only have a unique value for some (x,t), the acceptance of multi-valued solution is impossible and does not represent physical reality.

Plots of u versus x are shown for different times are shown in Fig. 2.13. These are obtained by plotting $x(x_0) = x_0 + e^{-x_0^2}t$ versus $u = e^{-x_0^2}$ for the times indicated. One can see the initial profile propagating to the right and the nonlinear behaviour is evident as larger value of u propagate faster than lower values. The profile becomes increasingly distorted in the forward direction and eventually acquiring a multi-valued solution as can be clearly seen at $t = 2$. The onset of this multi-valued solution occurs as u develops a vertical profile, that is, $\partial u / \partial x \to \infty$ at a specific time close to $t = 1$ as can be observed in Fig. 2.13. One can determine the time that the

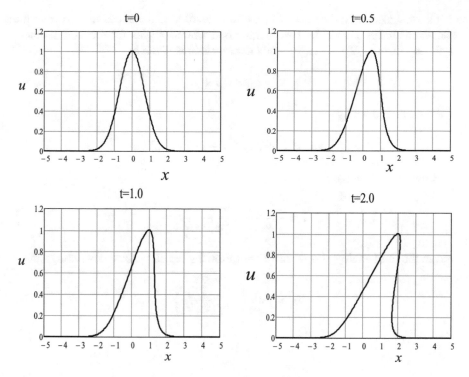

Fig. 2.13 Plots of u versus x for Eq. (2.55) with initial condition given by $u(x,0) = e^{-x^2}$. Note the change in profile at different times and the development of a multi-valued solution (see text)

profile develops an infinite slope by returning to some previous equations, that is, Eqs. (2.53) and (2.54), repeated below;

$$\frac{\partial u}{\partial x} = \frac{u'(x_0, 0)}{1 + c'(u(x_0, 0))t}$$

and

$$\frac{\partial u}{\partial t} = -\frac{u'(x_0, 0)c(u(x_0, 0))}{1 + c'(u(x_0, 0))t}.$$

The slopes, $\partial u/\partial t$ and $\partial u/\partial x$ become infinite when the denominator in the above equations goes to zero. The earliest time that this occurs is called the *breaking time* t_B and it occurs when $c'(u(x_0, 0))$ has the largest negative value, hence,

$$t_B = \min\left\{-\frac{1}{c'(u(x_0, 0))}\right\}. \tag{2.57}$$

One can also reach the same conclusion in relation to the breaking time from the characteristic lines: consider two neighboring characteristics for the example here with slope, $c(u(x_0, 0)) = u(x_0, 0)$; one characteristic is given by

$$x = u(x_0, 0)t + x_0 \tag{2.58}$$

and the other given by

$$x = u(x_0 + dx_0, 0)t + (x_0 + dx_0). \tag{2.59}$$

Solving for t we have

$$t = -\frac{dx_0}{u(x_0 + dx_0) - u(x_0)}$$

and in the limit as $dx_0 \to 0$ we have the following equation for the breaking time,

$$t_B = \min\left\{-\frac{1}{u'(x_0, 0)}\right\}.$$

2.6.4 The Breaking Time

Let us now determine the breaking time for the example here; $c(u(x_0, 0))$ is given by

$$c(u(x_0, 0)) = e^{-x_0^2},$$

hence,

$$c'(u(x_0, 0)) = -2x_0 e^{-x_0^2}. \tag{2.60}$$

We now wish to determine the maximum value of this function; differentiating and equating the result to zero gives

$$\frac{d}{dx_0}[c'(u(x_0, 0))] = -2\left(-2x_0^2 e^{-x_0^2} + e^{-x_0^2}\right) = 0.$$

hence,

$$x_0 = \frac{1}{\sqrt{2}}$$

and substituting this back in Eq. (2.60) gives

$$c'(u(x_0,0)) = -\frac{2}{\sqrt{2}}e^{-\left(\frac{1}{\sqrt{2}}\right)^2} = -0.858$$

so that

$$t_B = 1.165.$$

Accordingly, the earliest time for the solution to become multi-valued is 1.165 and the profile develops a vertical slope at $x = \sqrt{2}$.

2.7   Application of Riemann Invariants to Simple Flow Problems

Returning to the Riemann invariants we found that $u + 2c/(\gamma - 1)$ is a constant on the positive characteristic C_+ given by $dx/dt = u + c$ and $u - 2c/(\gamma - 1)$ is constant on the negative characteristic C_- given by $dx/dt = u - c$. Let

$$R_+ = u + \frac{2c}{\gamma - 1} \tag{2.61}$$

and

$$R_- = u - \frac{2c}{\gamma - 1}. \tag{2.62}$$

Solving these equations for u and c yields,

$$u = \frac{R_+ + R_-}{2} \quad \text{and} \quad c = \frac{(\gamma - 1)}{4}(R_+ - R_-).$$

Hence, on C_+ we obtain,

$$\frac{dx}{dt} = \left(\frac{\gamma + 1}{4}\right)R_+ + \left(\frac{3 - \gamma}{4}\right)R_- \tag{2.63}$$

and on C_- we obtain,

$$\frac{dx}{dt} = \left(\frac{3 - \gamma}{4}\right)R_+ + \left(\frac{\gamma + 1}{4}\right)R_-. \tag{2.64}$$

However, on C_+ we know that R_+ is constant, so that the slope of the positive characteristic in the xt-plane also depends on R_-. Similarly, on C_- we know that R_- is constant, so that the slope of the negative characteristic in the xt-plane also depends on R_+.

Let us now turn our attention to the use of these relationships to determine the flow in a few simple cases involving piston motion. The first example deals with expansion rather than compression waves.

2.7.1   Piston Withdrawal

Consider a long tube with a tight fitting piston located at $x = 0$ as shown in Fig. 2.14, at $t = 0$ the piston begins to move to the left at speed $u_p(t)$ and the path taken by the piston is also shown. The air is initially at rest to the right of the piston and it exhibits the normal ambient sound speed c_0. Let us now consider the region in the xt-plane that is penetrated by both positive and negative characteristics that cut the x-axis at $t = 0$. On a positive characteristic $R_+ = u + 2c/(\gamma - 1)$ is a constant and as this characteristic intersects the x-axis R_+ "picks up" the value of R_+ at $t = 0$, hence,

$$u + \frac{2c}{\gamma - 1} = \frac{2c_0}{\gamma - 1} \tag{2.65}$$

Similarly, for the negative characteristic we have

$$u - \frac{2c}{\gamma - 1} = -\frac{2c_0}{\gamma - 1}. \tag{2.66}$$

Adding and subtracting Eqs. (2.65) and (2.66) implies that $u = 0$ and $c = c_0$. Hence, the positive and negative characteristics are given by

Fig. 2.14 Piston withdrawal and piston path in xt-plane is shown

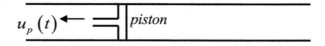

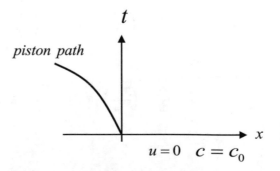

$$\frac{dx}{dt} = c_0 \text{ and } \frac{dx}{dt} = -c_0, \text{ respectively.}$$

These characteristics are straight lines, hence,

$$x = c_0 t + x0_+ \text{ and } x = -c_0 t + x0_-$$

where $x0_+$ and $x0_-$ are the intercepts on the x-axis for the C_+ and C_- characteristics, respectively. The region $\Re 1$ traversed by these characteristics (shown as dashed lines) is classified as a uniform region where $u = 0$ and $c = c_0$, and the region is bounded by the characteristic $x = c_0 t$ as shown in Fig. 2.15.

Let us now consider region $\Re 2$. In this region the positive characteristics start at the piston surface while the negative characteristics that start at $t = 0$ penetrate into this region, hence, Eq. (2.66) still applies to this region, that is,

$$u - \frac{2c}{\gamma - 1} = -\frac{2c_0}{\gamma - 1}$$

and solving this equation for c gives

$$c = c_0 + \left(\frac{\gamma - 1}{2}\right)u. \qquad (2.67)$$

However, $u + 2c/(\gamma - 1)$ is a constant on the positive characteristic and the constant it "picks up" is dependent on where the characteristic intersects the piston surface; call this constant K_s, then on C_+ we have

$$u + \frac{2c}{\gamma - 1} = K_s$$

and using Eq. (2.67) in this latter equation gives

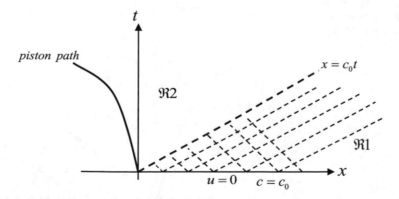

Fig. 2.15 Piston withdrawal showing the characteristics (broken lines) in the uniform region (see text)

$$u = \frac{K_s}{2} - \frac{c_0}{\gamma - 1}. \tag{2.68}$$

Hence, we deduce that both u and c are constant on the positive characteristic and these characteristics are straight lines with slopes dependent on the particular value of K_s. Typical characteristics in region $\mathfrak{R}2$ are sketched in Fig. 2.16.

Suppose one of these positive characteristics emanates from the piston surface at $t = t_0$, then the characteristics intersects the piston surface at $x_p(t_0)$ and, accordingly, the equation of this characteristic is

$$x(t) = x_p(t_0) + S(t_0)(t - t_0) \tag{2.69}$$

since it is a straight line and where the slope $S(t_0)$ is clearly dependent on t_0 as previously indicated. However, the slope, in general, is given by $u + c$ and using Eq. (2.67), the slope becomes

$$S(t_0) = c_0 + \left(\frac{\gamma + 1}{2}\right) u_p(t_0)$$

where

$$u_p(t_0) = \left(\frac{dx_p(t)}{dt}\right)_{t=t_0}.$$

Accordingly, Eq. (2.69) becomes

$$x(t) = x_p(t_0) + \left[c_0 + \left(\frac{\gamma + 1}{2}\right) u_p(t_0)\right](t - t_0). \tag{2.70}$$

Fig. 2.16 The positive characteristics are shown emanating from the piston surface (see text)

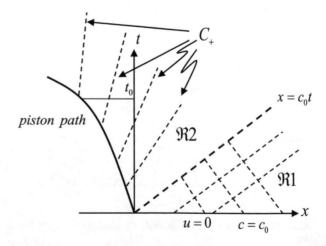

2.7.2 Piston Withdrawal at Constant Speed

Let us now apply the above considerations to the case where the piston is withdrawn at a constant speed u_0, hence, $x_p(t_0) = -u_0 t_0$ and $u_p(t_0) = -u_0$ so that Eq. (2.70) becomes

$$x(t) = -u_0 t_0 + \left[c_0 - \left(\frac{\gamma + 1}{2} \right) u_0 \right] (t - t_0), \qquad (2.71)$$

which applies to all positive characteristics leaving the piston and all of these characteristics have

$$u = -u_0 \quad \text{and} \quad c = c_0 - \left(\frac{\gamma - 1}{2} \right) u_0 \qquad (2.72)$$

with identical slopes of

$$c_0 - \left(\frac{\gamma + 1}{2} \right) u_0.$$

The first characteristic leaves the piston at $t_0 = 0$ and Eq. (2.71) gives

$$x(t) = \left[c_0 - \left(\frac{\gamma + 1}{2} \right) u_0 \right] t \qquad (2.73)$$

and the region $\mathfrak{R}3$ to the left of this characteristic as shown in Fig. 2.17 is a uniform region according to Eq. (2.72) and the negative characteristics are also straight lines.

One can observe from Fig. 2.17 that there is a region $\mathfrak{R}2$ with no positive characteristics as Eq. (2.71) only applies to positive characteristics leaving the piston; however, negative characteristics penetrate this region as Eq. (2.66) still

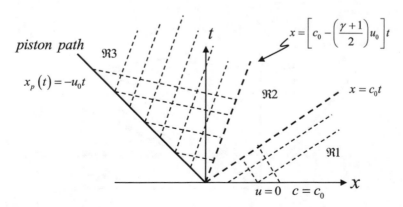

Fig. 2.17 Characteristics are sketched in the case where the piston is withdrawn at constant speed (see text)

applies. The only solution is to have the positive characteristics emanating from the origin so that they do not intersect the characteristics, $x = c_0 t$ and $x = [c_0 - ((\gamma + 1)/2)u_0]t$. Consequently, these characteristics are given by

$$\frac{x}{t} = u + c$$

and as $c = c_0 + [(\gamma - 1)/2]u$ from Eq. (2.67), we can solve for u and c, giving,

$$u = \frac{2}{\gamma + 1}\left[\frac{x}{t} - c_0\right] \tag{2.74}$$

and

$$c = \left(\frac{\gamma - 1}{\gamma + 1}\right)\frac{x}{t} + \frac{2c_0}{\gamma + 1}. \tag{2.75}$$

The region $\mathfrak{R}2$ is called an *expansion fan* or *centred expansion* and the characteristics are sketched in Fig. 2.18.

An infinite number of straight line characteristics can be drawn starting at $x = c_0 t$ which is the *head* of the expansion fan and ending at $x = [c_0 - ((\gamma + 1)/2)u_0]t$ which is the *tail* of the expansion fan.

As regards the negative characteristics in this region, we note that they are determined by the equation

$$\frac{dx}{dt} = u - c.$$

Substituting for u and c from Eqs. (2.74) and (2.75) gives

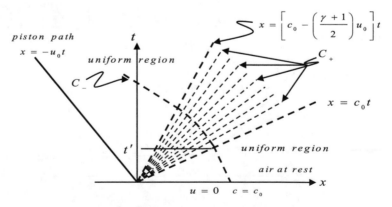

Fig. 2.18 A piston withdrawn at constant speed generates an expansion fan with the positive characteristics emanating from the origin

$$\frac{dx}{dt} = \left(\frac{3-\gamma}{1+\gamma}\right)\frac{x}{t} - \frac{4c_0}{1+\gamma} \qquad (2.76)$$

which determines the negative characteristics. In order to integrate this latter equation we let $A = (3 - \gamma)/(1 + \gamma)$ and $B = 4c_0/(1 + \gamma)$, hence,

$$\frac{dx}{dt} = A\frac{x}{t} - B \qquad (2.77)$$

and making the following change in variable; $z = x/t$, it is easy to show that

$$\frac{dx}{dt} = t\frac{dz}{dt} + z.$$

Substituting this result in Eq. (2.77) gives

$$t\frac{dz}{dt} = (A - 1)z - B,$$

hence,

$$\frac{dz}{(A - 1)z - B} = \frac{dt}{t} \qquad (2.78)$$

and by carrying out the integration we have

$$\ln\left[(A - 1)z - B\right] = (A - 1)\ln t + \text{ constant}.$$

Substituting back for z, yields,

$$\ln\left[(A - 1)x - Bt\right] = A \ln t + \text{ constant}. \qquad (2.79)$$

The constant of integration is obtained by noting that the negative characteristic cuts the positive characteristic, $x = c_0 t$, which is the boundary between the centred expansion region and the air at rest. Suppose the negative characteristic crosses $x = c_0 t$ at $t = t'$ as shown in Fig. 2.18, then the following equation is obtained;

$$\ln\left[(A - 1)c_0 t' - Bt'\right] = A \ln t' + \text{ constant},$$

which determines the constant of integration and substituting back, we obtain

$$\ln\left[\frac{(A - 1)x - Bt}{(A - 1)c_0 t' - Bt'}\right] = \ln\left(\frac{t}{t'}\right)^{\frac{3-\gamma}{1+\gamma}}. \qquad (2.80)$$

Finally, tidying up and substituting back the values for A and B, we obtain the following equation for the family of negative characteristics,

$$x(t) = -\frac{2c_0 t}{\gamma - 1} + \left(\frac{\gamma + 1}{\gamma - 1}\right) c_0 t' \left(\frac{t}{t'}\right)^{\frac{3-\gamma}{1+\gamma}}. \tag{2.81}$$

We can obtain expressions for the pressure, density and temperature in the expansion fan by noting that isentropic conditions apply throughout. The isentropic relation,

$$\frac{p}{\rho^\gamma} = \text{ constant}$$

and this taken in conjunction with the equations, $p = \rho R T$ and $c \propto \sqrt{T}$, it follows that

$$\frac{T}{T_0} = \left(\frac{c}{c_0}\right)^2, \frac{p}{p_0} = \left(\frac{c}{c_0}\right)^{\frac{2\gamma}{\gamma-1}}, \frac{\rho}{\rho_0} = \left(\frac{c}{c_0}\right)^{\frac{2}{\gamma-1}}.$$

Taking Eq. (2.75) and dividing by c_0 gives

$$\frac{c}{c_0} = \left(\frac{\gamma - 1}{\gamma + 1}\right) \frac{x}{c_0 t} + \frac{2}{\gamma + 1},$$

hence,

$$\frac{T}{T_0} = \left[\left(\frac{\gamma - 1}{\gamma + 1}\right) \frac{x}{c_0 t} + \frac{2}{\gamma + 1}\right]^2, \tag{2.82}$$

$$\frac{p}{p_0} = \left[\left(\frac{\gamma - 1}{\gamma + 1}\right) \frac{x}{c_0 t} + \frac{2}{\gamma + 1}\right]^{\frac{2\gamma}{\gamma-1}}. \tag{2.83}$$

$$\frac{\rho}{\rho_0} = \left[\left(\frac{\gamma - 1}{\gamma + 1}\right) \frac{x}{c_0 t} + \frac{2}{\gamma + 1}\right]^{\frac{2}{\gamma-1}}. \tag{2.84}$$

(see Appendix A for more details)

2.7.3   Piston Moving into a Tube

Suppose, on the other hand, the piston moves into the tube with uniform acceleration a, such that, its velocity is $u_p(t) = at$ and its displacement at time t is given by $x_p(t) = at^2/2$ as shown in Fig. 2.19.

By assuming that the air is at rest in the tube at $t = 0$ we can easy verify, as in the case where we considered the piston being withdrawn, that we obtain a uniform region bounded by the characteristic $x = c_0 t$ as shown in Fig. 2.20. The characteristics are shown as broken lines.

Taking a point A in the air adjacent to the piston as shown in Fig. 2.20 and let us consider a negative characteristic from the uniform region to A; the Riemann invariant for this negative characteristic is

$$u_A - \frac{2c_A}{\gamma - 1} = -\frac{2c_0}{\gamma - 1}$$

and solving for c_A gives

Fig. 2.19 Piston moving into a tube with uniform acceleration and the associated plot of the piston path on the xt-plane

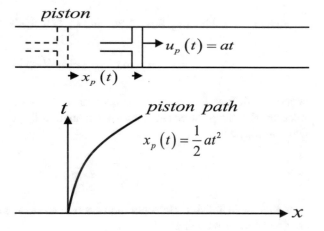

Fig. 2.20 Characteristics are sketched for the uniform region when the piston moves into the tube (see text)

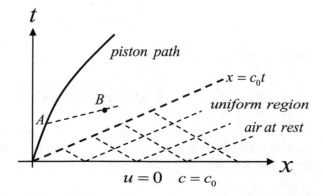

$$c_A = c_0 + \frac{(\gamma - 1)}{2} u_A. \tag{2.85}$$

Now consider a point B on the same positive characteristic emanating from point A, the Riemann invariants gives

$$u_B + \frac{2c_B}{\gamma - 1} = u_A + \frac{2c_A}{\gamma - 1},$$

and substituting Eq. (2.85) in this latter equation gives

$$u_B + \frac{2c_B}{\gamma - 1} = 2u_A + \frac{2c_0}{\gamma - 1}. \tag{2.86}$$

A negative characteristic from the uniform region and passing through point B gives the following Riemann invariant

$$u_B - \frac{2c_B}{\gamma - 1} = -\frac{2c_0}{\gamma - 1}. \tag{2.87}$$

Adding and subtracting Eqs. (2.86) and (2.87), yields

$$u_B = u_A \text{ and } c_B = c_0 + \frac{(\gamma - 1)}{2} u_A,$$

hence, we deduce that $c_B = c_A$ by comparing the latter equation with Eq. (2.85). Accordingly, the positive characteristic emanating from the piston at A is a straight line with slope

$$\frac{dx}{dt} = u_A + c_A. \tag{2.88}$$

If this characteristic intercepts the piston surface at, say, $t = t_0$, then this latter equation can be written as

$$\frac{dx}{dt} = u_p(t_0) + c_p(t_0), \tag{2.89}$$

where

$$c_p(t_0) = c_0 + \frac{(\gamma - 1)}{2} u_p(t_0) \tag{2.90}$$

is the speed of sound in the air at the piston surface when the piston is moving with velocity $u_p(t_0)$. Since $u_p(t_0) > 0$ and $c_p(t_0) > c_0$ the disturbance $u_p(t_0) + c_p(t_0)$ travels faster than the initial disturbance given by $(dx/dt) = c_0$, and as the slope of the

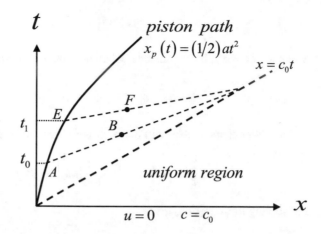

Fig. 2.21 Positive characteristics are shown converging to form a shock

positive characteristic *AB* as depicted in Fig. 2.20 appears as the reciprocal of *dx/dt* it converges towards the characteristic $x = c_0 t$.

If we consider points *E* and *F* on another positive characteristic leaving the piston's surface at, say, $t = t_1$ as shown in Fig. 2.21, we can draw similar conclusions to those drawn for the characteristic *AB*; namely, this characteristic is also a straight line with slope

$$\frac{dx}{dt} = u_p(t_1) + c_p(t_1) \tag{2.91}$$

Since $u_p(t_1) > u_p(t_0)$ and $c_p(t_1) > c_p(t_0)$, it follows that the characteristic *EF* has a larger slope than the characteristic *AB*, but in terms of the *tx*-plots shown in Fig. 2.21 the reverse applies. Consequently, each succeeding compression wave will travel faster than is predecessors and they eventually catch up with each other. When this occurs the characteristics coalesce to produce a shock wave moving at a speed greater than c_0 and in these circumstances the isentropic conditions break down.

In order to determine when the positive characteristics intersect let us consider one of these characteristics starting at the piston surface at time t_0. Since the characteristic is a straight line its equation is

$$x(t) = x_p(t_0) + S(t_0)(t - t_0) \tag{2.92}$$

where

$$S(t_0) = u_p(t_0) + c_p(t_0)$$

$$= at_0 + \left[c_0 + \left(\frac{\gamma - 1}{2} \right) at_0 \right]$$

$$= c_0 + \left(\frac{\gamma + 1}{2}\right) a t_0. \tag{2.93}$$

where $u_p(t_0) = a t_0$, hence, the characteristic equation becomes

$$x(t) = \frac{1}{2} a t_0^2 + \left[c_0 + \left(\frac{\gamma + 1}{2}\right) a t_0\right](t - t_0). \tag{2.94}$$

This latter characteristic will intersect with the characteristic $x = c_0 t$ when

$$\frac{1}{2} a t_0^2 + \left[c_0 + \left(\frac{\gamma + 1}{2}\right) a t_0\right](t - t_0) = c_0 t \tag{2.95}$$

and solving this equation, gives

$$\left(\frac{\gamma + 1}{2}\right) a t - c_0 - \frac{\gamma}{2} a t_0 = 0,$$

hence, the characteristics meet when

$$t = \frac{c_0 + \frac{\gamma}{2} a t_0}{\left(\frac{\gamma + 1}{2}\right) a} \tag{2.96}$$

and the earliest time for the shock wave to form is at $t_0 = 0$, hence,

$$t_{shock} = \frac{2 c_0}{(\gamma + 1) a}, \tag{2.97}$$

which is in agreement with Eq. (2.26).

2.7.4   Numerically Integrating the Equations of Motion Based Riemann's Method

Before we conclude Sect. 2.7 on Riemann invariants we will take a brief look at how the method of characteristics can be used to numerically integrate the equations of motion. It is important to bear in-mind, however, that the characteristic relationships only apply if the flow is isentropic. In relation to Fig. 2.22, suppose that the particle velocity u_i and speed of sound c_i are known at a set of points x_i; $i = 1, 2, 3. \ldots .n$ at $t = 0$. Then the various Riemann invariants at $t = 0$ are given by

$$R_+^{(i)} = u_i + \frac{2 c_i}{\gamma - 1} \tag{2.98}$$

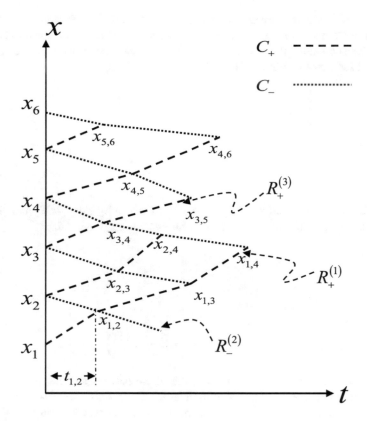

Fig. 2.22 Mesh structure for integrating along characteristics (see text)

and

$$R_-^{(i)} = u_i - \frac{2c_i}{\gamma - 1},\tag{2.99}$$

where $u_i = u(x_i, 0)$ and $c_i = c(x_i, 0)$ for $i = 1, 2, 3. \ldots .n$ and all Riemann invariants are known at $t = 0$.

The numerical procedure is as follows: for each point x_i starting at $t = 0$ two characteristic lines are drawn; one with slope,

$$\left(\frac{dx}{dt}\right)_{C_+} = u_i + c_i$$

and the other with slope,

$$\left(\frac{dx}{dt}\right)_{C_-} = u_i - c_i.$$

The positive characteristic emanating from x_i and the negative characteristic emanating from x_{i+1} intersect at the point (x_i, x_{i+1}). Since the Riemann invariants remain constant along the characteristics we can determine the quantities $u_{i,i+1}$ and $c_{i,i+1}$ at the first intersection points from the equations,

$$u_{i,i+1} + \frac{2c_{i,i+1}}{\gamma - 1} = R_+^{(i)} \tag{2.100}$$

$$u_{i,i+1} - \frac{2c_{i,i+1}}{\gamma - 1} = R_-^{(i+1)}. \tag{2.101}$$

where the Riemann invariants $R_+^{(i)}$ and $R_+^{(i+1)}$ are given by Eqs. (2.98) and (2.99).

By adding and subtracting these latter two equations, we obtain,

$$u_{i,i+1} = \frac{1}{2}\left(R_+^{(i)} + R_-^{(i+1)}\right) \tag{2.102}$$

and

$$c_{i,i+1} = \frac{(\gamma - 1)}{4}\left(R_+^{(i)} - R_-^{(i+1)}\right), \tag{2.103}$$

hence, $u_{i,i+1}$ and $c_{i,i+1}$ are now known at these first intersection points. Having determined the values of u and c at the intersection points, we must now find the location of these points and we know, in general, that the characteristics are not straight lines; hence, an approximate method must be applied. A good approximation[1] for the coordinates of the intersection points can be obtained by averaging the slope of the positive characteristic at $(x_i, 0)$ and $(x_{i, i+1}, t_{i, i+1})$, and, similarly by averaging the slope of the negative characteristic at $(x_{i+1}, 0)$ and $(x_{i, i+1}, t_{i, i+1})$. Hence, the average slope of the positive characteristic is

$$\left\langle\frac{dx}{dt}\right\rangle_{C_+} = \frac{1}{2}\left[(u_i + c_i) + (u_{i,i+1} + c_{i,i+1})\right] \tag{2.104}$$

and the average slope of the negative characteristic is

$$\left\langle\frac{dx}{dt}\right\rangle_{C_-} = \frac{1}{2}\left[(u_{i+1} - c_{i+1}) + (u_{i,i+1} - c_{i,i+1})\right], \tag{2.105}$$

hence, we can write

---

[1]The approximation will improve if the spacing between the initial set of data points is reduced.

$$\frac{x_{i,i+1} - x_i}{t_{i,i+1}} = \frac{1}{2}[(u_i + c_i) + (u_{i,i+1} + c_{i,i+1})] \tag{2.106}$$

and

$$\frac{x_{i,i+1} - x_{i+1}}{t_{i,i+1}} = \frac{1}{2}[(u_{i+1} - c_{i+1}) + (u_{i,i+1} - c_{i,i+1})]. \tag{2.107}$$

These latter two equations can be solved to give the coordinates $(x_{i, i+1}, t_{i, i+1})$ at the points of first intersection. It should be noted from Fig. 2.22 that, in general, these points of intersection occur at irregular intervals in both space and time. Apart from the new values of u and c at the intersection points, one can determine the other variables characterising the flow by using the isentropic equation for a perfect gas according to Eq. (2.21); hence, the pressure, density and temperature at the first intersection points are given by the following equations,

$$\frac{p_{i,i+1}}{p_i} = \left(\frac{c_{i,i+1}}{c_i}\right)^{\frac{2\gamma}{\gamma-1}},$$

$$\frac{\rho_{i,i+1}}{\rho_i} = \left(\frac{c_{i,i+1}}{c_i}\right)^{\frac{2}{\gamma-1}}$$

and

$$\frac{T_{i,i+1}}{T_i} = \left(\frac{c_{i,i+1}}{c_i}\right)^{2}.$$

Knowing the new values of u and c one can draw two other characteristics starting at these intersection points, one with slope

$$\frac{dx}{dt} = u_{i,i+1} + c_{i,i+1}$$

and the other with slope

$$\frac{dx}{dt} = u_{i,i+1} - c_{i,i+1}$$

and we can proceed to determine new intersection points as illustrated in Fig. 2.22. By continuing with this procedure, one can perform the entire numerical integration.

References

1. W. Band, *Introduction to Mathematical Physics* (Van Nostrand Company, Inc., Princeton, New Jersey, 1959)
2. S. Temkin, *Elements of Acoustics*, (John Wiley & Sons, New York, 1981), Section 3.7
3. J.D. Anderson, *Modern Compressible Flow with Historical Perspective*, 3rd edn. (McGraw-Hill, New York, 2003) Chapter 7
4. L.D. Landau, E.M. Lifshitz, *Fluid Mechanics* (Pergamon Press, London, 1966), p. 366
5. N. Curle, H.J. Davies, *Modern Fluid Dynamics*, vol 2 (Van Nostrand Reinhold Company, London, 1971), p. 68
6. O. V. Rudenko, S. I. Soluyan, *Theoretical Foundations of Nonlinear Acoustics*, (Consultants Bureau, New York, A Division of Plenum Publishing Company, 1977), Chapter 1
7. D. Mihalas, B. Weibel-Mihalas, *Foundations of Radiation Hydrodynamics* (Dover Publications, Inc., New York, 1999), p. 227
8. W. Band, G.E. Duvall, Physical nature of shock propagation. Am. J. Physiol. **29**, 780–785 (1961)
9. W.C. Griffith, W. Bleakney, Shock waves in gases. Am. J. Physiol. **22**, 597 (1954)
10. N. Curle, H.J. Davies, *Modern Fluid Dynamics*, vol 2 (Van Nostrand Reinhold Company, London, 1971) Section 3.3.2
11. L.D. Landau, E.M. Lifshitz, *Fluid Mechanics* (Pergamon Press, London, 1966) Chapter 9
12. H.W. Liepmann, A. Roshko, *Elements of Gasdynamics* (Dover Publications Inc., Mineola, New York, 1956) Section 3.9
13. G.B. Whitham, *Linear and Nonlinear Waves* (John Wiley & Sons, Inc., New York, 1999)
14. R. Courant, K.O. Friedrichs, *Supersonic Flow and Shock Waves* (Interscience Publishers, Inc., New York, 1956)
15. A.R. Paterson, *A First Course in Fluid Dynamics* (Cambridge University Press, London, 1983) Chapter 14
16. J.D. Logan, *Applied Mathematics*, 2nd edn. (Wiley & Sons, Inc., New York, 1977) Chapter 6
17. A.J. Chorin, J.E. Marsden, *A Mathematical Introduction to Fluid Mechanics* (Springer-Verlag, New York, 1979) Chapter 3
18. F. H. Harlow, LA-2412 Report (Los Alamos Scientific Laboratory of the University of California, Los Alamos, New Mexico, November 1960), Chapter 3
19. M. A. Saad, *Compressible Fluid Flow*, (Prentice-Hall, Inc., Englewood Cliffs, New Jersey 1985), Chapter 9
20. J. Billingham, A.C. King, *Wave Motion* (Cambridge University Press, New York, 2000) Chapter 7
21. O. Regev, O.M. Umurhan, P.A. Yecko, *Modern Fluid Dynamics for Physics and Astrophysics* (Springer, New York, 2016) Chapter 6
22. J.H.S. Lee, *The Gas Dynamics of Explosions* (Cambridge University Press, New York, 2016) Chapter 1
23. W.A. Strauss, *Partial Differential Equations: An Introduction* (John Wiley & Sons, Inc., New York, 1992) Chapter 14
24. P.J. Olver, *Introduction to Partial Differential Equations* (Springer, 2014)
25. S. Salsa, *Partial Differential Equations in Action: From Modelling to Theory* (Springer, Milan, Italy, 2008) Chapter 4

Chapter 3
Conditions Across the Shock: The Rankine-Hugoniot Equations

3.1 Introduction to Normal Shock Waves

In this chapter we will investigate the relationship between the states on both sides of a normal shock wave. These relationships are known as the Rankine-Hugoniot equations and they can be derived by applying the laws of mass, momentum and energy conservation. The relationships derived will be used in the subsequent chapters and, in particular, they will be used to ascertain the accuracy of the solutions obtained numerically.

3.2 Conservation Equations

The laws of mass, momentum and energy conservation are applied to a fluid traversing the shock front [1–6]. Let us consider a shock wave propagating to the right into a stationary fluid with velocity U_s as shown in Fig. 3.1a. The pressure and density of the fluid ahead of the shock front are assumed to be p_0 and ρ_0, respectively, while the compressed fluid behind the shock front is moving with velocity u_p and has pressure p and density ρ. In a reference system in which the shock is stationary as illustrated in Fig. 3.1b, the velocity of the fluid entering the shock is U_s and the velocity of the fluid leaving the shock is $U_s - u_p$.

© The Author(s), under exclusive license to Springer Nature Switzerland AG 2021
S. Prunty, *Introduction to Simple Shock Waves in Air*, Shock Wave and High
Pressure Phenomena, https://doi.org/10.1007/978-3-030-63606-7_3

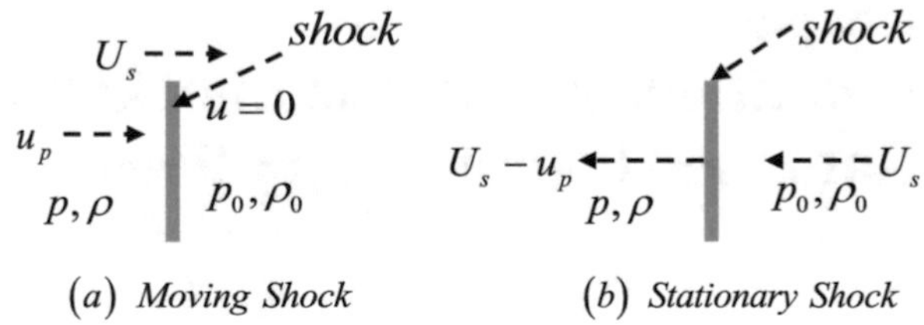

Fig. 3.1 Fluid velocities are
shown for (**a**) a moving
shock, and in a frame of
reference in which (**b**) the
shock is stationary (see text)

(*a*) *Moving Shock*   (*b*) *Stationary Shock*

3.2.1 Conservation of Mass

Mass conservation implies that the mass of fluid entering per unit area per unit time
is equal to the mass leaving per unit area per unit time, so that

$$\rho_0 U_s = \rho (U_s - u_p). \tag{3.1}$$

3.2.2 Conservation of Momentum

Conservation of momentum states that the difference between the rate of momentum
arriving at the shock front and the rate of momentum leaving the shock front is equal
to the net force per unit area acting on the material as it crosses the shock front;
hence,

$$\rho_0 U_s^2 - \rho (U_s - u_p)^2 = p - p_0 \tag{3.2}$$

3.2.3 Conservation of Energy

Conservation of energy requires that the total energy arriving at the shock front per
unit area per unit time minus the energy leaving the front per unit area per unit time is
equal to the rate at which work is done by the pressure difference across the front. In
general, the rate of increase of energy (kinetic energy plus internal energy) per unit
time is equal to

$$\frac{d}{dt}\left(\frac{1}{2}\rho v^2 V + e\rho V\right)$$

where ρ is the density, V is the volume, v is the velocity and e is the internal energy
per unit mass; hence, the rate of increase of energy per unit area per unit time
becomes

$$\frac{1}{2}\rho v^3 + e\rho v,$$

and, in general, the work (pressure × area × distance) done per unit area per unit time by the pressure forces is

$$pv.$$

Applying these equations to the energy conservation condition as expressed above gives

$$\left[\frac{1}{2}\rho(U_s - u_p)^3 + \rho e(U_s - u_p)\right] - \left[\frac{1}{2}\rho_0 U_s^3 + \rho_0 e_0 U_s\right] = p_0 U_s - p(U_s - u_p),$$

where e and e_0 are the internal energies per unit mass leaving and entering the shock front, respectively. The latter equation can be written in the form;

$$\frac{1}{2}\rho_0 U_s^3 + \rho_0 e_0 U_s + p_0 U_s = \frac{1}{2}\rho(U_s - u_p)^3 + \rho e(U_s - u_p) + p(U_s - u_p) \quad (3.3)$$

Let us now spend some time manipulating these equations in order to establish some useful relationships that can be used at a later stage. Eq. (3.1) gives

$$u_p = \left(1 - \frac{\rho_0}{\rho}\right)U_s, \quad (3.4)$$

and if we write Eq. (3.2) as

$$\rho_0 U_s^2 - \rho(U_s - u_p)(U_s - u_p) = p - p_0$$

and substituting Eq. (3.1) in this latter equation we have

$$\rho_0 U_s^2 - \rho_0 U_s(U_s - u_p) = p - p_0,$$

hence,

$$\rho_0 U_s u_p = p - p_0. \quad (3.5)$$

Multiplying Eq. (3.5) by Eq. (3.4) gives

$$\rho_0 U_s u_p^2 = (p - p_0)\left(1 - \frac{\rho_0}{\rho}\right)U_s,$$

so that the fluid velocity is

$$u_p = \sqrt{(p - p_0)\left(\frac{1}{\rho_0} - \frac{1}{\rho}\right)}. \tag{3.6}$$

Similarly, if Eq. (3.4) is used in the latter equation we obtain the shock velocity in terms of pressure and density according to

$$U_s = \sqrt{\left(\frac{\rho}{\rho_0}\right)\left(\frac{p - p_0}{\rho - \rho_0}\right)}. \tag{3.7}$$

By writing Eq. (3.3) in the form;

$$\frac{1}{2}\rho_0 U_s^3 + \rho_0 e_0 U_s + p_0 U_s = \frac{1}{2}\rho(U_s - u_p)^2(U_s - u_p) + \rho e(U_s - u_p) + p(U_s - u_p)$$

and using Eq. (3.1) we have

$$\frac{1}{2}\rho_0 U_s^3 + \rho_0 e_0 U_s + p_0 U_s = \frac{1}{2}\rho_0 U_s(U_s - u_p)^2 + \rho_0 U_s e + \frac{p\rho_0}{\rho}U_s.$$

Cancelling $\rho_0 U_s$ across gives

$$\frac{1}{2}U_s^2 + e_0 + \frac{p_0}{\rho_0} = \frac{1}{2}(U_s - u_p)^2 + e + \frac{p}{\rho}, \tag{3.8}$$

and re-arranging this latter equation we obtain

$$e - e_0 = \frac{p_0}{\rho_0} - \frac{p}{\rho} + U_s u_p - \frac{1}{2}u_p^2.$$

Substituting Eqs. (3.5) and (3.6) in this latter equation gives

$$e - e_0 = \frac{p_0}{\rho_0} - \frac{p}{\rho} + \left(\frac{p}{\rho_0} - \frac{p_0}{\rho_0}\right) - \frac{1}{2}(p - p_0)\left(\frac{1}{\rho_0} - \frac{1}{\rho}\right)$$

so that the increase in internal energy as a result of the shock is

$$e - e_0 = \frac{1}{2}(p + p_0)\left(\frac{1}{\rho_0} - \frac{1}{\rho}\right), \tag{3.9}$$

which is called the *Hugoniot equation*. Other equations based on the conservation laws are known as the *Rankine-Hugoniot equations* and they will be considered in due course.

3.3 Thermodynamic Relations

The energy equation above can be simplified by introducing the enthalpy, H which we have previously met in Chap. 1. Noting that

$$H = E + pV$$

where E is the internal energy and in terms of unit mass of material we have

$$\frac{H}{m} = \frac{E}{m} + p\frac{V}{m}$$

so that

$$h = e + \frac{p}{\rho}$$

where, as before, h is the enthalpy per unit mass, e is the usual internal energy per unit mass and ρ is the density and, as we have already seen, the following relationship applies;

$$h = c_P T$$

where c_P is the specific heat at constant pressure and T is the temperature. The conservation of energy equation, namely, Eq. (3.8), now becomes

$$\frac{1}{2}U_s^2 + c_P T_0 = \frac{1}{2}(U_s - u_p)^2 + c_P T.$$

Collecting all three conservation equations we can write them as

$$\rho_0 U_s = \rho(U_s - u_p) \tag{3.10a}$$

$$\rho_0 U_s^2 + p_0 = \rho(U_s - u_p)^2 + p \tag{3.10b}$$

$$\frac{1}{2}U_s^2 + c_P T_0 = \frac{1}{2}(U_s - u_p)^2 + c_P T \tag{3.10c}$$

These three equations contain four variables; velocity, density, pressure and temperature. Assuming conditions are known on one side of the shock front then another equation is required to find the conditions on the other side; this equation is the equation of state and in the case of an ideal gas it is given by; $p = \rho RT$, where R is a constant depending on the gas as discussed in Chap. 1 ($R = 287 Jkg^{-1}K^{-1}$ for air).

3.4 Alternative Notation for the Conservation Equations Across the Shock

It is convenient at this stage to introduce new variables to define the conditions on either side of the shock front; on one side let us define;

$$U_s \rightarrow U_1, \rho_0 \rightarrow \rho_1, p_0 \rightarrow p_1 \text{ and } T_0 \rightarrow T_1$$

and on the other side of the shock front we define;

$$\left(U_s - u_p\right) \rightarrow U_2, \ \rho \rightarrow \rho_2, \ p \rightarrow p_2 \text{ and } T \rightarrow T_2,$$

so that

$$u_p = U_1 - U_2 \tag{3.11}$$

where U_1 is the fluid velocity entering the shock and U_2 is the fluid velocity leaving the shock as seen by an observer that moves with the shock.

Accordingly, the conservation equations now become

$$\rho_1 U_1 = \rho_2 U_2 \tag{3.12a}$$

$$\rho_1 U_1^2 + p_1 = \rho_2 U_2^2 + p_2 \tag{3.12b}$$

$$\frac{1}{2} U_1^2 + c_P T_1 = \frac{1}{2} U_2^2 + c_P T_2 \tag{3.12c}$$

3.5 A Very Weak Shock

In the case of a very weak shock let us follow the analysis in the article by Blum [7]. Suppose velocity, temperature and pressure are known on either side of the shock and if we let

$$\rho_1 U_1 = \rho_2 U_2 \equiv K_1 \tag{3.12d}$$

$$\rho_1 U_1^2 + p_1 = \rho_2 U_2^2 + p_2 \equiv K_2 \tag{3.12e}$$

$$\frac{1}{2}U_1^2 + c_P T_1 = \frac{1}{2}U_2^2 + c_P T_2 \equiv K_3 \qquad (3.12\text{f})$$

with the equation of state for an ideal gas.

$$p_1 = \rho_1 R T_1 \text{ and } p_2 = \rho_2 R T_2,$$

then the system of Eqs. (3.12d, 3.12e and 3.12f) can be solved in terms of the constants K_1, K_2 and K_3. Solving for U_1 or U_2 by eliminating ρ_1, p_1 and T_1 (or ρ_2, p_2 and T_2) we have

$$\rho_1 = \frac{K_1}{U_1}$$

and Eq. (3.12e) gives

$$\rho_1 U_1^2 + \rho_1 R T_1 = K_2.$$

Dividing across by ρ_1 and rearranging the latter equation gives

$$RT_1 = \frac{K_2}{\rho_1} - U_1^2.$$

and using $\rho_1 = K_1/U_1$ in this latter equation yields

$$T_1 = \frac{K_2}{K_1 R} U_1 - \frac{U_1^2}{R}$$

and substituting this result in Eq. (3.12f) we obtain the following quadratic equation for U_1;

$$\left(\frac{c_P}{R} - \frac{1}{2}\right)U_1^2 - \frac{c_P}{R}\frac{K_2}{K_1}U_1 + K_3 = 0; \qquad (3.13)$$

clearly, the same equation is obtained with U_2 in place of U_1. The solution of the latter equation is given by

$$U_1, U_2 = \frac{c_P(K_2/K_1)}{c_P + c_V}\left[1 \pm \sqrt{1 - \frac{4K_3 K_1^2 R^2}{c_P^2 K_2^2}\left(\frac{c_P}{R} - \frac{1}{2}\right)}\right]$$

$$= \frac{c_P(K_2/K_1)}{c_P + c_V}[1 \pm \Delta] \qquad (3.14)$$

where

$$\Delta^2 = 1 - \frac{4K_3 K_1^2 R^2}{c_P^2 K_2^2}\left(\frac{c_P}{R} - \frac{1}{2}\right)$$

and we have used $R = c_P - c_V$. It is straightforward to show that

$$\Delta^2 = 1 - \frac{2K_3 K_1^2}{K_2^2}\frac{(\gamma^2 - 1)}{\gamma^2},$$

where $\gamma = c_P/c_V$. Substituting for K_1, K_2 and K_3 on one side of the shock; namely, $K_1 = \rho_1 U_1, K_2 = \rho_1 U_1^2 + p_1$ and $K_3 = \frac{1}{2}U_1^2 + c_P T_1$ (and using $p_1 = \rho_1 R T_1$), we have

$$\Delta^2 = 1 - \frac{(U_1^2 + 2c_P T_1)(\rho_1 U_1)^2}{(\rho_1 U_1^2 + p_1)^2}\frac{(\gamma^2 - 1)}{\gamma^2}. \tag{3.15}$$

It is easy to show that $\Delta = 0$ in the special case where

$$U_1^2 = \gamma R T_1$$

and it also follows that $U_2^2 = \gamma R T_1$ when $\Delta = 0$, so that $T_2 = T_1$. These latter equations represent the speed of sound ($c_0 = \sqrt{\gamma R T_1}$) so that an ideal sound wave can be viewed as the limiting case of a very weak shock and it is easy to verify that all parameters are unaffected by the passage of a weak shock.

3.6   Rankine-Hugoniot Equations

These conservation equations, expressing the changes in the physical parameters across the shock front, can be used to derive the *Rankine-Hugoniot equations* as outlined below. As we shall see in due course, these equations can be combined to produce some useful relationships between the parameters on either side of the shock.

Dividing Eq. (3.12b) by Eq. (3.12a) gives

$$U_2 - U_1 = \frac{p_1}{\rho_1 U_1} - \frac{p_2}{\rho_2 U_2}$$

and multiply both sides by $U_2 + U_1$, yields

$$U_2^2 - U_1^2 = \frac{(U_2 + U_1)}{\rho_1 U_1}(p_1 - p_2)$$

after using Eq. (3.12a) on the right-hand side. Expanding the latter equation gives

$$U_2^2 - U_1^2 = \left(\frac{U_2}{\rho_1 U_1} + \frac{1}{\rho_1}\right)(p_1 - p_2)$$

$$= \left(\frac{1}{\rho_2} + \frac{1}{\rho_1}\right)(p_1 - p_2), \tag{3.16}$$

where we have used Eq. (3.12a) again. However, Eq. (3.12c) gives

$$\frac{1}{2}\left(U_2^2 - U_1^2\right) = c_P(T_1 - T_2)$$

$$= c_P\left(\frac{p_1}{R\rho_1} - \frac{p_2}{R\rho_2}\right)$$

$$= \frac{\gamma}{\gamma - 1}\left(\frac{p_1}{\rho_1} - \frac{p_2}{\rho_2}\right),$$

where we have used the fact that $p_{1,\,2} = \rho_{1,\,2}RT_{1,\,2}$, $c_P - c_V = R$ and $\gamma = c_P/c_V$. Hence,

$$U_2^2 - U_1^2 = \frac{2\gamma}{\gamma - 1}\left(\frac{p_1}{\rho_1} - \frac{p_2}{\rho_2}\right)$$

and comparing this equation with Eq. (3.16) above gives

$$\left(\frac{1}{\rho_1} + \frac{1}{\rho_2}\right)(p_1 - p_2) = \frac{2\gamma}{\gamma - 1}\left(\frac{p_1}{\rho_1} - \frac{p_2}{\rho_2}\right).$$

By re-arranging this latter equation it is straightforward to show that

$$\frac{\rho_2}{\rho_1} = \frac{(\gamma - 1) + (\gamma + 1)(p_2/p_1)}{(\gamma + 1) + (\gamma - 1)(p_2/p_1)} \tag{3.17a}$$

which is one of the Rankine-Hugoniot relations.

3.6.1   Pressure and Density Changes for a Weak Shock

Let us now investigate the above Rankine-Hugoniot relationship as given by Eq. (3.17a) for very weak shocks. Writing Eq. (3.17a) in the form,

$$\frac{\rho_2}{\rho_1} = \frac{p_2 + \frac{(\gamma - 1)}{(\gamma + 1)}p_1}{p_1 + \frac{(\gamma - 1)}{(\gamma + 1)}p_2},$$

and replacing p_2 by $p + dp$ and p_1 by p to indicate small changes in pressure, similarly let $\rho_2 \rightarrow \rho + d\rho$ and $\rho_1 \rightarrow \rho$, and define $\xi = (\gamma - 1)/(\gamma + 1)$, then the above equation becomes,

$$\frac{\rho + d\rho}{\rho} = \frac{p + dp + \xi p}{p + \xi(p + dp)}$$
$$= \frac{p(1 + \xi) + dp}{p(1 + \xi) + \xi dp}$$
$$= \left[1 + \frac{dp}{(1 + \xi)p}\right]\left[1 + \frac{\xi dp}{(1 + \xi)p}\right]^{-1}.$$

Hence,

$$1 + \frac{d\rho}{\rho} \approx 1 - \frac{\xi dp}{(1 + \xi)p} + \frac{dp}{(1 + \xi)p}$$
$$= 1 + \left(\frac{1 - \xi}{1 + \xi}\right)\frac{dp}{p}$$

and it is easy to show that $\gamma = (1 + \xi)/(1 - \xi)$ so that the latter equation becomes

$$\frac{dp}{p} = \gamma \frac{d\rho}{\rho}$$

and integrating we have the isentropic relation

$$p\rho^{-\gamma} = \text{constant}$$

or in terms of the specific volume v, we have

$$pv^\gamma = \text{constant}.$$

Consequently, the Rankine-Hugoniot relationship as given by Eq. (3.14) reduces to the equation for an isentropic process in the case of a very weak shock and plots of these relationships are shown in Fig. 3.2.

In the case of a very strong shock; $p_2/p_1 \gg 1$, it can be seen from Eq. (3.17a) that

$$\frac{\rho_2}{\rho_1} \rightarrow \frac{(\gamma + 1)}{(\gamma - 1)} \tag{3.17b}$$

so that the density ratio tends to a limiting value (for example, $\rho_2/\rho_1 \rightarrow 6$ when $\gamma = 1.4$). The relationship between the densities as expressed in Eq. (3.17a) in conjunction with the continuity equation implies that

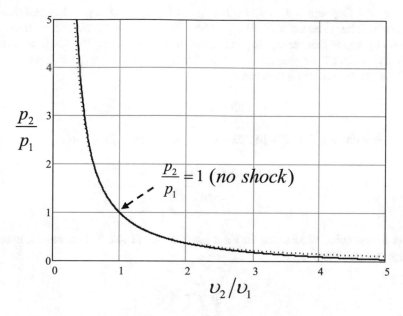

Fig. 3.2 Pressure ratio versus specific volume ratio across the shock (solid line) is shown. The isentropic relation, $pv^{\gamma} =$ constant, is also shown (broken line), (see text)

$$\frac{U_1}{U_2} = \frac{(\gamma - 1) + (\gamma + 1)(p_2/p_1)}{(\gamma + 1) + (\gamma - 1)(p_2/p_1)} \tag{3.18}$$

and from the equation of state we obtain

$$\frac{T_2}{T_1} = \left(\frac{p_2}{p_1}\right) \frac{(\gamma + 1) + (\gamma - 1)(p_2/p_1)}{(\gamma - 1) + (\gamma + 1)(p_2/p_1)}. \tag{3.19}$$

3.7   Entropy Change of the Gas on Its Passage Through a Shock

By returning again to the conservation equations;

$$\rho_1 U_1 = \rho_2 U_2$$

$$\rho_1 U_1^2 + p_1 = \rho_2 U_2^2 + p_2$$

$$\frac{1}{2} U_1^2 + c_P T_1 = \frac{1}{2} U_2^2 + c_P T_2$$

we note that they are satisfied whether $U_1 < U_2$ or $U_1 > U_2$, although intuition expects the latter to be the case. Blum [7] has also presented an analysis to show that $U_1 > U_2$ by noting that the entropy of the gas must increase on its passage across the shock as it is an irreversible process. Here we will follow Blum's analysis.

The entropy change dS is given by

$$TdS = dE + pdV$$

and as the enthalpy, $H = E + pV$, than $dH = dE + pdV + Vdp$, hence,

$$TdS = dH - Vdp$$
$$= mc_PdT - \frac{mRT}{p}dp,$$

where the equation of state for an ideal gas has been used. Therefore, the entropy change per unit mass is

$$\frac{dS}{m} = c_P\frac{dT}{T} - R\frac{dp}{p}$$

and integrating we have

$$\frac{S_2 - S_1}{m} = c_P\ln\frac{T_2}{T_1} - R\ln\frac{p_2}{p_1}. \tag{3.20}$$

In order to determine the entropy change we need expressions for the pressure and temperature ratios, p_2/p_1 and T_2/T_1, respectively. If we let $x = U_2/U_1$, we can write Eq. (3.18) for the pressure ratio as

$$\frac{p_2}{p_1} = \frac{(\gamma + 1) - x(\gamma - 1)}{(\gamma + 1)x - (\gamma - 1)}.$$

and it is easy to show that Eq. (3.19) for the temperature ratio can also be written in terms of x as

$$\frac{T_2}{T_1} = x\frac{(\gamma + 1) - x(\gamma - 1)}{(\gamma + 1)x - (\gamma - 1)}.$$

Consequently, the entropy change per unit mass is

$$\frac{S_2 - S_1}{m} = c_P\ln x + c_P\ln\frac{(\gamma + 1) - x(\gamma - 1)}{(\gamma + 1)x - (\gamma - 1)} - R\ln\frac{(\gamma + 1) - x(\gamma - 1)}{(\gamma + 1)x - (\gamma - 1)},$$

and noting that $R = c_P - c_V$, we have

$$\frac{S_2 - S_1}{m} = c_P \ln x + c_V \ln \frac{(\gamma + 1) - x(\gamma - 1)}{(\gamma + 1)x - (\gamma - 1)}$$

or

$$\frac{S_2 - S_1}{mc_V} = \gamma \ln x + \ln \frac{(\gamma + 1) - x(\gamma - 1)}{(\gamma + 1)x - (\gamma - 1)}. \tag{3.21}$$

Let $(S_2 - S_1)/mc_V = F(x)$ then

$$F(x) = \gamma \ln x + \ln \frac{(\gamma + 1) - x(\gamma - 1)}{(\gamma + 1)x - (\gamma - 1)}. \tag{3.22}$$

In the case of a sound wave we have already shown that $U_1 = U_2 = \sqrt{\gamma RT}$, so that $x = 1$ and hence, $F(1) = 0$; implying no entropy change. By differentiating $F(x)$, we obtain

$$\begin{aligned}
\frac{dF}{dx} &= \frac{\gamma}{x} - \frac{(\gamma - 1)}{[(\gamma + 1) - x(\gamma - 1)]} - \frac{(\gamma + 1)}{[(\gamma + 1)x - (\gamma - 1)]} \\
&= \frac{\gamma[2\gamma^2 x + 2x - \gamma^2 + 1 - \gamma^2 x^2 + x^2] - 4\gamma x}{x[(\gamma + 1) - (\gamma - 1)x][(\gamma + 1)x - (\gamma - 1)]} \\
&= \frac{\gamma[2x(\gamma^2 - 1) - (\gamma^2 - 1) - x^2(\gamma^2 - 1)]}{x[(\gamma + 1) - (\gamma - 1)x][(\gamma + 1)x - (\gamma - 1)]} \tag{3.23} \\
&= -\frac{\gamma(\gamma^2 - 1)(x - 1)^2}{x[(\gamma + 1) - (\gamma - 1)x][(\gamma + 1)x - (\gamma - 1)]} \\
&= -\frac{\gamma(x - 1)^2}{x\left[\left(\frac{\gamma + 1}{\gamma - 1}\right) - x\right]\left[x - \left(\frac{\gamma - 1}{\gamma + 1}\right)\right]}
\end{aligned}$$

One can see that $(dF/dx)_{x=1} = 0$ is the only extremum and $dF/dx \leq 0$ in the range

$$\frac{\gamma - 1}{\gamma + 1} < x < \frac{\gamma + 1}{\gamma - 1}$$

and the only allowed values of x that satisfy the entropy condition, that is, $0 \leq F(x) < \infty$, are

$$\frac{\gamma - 1}{\gamma + 1} < x \leq 1.$$

These assertions are shown plotted in Fig. 3.3 for $\gamma = 1.4$.

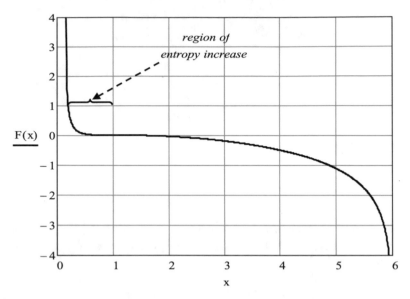

Fig. 3.3 Entropy change across the shock as a function of x for $\gamma = 1.4$ (see text)

Hence, we obtain the important result that $U_1 \geq U_2$ and not the other way round, while the lower limit corresponds to the case of a very strong shock. Accordingly, the second law of thermodynamics demands that $S_2 - S_1 \geq 0$ so that $U_1 \geq U_2$; the equality sign corresponds to an ordinary sound wave as previously discussed.

3.8 Other Useful Relationships in Terms of Mach Number

Let us now establish some other useful relationships in terms of the Mach number [8, 9]. The Mach number M on either side of the shock is defined as

$$M_{1,2} = \frac{U_{1,2}}{c_{1,2}} = \frac{U_{1,2}}{\sqrt{\gamma R T_{1,2}}}. \tag{3.24}$$

where $c_{1,2}$ is the acoustic speed on either side of the shock. Eq. (3.12b) gives

$$
\begin{aligned}
p_2 - p_1 &= \rho_1 U_1^2 - \rho_2 U_2^2 \\
&= \rho_1 U_1^2 \left(1 - \frac{U_2}{U_1} \right)
\end{aligned}
$$

after using Eq. (3.12a). Dividing both sides by p_1 and using the equation of state, namely, $p_1 = \rho_1 R T_1$, we have

$$\frac{p_2}{p_1} - 1 = \frac{\rho_1 U_1^2}{\rho_1 RT_1}\left(1 - \frac{U_2}{U_1}\right)$$

$$= \frac{\gamma U_1^2}{\gamma RT_1}\left(1 - \frac{U_2}{U_1}\right)$$

$$= \gamma M_1^2\left(1 - \frac{U_2}{U_1}\right).$$

However, the continuity equation implies that $U_2/U_1 = \rho_1/\rho_2$, hence, using Eq. (3.17a) for the density ratio we obtain

$$\frac{p_2}{p_1} - 1 = \gamma M_1^2\left[1 - \frac{(\gamma + 1) + (\gamma - 1)(p_2/p_1)}{(\gamma - 1) + (\gamma + 1)(p_2/p_1)}\right]$$

and solving for the ratio p_2/p_1 we have

$$\frac{p_2}{p_1} = 1 + \frac{2\gamma}{\gamma + 1}\left(M_1^2 - 1\right) \tag{3.25}$$

which gives the pressure ratio in terms of the Mach number M_1. Solving this latter equation for the Mach number in terms of the pressure ratio we obtain the following Rankine-Hugoniot equation,

$$\frac{U_1^2}{\gamma RT_1} = \frac{1}{2\gamma}\left[(\gamma - 1) + (\gamma + 1)\frac{p_2}{p_1}\right], \tag{3.26a}$$

and in the limit of a very strong shock ($p_2/p_1 >> 1$) we obtain,

$$\frac{U_1^2}{\gamma RT_1} = \frac{(\gamma + 1)}{2\gamma}\frac{p_2}{p_1},$$

and as $p_1 = \rho_1 RT_1$, we find that

$$p_2 \rightarrow \frac{2}{\gamma + 1}\rho_1 U_1^2. \tag{3.26b}$$

Returning to Eq. (3.26a) we can obtain the following expression for the velocity of the moving shock wave U_s (noting that $U_1 = U_s$) in terms of the pressure ratio across the shock and the acoustic speed c_1 of the air, for example, into which the shock is propagating,

$$U_s = c_1\left[\frac{(\gamma - 1) + (\gamma + 1)\frac{p_2}{p_1}}{2\gamma}\right]^{1/2} \tag{3.27}$$

Similarly, using Eq. (3.17a) that relates the pressure ratio to the density ratio we can establish the following relationship;

$$\frac{\rho_2}{\rho_1} = \frac{(\gamma + 1)M_1^2}{\left[(\gamma - 1)M_1^2 + 2\right]} \tag{3.28}$$

and, in addition, one can also show that

$$\frac{T_2}{T_1} = \frac{\left[2\gamma M_1^2 - (\gamma - 1)\right]\left[(\gamma - 1)M_1^2 + 2\right]}{(\gamma + 1)^2 M_1^2} \tag{3.29}$$

by noting that $p_2/p_1 = (\rho_2 T_2 / \rho_1 T_1)$. Clearly, $p_2/p_1 = 1$, $\rho_2/\rho_1 = 1$ and $T_2/T_1 = 1$ when $M_1 = 1$, this corresponds to the propagation of an ordinary sound wave which we have already seen as the limiting case of a very weak shock. Plots of pressure, density and temperature ratios are shown in Fig. 3.4.

Writing again the momentum equation as

$$\rho_1 U_1^2 + p_1 = \rho_2 U_2^2 + p_2$$

and using $p_{1,2} = \rho_{1,2} R T_{1,2}$ and $M_{1,2} = U_{1,2}/\sqrt{\gamma R T_{1,2}}$, while eliminating ρ_1 and ρ_2 one can verify that

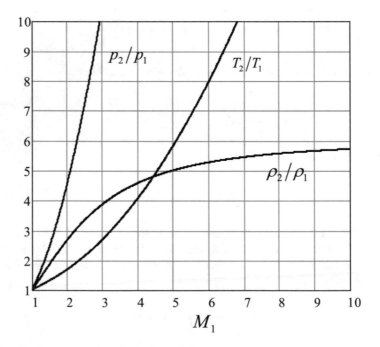

Fig. 3.4 Plots of pressure, density and temperature ratios as a function of Mach number M_1 across a normal shock wave. ($\gamma = 1.4$)

$$\frac{p_2}{p_1} = \frac{1 + \gamma M_1^2}{1 + \gamma M_2^2} \tag{3.30}$$

Similarly, by using the equation of state and the continuity equation it is easy to show that

$$\frac{\rho_2}{\rho_1} = \frac{M_1}{M_2}\sqrt{\frac{T_1}{T_2}}$$

and

$$\frac{p_2}{p_1} = \frac{M_1}{M_2}\sqrt{\frac{T_2}{T_1}},$$

while the conservation of energy equation gives

$$\frac{T_2}{T_1} = \frac{1 + \frac{1}{2}(\gamma - 1)M_1^2}{1 + \frac{1}{2}(\gamma - 1)M_2^2}. \tag{3.31}$$

Let us now establish a relationship between M_1 and M_2. Using Eqs. (3.25) and (3.30) we obtain

$$\frac{1 + \gamma M_2^2}{1 + \gamma M_1^2} = \frac{\gamma + 1}{(1 - \gamma) + 2\gamma M_1^2}$$

and solving for M_2 we have

$$M_2 = \sqrt{\frac{2 + (\gamma - 1)M_1^2}{2\gamma M_1^2 - (\gamma - 1)}}, \tag{3.32}$$

which give the Mach number M_2 in terms of the Mach number M_1. It can be seen that $M_2 = 1$ when $M_1 = 1$ and $M_2 < 1$ when $M_1 > 1$ and vice versa.

3.9 Entropy Change Across the Shock in Terms of Mach Number

Equation (3.32) is symmetric in M_1 and M_2 as can be seen by writing it in the form,

$$2\gamma M_1^2 M_2^2 - (\gamma - 1)(M_1^2 + M_2^2) - 2 = 0.$$

Accordingly, the conservation equations allow the existence of a compressive solution ($M_1 > 1$, $M_2 < 1$) and an expansive solution ($M_1 < 1$, $M_2 > 1$). However, the second law of thermodynamics requires that the fluid flow through a stationary shock wave always requires a supersonic-to-subsonic transition [7, 10]. To show this, we invoke the second law of thermodynamics by noting that the entropy of the fluid must increase as it moves across the shock. The entropy change dS is given by

$$TdS = dE + pdV$$

and as enthalpy is $H = E + pV$, so that $dH = dE + pdV + Vdp$, hence,

$$TdS = dH - Vdp$$
$$= mc_p dT - \frac{mRT}{p}dp,$$

therefore, the entropy change per unit mass is

$$\frac{dS}{m} = c_P \frac{dT}{T} - R \frac{dp}{p}$$

and integrating we have

$$\frac{S_2 - S_1}{m} = c_P \ln \frac{T_2}{T_1} - R \ln \frac{p_2}{p_1}.$$

Eliminating the ratio T_2/T_1 by using the equation of state for a perfect gas, namely, $p_{1,\,2} = \rho_{1,\,2} R T_{1,\,2}$, we have

$$\frac{S_2 - S_1}{m} = c_P \ln \left(\frac{p_2}{p_1}\right)\left(\frac{\rho_1}{\rho_2}\right) - R \ln \frac{p_2}{p_1}$$
$$= c_P \ln \left(\frac{p_2}{p_1}\right) + c_P \ln \left(\frac{\rho_1}{\rho_2}\right) - R \ln \frac{p_2}{p_1}$$
$$= c_V \ln \left(\frac{p_2}{p_1}\right) + c_P \ln \left(\frac{\rho_1}{\rho_2}\right),$$

hence,

$$\frac{S_2 - S_1}{mc_V} = \ln \left(\frac{p_2}{p_1}\right)\left(\frac{\rho_1}{\rho_2}\right)^{\gamma}.$$

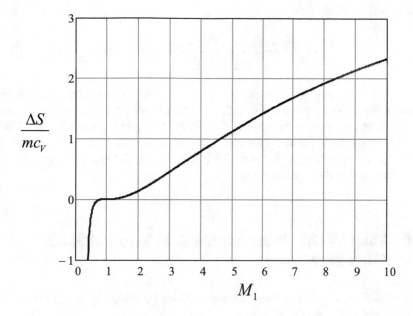

Fig. 3.5   The change in entropy across a normal shock is shown ($\gamma = 1.4$)

Substituting Eqs. (3.25) and (3.28) in this latter equation yields

$$\frac{\Delta S}{mc_V} = \ln\left(\frac{1 - \gamma + 2\gamma M_1^2}{\gamma + 1}\right)\left[\frac{2 + (\gamma - 1)M_1^2}{(\gamma + 1)M_1^2}\right]^{\gamma}, \tag{3.33}$$

where $\Delta S = S_2 - S_1$ and Eq. (3.33) is shown plotted in Fig. 3.5. Defining,

$$\frac{\Delta S}{mc_V} \equiv \ln\left[f(x)\right],$$

where

$$f(x) = \left(\frac{1 - \gamma + 2\gamma x}{\gamma + 1}\right)\left[\frac{2 + (\gamma - 1)x}{(\gamma + 1)x}\right]^{\gamma} \tag{3.34}$$

and $x \equiv M_1^2$. It can be seen that $f(1) = 1$, so that $S_2 - S_1 = 0$, hence, there is no entropy increase when the upstream Mach number is unity and we have noted that $M_2 = 1$ when $M_1 = 1$, so there is no shock and no discontinuity.

It is easy to verify that

$$\frac{df}{dx} = \frac{2\gamma(\gamma - 1)(x - 1)^2[2 + (\gamma - 1)x]^{\gamma-1}}{[(\gamma + 1)x]^{\gamma+1}} \geq 0, \tag{3.35}$$

so that $f(x)$ is increasing in the range, $x \geq 1$. The *second law of thermodynamics* demands that $S_2 - S_1 \geq 0$ (see Fig. 3.5), so it follows that $M_1 \geq 1$ and we conclude that the upstream flow is supersonic, that is, $M_1 > 1$ and the downstream flow is subsonic ($M_2 < 1$). In these circumstances it is evident from Eqs. (3.25), (3.28) and (3.29) that $p_2/p_1 > 1$, $\rho_2/\rho_1 > 1$ and $T_2/T_1 > 1$, that is, the gas is compressed and heated on passing through the shock.

3.10   Fluid Motion Behind the Shock in Terms of Shock Wave Parameters

Let us establish another Rankine-Hugoniot relation by using Eq. (3.11) for the flow velocity behind the shock, namely,

$$\begin{aligned} u_p &= U_1 - U_2 \\ &= \left(1 - \frac{\rho_1}{\rho_2}\right)U_s \end{aligned} \tag{3.36}$$

after using the continuity equation and noting that $U_1 = U_s$, where U_s is the velocity of the shock wave. Using Eq. (3.17a) in this latter equation gives

$$u_p = \left[1 - \frac{(\gamma + 1) + (\gamma - 1)(p_2/p_1)}{(\gamma - 1) + (\gamma + 1)(p_2/p_1)}\right]U_s, \tag{3.37}$$

hence,

$$u_p = \frac{2\left(\frac{p_2}{p_1} - 1\right)U_s}{(\gamma - 1) + (\gamma + 1)\frac{p_2}{p_1}}, \tag{3.38a}$$

which is another Rankine-Hugoniot relationship. In the limit of a very strong shock ($p_2/p_1 \gg 1$) it follows that

$$u_p \rightarrow \frac{2}{\gamma + 1}U_s. \tag{3.38b}$$

However, in the general case, we substitute Eq. (3.25) in Eq. (3.38a), yielding,

$$u_p = \frac{2\frac{2\gamma}{\gamma+1}\left(\frac{U_s^2}{c_1^2} - 1\right)U_s}{(\gamma-1) + (\gamma+1)\left[1 + \frac{2\gamma}{\gamma+1}\left(\frac{U_s^2}{c_1^2} - 1\right)\right]},$$

where $c_1 = \sqrt{\gamma R T_1}$ is the sonic velocity upstream of the shock. Simplifying the latter equation we obtain,

$$u_p = \frac{2c_1^2\left(\frac{U_s^2}{c_1^2} - 1\right)}{(\gamma+1)U_s} \tag{3.39a}$$

or alternatively as,

$$u_p = \frac{2c_1}{(\gamma+1)}\left(M_1 - \frac{1}{M_1}\right) \tag{3.39b}$$

and solving Eq. (3.39a) for U_s we obtain,

$$U_s = \frac{(\gamma+1)}{4}u_p + \sqrt{\frac{1}{16}(\gamma+1)^2 u_p^2 + c_1^2}, \tag{3.40}$$

which give the velocity of the shock wave in terms of the air flow or particle velocity u_p behind the shock. We can also obtain an expression for the air flow velocity behind the shock wave in terms of the pressure ratio p_2/p_1 and the speed of sound ahead of the shock by using Eqs. (3.38a) and (3.27); after substituting we obtain,

$$u_p = \frac{2c_1\left(\frac{p_2}{p_1} - 1\right)\left[\frac{(\gamma-1)+(\gamma+1)\frac{p_2}{p_1}}{2\gamma}\right]^{1/2}}{(\gamma-1) + (\gamma+1)\frac{p_2}{p_1}}$$

and simplifying this expression, yields,

$$u_p = \frac{c_1}{\gamma}\left(\frac{p_2}{p_1} - 1\right)\left(\frac{\frac{2\gamma}{\gamma+1}}{\frac{\gamma-1}{\gamma+1} + \frac{p_2}{p_1}}\right)^{1/2}, \tag{3.41}$$

which gives the particle or mass motion velocity u_p behind the shock in terms of the pressure ratio across the shock and the sonic velocity c_1 in the gas ahead of the shock.

3.11   Reflection of a Plane Shock from a Rigid Plane Surface

Let us now consider a plane shock wave propagating to the right down a tube with velocity U_i and incident on a plane rigid end-wall of the tube as shown schematically in Fig. 3.6. When the shock front reaches the end-wall the gas or particle motion u_p behind the incident shock is suddenly stopped by the immediate generation of a reflected shock wave propagating to the left with velocity U_r, whose magnitude ensures that the gas behind the reflected shock is at zero velocity and, accordingly, is at rest relative to the tube.

The conservation equations for the incident shock wave with respect to regions 1 and 2 are

$$\rho_1 U_i = \rho_2 \left(U_i - u_p \right)$$
$$p_1 + \rho_1 U_i^2 = p_2 + \rho_2 \left(U_i - u_p \right)^2$$
$$h_1 + \frac{U_i^2}{2} = h_2 + \frac{\left(U_i - u_p \right)^2}{2}$$

and corresponding conservation equations for the reflected shock wave with respect to regions 2 and 3 are

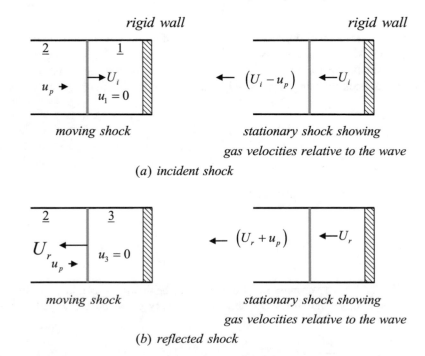

(a) incident shock

(b) reflected shock

Fig. 3.6 Schematic diagram showing the moving and stationary shocks for both the incident and reflected shock waves

$$\rho_3 U_r = \rho_2 \left(U_r + u_p \right)$$
$$p_3 + \rho_3 U_r^2 = p_2 + \rho_2 \left(U_r + u_p \right)^2$$
$$h_3 + \frac{U_r^2}{2} = h_2 + \frac{\left(U_r + u_p \right)^2}{2}.$$

One can see that the relative velocity of the gas on both sides of the discontinuity is the same for both the incident and reflected shock waves and equal to u_p [8]. Using the equation for the relative velocity, namely, Eq. (3.41), we have for the incident wave,

$$u_p = \frac{c_1}{\gamma} \left(\frac{p_2}{p_1} - 1 \right) \left(\frac{\frac{2\gamma}{\gamma+1}}{\frac{\gamma-1}{\gamma+1} + \frac{p_2}{p_1}} \right)^{1/2}$$

and similarly for the reflected wave we can also write,

$$u_p = \frac{c_2}{\gamma} \left(\frac{p_3}{p_2} - 1 \right) \left(\frac{\frac{2\gamma}{\gamma+1}}{\frac{\gamma-1}{\gamma+1} + \frac{p_3}{p_2}} \right)^{1/2}. \tag{3.42}$$

Equating these two latter equations permits one to determine the pressure ratio, p_3/p_2, across the reflected shock in terms of the pressure ratio, p_2/p_1, across the incident shock. For this we need the sonic speed ratio, c_1/c_2, and noting that the speed of sound in an ideal gas is given by the equation, $c^2 = \gamma p/\rho$, hence,

$$\frac{c_1^2}{c_2^2} = \frac{p_1 \rho_2}{p_2 \rho_1},$$

and according to Eq. (3.17a) the density ratio is given by

$$\frac{\rho_2}{\rho_1} = \frac{(\gamma - 1) + (\gamma + 1)(p_2/p_1)}{(\gamma + 1) + (\gamma - 1)(p_2/p_1)},$$

hence,

$$\frac{c_1^2}{c_2^2} = \frac{p_1}{p_2} \frac{(\gamma - 1) + (\gamma + 1)(p_2/p_1)}{(\gamma + 1) + (\gamma - 1)(p_2/p_1)}. \tag{3.43}$$

Before proceeding further to determine the pressure ratio p_3/p_2 in terms of the pressure ratio p_2/p_1, it is convenient to define the following parameters in order to avoid carrying terms involving $(\gamma - 1)$ and $(\gamma + 1)$; letting

$$\xi = \frac{\gamma - 1}{\gamma + 1}; \text{hence,} \quad \frac{2\gamma}{\gamma + 1} = 1 + \xi, \text{similarly let } x \equiv p_2/p_1 \text{ and } y \equiv p_3/p_2.$$

Combining Eqs. (3.41), (3.42) and (3.43) yields,

$$(y - 1)^2 \left(\frac{1 + \xi}{y + \xi}\right) = (x - 1)^2 \left(\frac{1 + \xi}{x + \xi}\right) \frac{1}{x} \frac{\xi + x}{1 + \xi x}.$$

Simplifying the latter equation by eliminating common factors the following quadratic equation for y in terms of x and the parameter ξ is obtained,

$$x(1 + \xi x)(y - 1)^2 = (x - 1)^2 (y + \xi).$$

Writing this latter equation in the following form,

$$x(1 + \xi x)(y - 1)^2 = (x - 1)^2 [(y - 1) + (1 + \xi)],$$

or, alternatively, in the form,

$$x(1 + \xi x)(y - 1)^2 - (x - 1)^2 (y - 1) - (x - 1)^2 (1 + \xi) = 0, \qquad (3.44)$$

and solving for $(y - 1)$ gives,

$$y - 1 = \frac{(x - 1)^2 \pm (x - 1)\sqrt{(x - 1)^2 + 4x(1 + \xi)(1 + \xi x)}}{2x(1 + \xi x)}.$$

The quantity under the square-root sign is a perfect square, which is equal to $[x(1 + 2\xi) + 1]^2$, hence,

$$y - 1 = \frac{(x - 1)^2 \pm (x - 1)[x(1 + 2\xi) + 1]}{2x(1 + \xi x)}. \qquad (3.45)$$

Taking the positive sign and solving for y gives,

$$y = \frac{x(1 + 2\xi) - \xi}{1 + \xi x}; \qquad (3.46)$$

on the other hand, by taking the negative sign it yields the trivial root $p_3 = p_1$ [8]. Substituting for x, y and ξ in Eq. (3.46) gives,

$$\frac{p_3}{p_2} = \frac{(3\gamma - 1)p_2 - (\gamma - 1)p_1}{(\gamma - 1)p_2 + (\gamma + 1)p_1}, \qquad (3.47)$$

which is our final result for the pressure ratio across the reflected shock. For a very strong incident shock, $(p_2/p_1) \gg 1$, Eq. (3.47) gives

$$\frac{p_3}{p_2} \rightarrow \frac{\left(\frac{3\gamma-1}{\gamma+1}\right)\frac{p_2}{p_1}}{\left(\frac{\gamma-1}{\gamma+1}\right)\frac{p_2}{p_1}} = \frac{3\gamma-1}{\gamma-1}$$

and with $\gamma = 1.4$ the reflected shock pressure ratio, $p_3/p_2 \rightarrow 8$, so that the strength of the reflected shock is magnified by a factor of 8 and in the case of a weak incident shock, $p_2/p_1 \rightarrow 1$, one finds that $p_3/p_2 \rightarrow 1$ as expected.

In an analogous fashion for the density ratio across the incident shock, the density ratio across the reflected shock can be written as,

$$\frac{\rho_3}{\rho_2} = \frac{(\gamma-1) + (\gamma+1)(p_3/p_2)}{(\gamma+1) + (\gamma-1)(p_3/p_2)} \tag{3.48}$$

and the corresponding temperature ratio is

$$\frac{T_3}{T_2} = \left(\frac{p_3}{p_2}\right)\frac{(\gamma+1) + (\gamma-1)(p_3/p_2)}{(\gamma-1) + (\gamma+1)(p_3/p_2)}. \tag{3.49}$$

In an analogous fashion to Eq. (3.25) one can also write,

$$\frac{p_3}{p_2} = 1 + \frac{2\gamma}{\gamma+1}(M_r^2 - 1), \tag{3.50}$$

where $M_r = (U_r + u_p)/c_2$. Similarly, by using Eq. (3.39b) for both the incident and reflected waves one can also write

$$\frac{2c_2}{(\gamma+1)}\left(M_r - \frac{1}{M_r}\right) = \frac{2c_1}{(\gamma+1)}\left(M_i - \frac{1}{M_i}\right),$$

hence,

$$\frac{M_r}{M_r^2 - 1} = \frac{M_i}{M_i^2 - 1}\frac{c_2}{c_1}. \tag{3.51}$$

Now using Eq. (3.43) in this latter equation we have

$$\frac{M_r}{M_r^2 - 1} = \frac{M_i}{M_i^2 - 1}\sqrt{\frac{p_2}{p_1}\frac{(\gamma+1) + (\gamma-1)(p_2/p_1)}{(\gamma-1) + (\gamma+1)(p_2/p_1)}} \tag{3.52}$$

and by substituting Eq. (3.25) for the pressure ratio p_2/p_1 in Eq. (3.52) we obtain

$$\frac{M_r}{M_r^2 - 1} = \frac{M_i}{M_i^2 - 1} \sqrt{\left[1 + \frac{2\gamma}{\gamma + 1}\left(M_i^2 - 1\right)\right] \frac{\gamma + 1 + \gamma - 1 + \frac{2\gamma(\gamma - 1)}{\gamma + 1}\left(M_i^2 - 1\right)}{\gamma - 1 + \gamma + 1 + 2\gamma\left(M_i^2 - 1\right)}}$$

and simplifying gives

$$\frac{M_r}{M_r^2 - 1} = \frac{M_i}{M_i^2 - 1} \sqrt{\frac{\left[1 + \frac{2\gamma}{\gamma + 1}\left(M_i^2 - 1\right)\right]\left[1 + \left(\frac{\gamma - 1}{\gamma + 1}\right)\left(M_i^2 - 1\right)\right]}{M_i^2}}. \tag{3.53}$$

Multiplying the quantities under the radical sign gives

$$\frac{1}{M_i^2(\gamma + 1)^2}$$

$$\times \left[(\gamma + 1)^2 + (\gamma^2 - 1)(M_i^2 - 1) + 2\gamma(\gamma + 1)(M_i^2 - 1) + 2\gamma(\gamma - 1)(M_i^2 - 1)^2\right].$$

Adding and subtracting $(\gamma + 1)^2 M_i^2$ in the numerator in the latter expression and simplifying yields

$$1 + 2\left(\frac{1 - \gamma - M_i^2 + 2\gamma M_i^2 - \gamma^2 M_i^2 + \gamma^2 M_i^4 - \gamma M_i^2}{(\gamma + 1)^2 M_i^2}\right)$$

and dividing above and below by M_i^2 we obtain

$$1 + 2\left(\frac{\frac{1}{M_i^2} - \frac{\gamma}{M_i^2} - 1 + 2\gamma - \gamma^2 + \gamma^2 M_i^2 - \gamma M_i^2}{(\gamma + 1)^2}\right)$$

$$= 1 + 2\left[\frac{\gamma M_i^2(\gamma - 1) - \gamma(\gamma - 1) + (\gamma - 1) - \frac{1}{M_i^2}(\gamma - 1)}{(\gamma + 1)^2}\right]$$

$$= 1 + \frac{2(\gamma - 1)}{(\gamma + 1)^2}\left(\gamma M_i^2 - \gamma + 1 - \frac{1}{M_i^2}\right)$$

$$= 1 + \frac{2(\gamma - 1)(M_i^2 - 1)\left(\gamma + \frac{1}{M_i^2}\right)}{(\gamma + 1)^2}$$

and with this substitution in Eq. (3.53) we have the following relationship between the Mach numbers;

$$\frac{M_r}{M_r^2 - 1} = \frac{M_i}{M_i^2 - 1} \sqrt{1 + \frac{2(\gamma - 1)(M_i^2 - 1)\left(\gamma + \frac{1}{M_i^2}\right)}{(\gamma + 1)^2}}, \tag{3.54}$$

as specified by Anderson [11], where $M_i = U_i/c_1$ is the Mach number of the incident shock. Referring back to Eq. (3.51) and noting that $c \propto \sqrt{T}$ one can conclude as a result of Eq. (3.54) that the temperature ratio across the incident shock wave can be written as

$$\frac{T_2}{T_1} = 1 + \frac{2(\gamma - 1)(M_i^2 - 1)\left(\gamma + \frac{1}{M_i^2}\right)}{(\gamma + 1)^2}, \tag{3.55}$$

which is an alternate form of Eq. (3.29).

We can also determine the reflected shock speed U_r in terms of the incident shock speed U_i by using Eq. (3.27), hence,

$$U_i = c_1 \left[\frac{(\gamma - 1) + (\gamma + 1)\frac{p_2}{p_1}}{2\gamma}\right]^{1/2} \quad \text{and} \quad U_r = c_2 \left[\frac{(\gamma - 1) + (\gamma + 1)\frac{p_3}{p_2}}{2\gamma}\right]^{1/2}$$

Using our previous substitutions, namely, $\xi = (\gamma - 1)/(\gamma + 1)$, $x = p_2/p_1$ and $y = p_3/p_2$, we can write the latter two equations as,

$$U_i = c_1 \left[\frac{\xi + x}{1 + \xi}\right]^{1/2} \quad \text{and} \quad U_r = c_2 \left[\frac{\xi + y}{1 + \xi}\right]^{1/2}.$$

Forming the ratio U_r/U_i we have

$$\left(\frac{U_r}{U_i}\right)^2 = \frac{(\xi + y)}{(\xi + x)}\frac{c_2^2}{c_1^2} = \frac{(\xi + y)}{(\xi + x)}\frac{x(1 + \xi x)}{(\xi + x)}.$$

Substituting Eq. (3.46) for y in this latter equation gives

$$\left(\frac{U_r}{U_i}\right)^2 = \frac{x(1 + \xi x)}{(\xi + x)^2}\left[\xi + \frac{x(1 + 2\xi) - \xi}{1 + \xi x}\right] = \frac{x^2}{(\xi + x)^2}(1 + \xi)^2,$$

hence,

$$\frac{U_r}{U_i} = \frac{x}{(\xi + x)}(1 + \xi).$$

$$= \frac{\dfrac{2\gamma}{\gamma + 1}\dfrac{p_2}{p_1}}{\dfrac{\gamma - 1}{\gamma + 1} + \dfrac{p_2}{p_1}}. \tag{3.56}$$

3.12   Approximate Analytical Expressions for Weak Shock Waves

Let us now consider some of the properties of weak shocks or shocks of moderate strength as this material will be useful for later consideration in Chap. 4. We can define the shock strength in terms of the excess pressure ratio $(p_2 - p_1)/p_1$ across the shock or in terms of the particle or material velocity u behind the shock front. Here, we will use the particle velocity as a measure of the shock strength and in the case of weak shocks we will assume that $(u/c_1) \ll 1$, where c_1 is the speed of sound in the still air ahead of the shock. Let p_1 and ρ_1 be the pressure and density, respectively, for the still air ahead of the shock and let p_2, ρ_2 and c_2 be the pressure, density and the speed of sound, respectively, in the air behind the shock front whose velocity is U.

3.12.1   Shock Velocity for Weak Shocks

Using Eq. (3.40) we can write the shock velocity U in the following form,

$$U = c_1 \left[\left(\frac{\gamma + 1}{4}\right)\frac{u}{c_1} + \sqrt{1 + \left(\frac{\gamma + 1}{4}\right)^2 \frac{u^2}{c_1^2}} \right].$$

Letting $x = \frac{(\gamma + 1)}{4}\frac{u}{c_1}$, the latter equation becomes

$$U = c_1\left[x + \sqrt{1 + x^2}\right]$$
$$= c_1\left[x + \left(1 + x^2\right)^{1/2}\right].$$

Using the following binomial expansion;

$$(1 + y)^{1/2} = 1 + (y/2) - (y^2/8) + (3y^3/48) + \ldots\ldots$$

and by retaining terms to second order we can show that the shock velocity can be written in the form,

$$U = c_1 \left[1 + \left(\frac{\gamma + 1}{4} \right) \frac{u}{c_1} + \frac{1}{2} \left(\frac{\gamma + 1}{4} \right)^2 \frac{u^2}{c_1^2} \right]. \tag{3.57}$$

3.12.2   Pressure Ratio for Weak Shocks

Using Eq. (3.25) for the pressure ratio we have

$$\frac{p_2}{p_1} = 1 + \frac{2\gamma}{\gamma + 1} \left(\frac{U^2}{c_1^2} - 1 \right).$$

Writing this in the form

$$\frac{p_2}{p_1} = 1 + \frac{2\gamma}{\gamma + 1} \left(\frac{U}{c_1} - 1 \right) \left(\frac{U}{c_1} + 1 \right)$$

and after using Eq. (3.57) in this latter equation we obtain

$$\frac{p_2}{p_1} = 1 + \frac{2\gamma}{\gamma + 1} \left[\left(\frac{\gamma + 1}{4} \right) \frac{u}{c_1} + \frac{1}{2} \left(\frac{\gamma + 1}{4} \right)^2 \frac{u^2}{c_1^2} \right] \left[2 + \left(\frac{\gamma + 1}{4} \right) \frac{u}{c_1} + \frac{1}{2} \left(\frac{\gamma + 1}{4} \right)^2 \frac{u^2}{c_1^2} \right].$$

Performing the multiplication in this latter equation and after retaining terms of the order of u^3/c_1^3, we have

$$\frac{p_2}{p_1} = 1 + \frac{2\gamma}{\gamma + 1} \left[2 \left(\frac{\gamma + 1}{4} \right) \frac{u}{c_1} + \frac{(\gamma + 1)^2}{8} \frac{u^2}{c_1^2} + \frac{(\gamma + 1)^3}{64} \frac{u^3}{c_1^3} \right]$$

and, on further simplification, we finally arrive at the following expression for the pressure ratio in the case of weak shocks,

$$\frac{p_2}{p_1} = 1 + \gamma \frac{u}{c_1} + \frac{\gamma(\gamma + 1)}{4} \frac{u^2}{c_1^2} + \frac{\gamma(\gamma + 1)^2}{32} \frac{u^3}{c_1^3} \tag{3.58}$$

The waves represented by Eqs. (2.14) to (2.19) of Chap. 2 are often described as *simple waves* [8] and according to Eqs. (2.18) and (2.21) we have

$$\frac{p_2}{p_1} = \left(\frac{c_2}{c_1}\right)^{\frac{2\gamma}{\gamma-1}} = \left[1 + \left(\frac{\gamma-1}{2}\right)\frac{u}{c_1}\right]^{\frac{2\gamma}{\gamma-1}}$$

By using the binomial expansion,

$$(1+x)^n = 1 + nx + \frac{n(n-1)}{2}x^2 + \frac{n(n-1)(n-2)}{6}x^3 + ..$$

we can write this pressure ratio as

$$\frac{p_2}{p_1} = 1 + \frac{2\gamma}{\gamma-1}\left(\frac{\gamma-1}{2}\right)\frac{u}{c_1} + \frac{1}{2}\left(\frac{2\gamma}{\gamma-1}\right)\left(\frac{2\gamma}{\gamma-1}-1\right)\left(\frac{\gamma-1}{2}\right)^2\frac{u^2}{c_1^2} + ..$$

and on further simplifying this equation and retaining terms to third order we arrive at the following result,

$$\frac{p_2}{p_1} = 1 + \gamma\frac{u}{c_1} + \frac{\gamma(\gamma+1)}{4}\frac{u^2}{c_1^2} + \frac{\gamma(\gamma+1)}{12}\frac{u^3}{c_1^3} + \ldots,$$

which shows that the pressure ratio for the so-called *simple wave* and the shock wave agree up to terms of second order in the shock strength.

3.12.3   Density Ratio for Weak Shocks

The equation for the density ratio according to Eq. (3.36) can be written as

$$\frac{\rho_2}{\rho_1} = \frac{U/c_1}{(U/c_1) - (u/c_1)}$$

and after substituting Eq. (3.57) in this latter equation we obtain

$$\frac{\rho_2}{\rho_1} = \frac{1 + \left(\frac{\gamma+1}{4}\right)\frac{u}{c_1} + \frac{1}{2}\left(\frac{\gamma+1}{4}\right)^2\frac{u^2}{c_1^2}}{1 + \left(\frac{\gamma+1}{4}\right)\frac{u}{c_1} + \frac{1}{2}\left(\frac{\gamma+1}{4}\right)^2\frac{u^2}{c_1^2} - \frac{u}{c_1}}$$

$$= \frac{1 + \left(\frac{\gamma+1}{4}\right)\frac{u}{c_1} + \frac{1}{2}\left(\frac{\gamma+1}{4}\right)^2\frac{u^2}{c_1^2}}{1 + \left(\frac{\gamma-3}{4}\right)\frac{u}{c_1} + \frac{1}{2}\left(\frac{\gamma+1}{4}\right)^2\frac{u^2}{c_1^2}}.$$

By using the binomial expansion; $(1 + x)^{-1} = 1 - x + x^2 - x^3 + ..$ for the denominator term and by retaining terms of the order of u^3/c_1^3, we have

$$\frac{\rho_2}{\rho_1} = \left[1 + \left(\frac{\gamma+1}{4}\right)\frac{u}{c_1} + \frac{1}{2}\left(\frac{\gamma+1}{4}\right)^2\frac{u^2}{c_1^2}\right] \times$$
$$\left[1 - \left(\frac{\gamma-3}{4}\right)\frac{u}{c_1} - \frac{1}{2}\left(\frac{\gamma+1}{4}\right)^2\frac{u^2}{c_1^2} + \left(\frac{\gamma-3}{4}\right)^2\frac{u^2}{c_1^2} + \frac{(\gamma-3)(\gamma+1)^2}{64}\frac{u^3}{c_1^3} - \left(\frac{\gamma-3}{4}\right)^3\frac{u^3}{c_1^3}\right]$$

Multiplying the terms on the right-hand side and by retaining terms of the order of u^3/c_1^3, we find that

$$\frac{\rho_2}{\rho_1} = 1 + \left\{\frac{\gamma+1}{4} - \left(\frac{\gamma-3}{4}\right)\right\}\frac{u}{c_1} + \left\{\frac{1}{32}(\gamma^2 - 14\gamma + 17) - \frac{(\gamma+1)(\gamma-3)}{16} + \frac{1}{2}\left(\frac{\gamma+1}{4}\right)^2\right\}\frac{u^2}{c_1^2} +$$
$$\left\{\frac{(\gamma-3)(\gamma-1)}{8} + \frac{(\gamma+1)(\gamma^2-14\gamma+17)}{32} - \frac{1}{2}\left(\frac{\gamma+1}{4}\right)^2\left(\frac{\gamma-3}{4}\right)\right\}\frac{u^3}{c_1^3}$$

and after simplifying the expression on the right-hand side, we finally obtain the following result for the density ratio for weak shocks,

$$\frac{\rho_2}{\rho_1} = 1 + \frac{u}{c_1} + \left(\frac{3-\gamma}{4}\right)\frac{u^2}{c_1^2} + \frac{(\gamma^2 - 14\gamma + 17)}{32}\frac{u^3}{c_1^3} \qquad (3.59)$$

Let us now compare this result with the density ratio in the case of *simple waves*. According to Eq. (2.21) we have

$$\frac{c_2}{c_1} = \left(\frac{\rho_2}{\rho_1}\right)^{\frac{\gamma-1}{2}}$$

and by using Eq. (2.18) we can write the latter equation as

$$\frac{\rho_2}{\rho_1} = \left[1 + \left(\frac{\gamma-1}{2}\right)\frac{u}{c_1}\right]^{\frac{2}{\gamma-1}}.$$

Using the binomial expansion,

$$(1 + x)^n = 1 + nx + \frac{n(n-1)}{2}x^2 + \frac{n(n-1)(n-2)}{6}x^3 + ..$$

we have

$$\frac{\rho_2}{\rho_1} = 1 + \left(\frac{2}{\gamma-1}\right)\left(\frac{\gamma-1}{2}\right)\frac{u}{c_1} + \frac{1}{2}\left(\frac{2}{\gamma-1}\right)\left(\frac{2}{\gamma-1}-1\right)\left(\frac{\gamma-1}{2}\right)^2\frac{u^2}{c_1^2} + ..$$

and after simplifying this equation by retaining terms to third order we obtain

$$\frac{\rho_2}{\rho_1} = 1 + \frac{u}{c_1} + \frac{(3-\gamma)}{4}\frac{u^2}{c_1^2} + \frac{(\gamma^2-5\gamma+6)}{12}\frac{u^3}{c_1^3} + ...$$

By comparing this result with Eq. (3.59) we see that the density ratio for the simple wave and the shock agree up to terms of second order in the shock strength.

3.12.4   Temperature Ratio for Weak Shocks

In the case of an ideal gas we have the equation, $p = \rho R T$, hence, the temperature ratio across the shock is given by

$$\frac{T_2}{T_1} = \frac{p_2}{p_1}\left(\frac{\rho_2}{\rho_1}\right)^{-1}$$

and by using the equation for the density ratio one can show that its inverted form is

$$\left(\frac{\rho_2}{\rho_1}\right)^{-1} = 1 - \frac{u}{c_1} + \left(\frac{\gamma+1}{4}\right)\frac{u^2}{c_1^2} - \frac{(\gamma+1)^2}{32}\frac{u^3}{c_1^3}$$

and taking this in conjunction with Eq. (3.58) for the pressure ratio, we have

$$\frac{T_2}{T_1} = \left[1 + \gamma\frac{u}{c_1} + \frac{\gamma(\gamma+1)}{4}\frac{u^2}{c_1^2} + \frac{\gamma(\gamma+1)^2}{32}\frac{u^3}{c_1^3}\right]$$
$$\times \left[1 - \frac{u}{c_1} + \left(\frac{\gamma+1}{4}\right)\frac{u^2}{c_1^2} - \frac{(\gamma+1)^2}{32}\frac{u^3}{c_1^3}\right].$$

Multiplying out on the right-hand side of this latter equation and retaining terms of the order of u^3/c_1^3, we find that the temperature ratio is given by

$$\frac{T_2}{T_1} = 1 + (\gamma-1)\frac{u}{c_1} + \frac{(\gamma-1)^2}{4}\frac{u^2}{c_1^2} + \frac{(\gamma-1)(\gamma+1)^2}{32}\frac{u^3}{c_1^3}, \tag{3.60}$$

and it is straightforward to show that the temperature ratio for simple waves agrees with this result for terms up to second order in the shock strength.

3.12.5 Sound Speed Ratio for Weak Shocks

The sound speed ratio is given by the equation

$$\frac{c_2}{c_1} = \sqrt{\frac{T_2}{T_1}}$$

and by using Eq. (3.60) for the temperature ratio one can show, after some algebra, that the sound speed ratio is given by

$$\frac{c_2}{c_1} = 1 + \left(\frac{\gamma - 1}{2}\right)\frac{u}{c_1} + \frac{(\gamma - 1)(\gamma + 1^2)}{64}\frac{u^3}{c_1^3}. \tag{3.61}$$

Note that the term of order u^2/c_1^2 drops out, hence, we deduce that the sound speed ratio in the case a weak shock and that of a simple wave (see, for example, Eq. (2.18)) agree up to terms of second order in the shock strength.

3.12.6 Entropy Change for Weak Shocks

Recalling our discussion in Sect. 1.6 of Chap. 1 and in Sects. 3.7 and 3.9 of Chap. 3, we found that the entropy change ($s_2 - s_1$) per unit mass of fluid crossing the shock can be written in the form,

$$\frac{s_2 - s_1}{c_V} = \ln\left(\frac{p_2}{p_1}\right) - \gamma \ln\left(\frac{\rho_2}{\rho_1}\right).$$

Using previous expressions for the ratios, p_2/p_1 and ρ_2/ρ_1, we have

$$\frac{s_2 - s_1}{c_V} = \ln\left[1 + \gamma\frac{u}{c_1} + \gamma\left(\frac{\gamma + 1}{4}\right)\frac{u^2}{c_1^2} + \gamma\frac{(\gamma + 1)^2}{32}\frac{u^3}{c_1^3}\right]$$
$$- \gamma \ln\left[1 + \frac{u}{c_1} + \left(\frac{3 - \gamma}{4}\right)\frac{u^2}{c_1^2} + \frac{(\gamma^2 - 14\gamma + 17)}{32}\frac{u^3}{c_1^3}\right]$$

Using the expansion, $\ln(1 + x) = x - \frac{x^2}{2} + \frac{x^3}{3} - \ldots$ for terms up to third order for small values of x, we can write the expression for the entropy change as

$$\frac{s_2 - s_1}{c_V} = \left\{\gamma\frac{u}{c_1} + \gamma\left(\frac{\gamma + 1}{4}\right)\frac{u^2}{c_1^2} + \gamma\frac{(\gamma + 1)^2}{32}\frac{u^3}{c_1^3}\right\} - \frac{1}{2}\left\{\gamma^2\frac{u^2}{c_1^2} + 2\gamma^2\left(\frac{\gamma + 1}{4}\right)\frac{u^3}{c_1^3}\right\} + \frac{\gamma^3}{3}\frac{u^3}{c_1^3}$$
$$- \gamma\left[\left\{\frac{u}{c_1} + \left(\frac{3 - \gamma}{4}\right)\frac{u^2}{c_1^2} + \frac{(\gamma^2 - 14\gamma + 17)}{32}\frac{u^3}{c_1^3}\right\} - \frac{1}{2}\left\{\frac{u^2}{c_1^2} + 2\left(\frac{3 - \gamma}{4}\right)\frac{u^3}{c_1^3}\right\} + \frac{1}{3}\frac{u^3}{c_1^3}\right]$$

One observes immediately that the terms of the order of u/c_1 drop out, hence, collecting terms of the order of u^2/c_1^2, we have

$$\frac{\gamma(\gamma+1)}{4} - \frac{\gamma^2}{2} - \frac{\gamma(3-\gamma)}{4} + \frac{\gamma}{2}$$
$$= \frac{1}{4}\left(\gamma^2 + \gamma - 2\gamma^2 - 3\gamma + \gamma^2 + 2\gamma\right)$$
$$= 0$$

and, therefore, terms of the order of u^2/c_1^2 also drop out. Collecting terms of the order of u^3/c_1^3, we have the following terms,

$$\frac{\gamma(\gamma+1)^2}{32} - \frac{\gamma^2(\gamma+1)}{4} + \frac{\gamma^3}{3} - \frac{\gamma(\gamma^2 - 14\gamma + 17)}{32} + \frac{\gamma(3-\gamma)}{4} - \frac{\gamma}{3}.$$

By simplifying this latter expression it is easy to show that it reduces to

$$\frac{\gamma(\gamma^2 - 1)}{12}$$

hence, the entropy change in the case of a weak shock can be written as

$$\frac{s_2 - s_1}{c_V} = \frac{\gamma(\gamma^2 - 1)}{12} \frac{u^3}{c_1^3}, \tag{3.62}$$

which is of third order when expanded in powers of the shock strength. Consequently, the increase in the entropy of the fluid as it crosses the shock is zero for terms up to second order. This implies that the isentropic equations can be used for determining the flow properties in a disturbance up to second order in the shock strength.

3.12.7   Change in the Riemann Invariant R_- for Weak Shocks

We already know that the Riemann invariant is constant across a simple wave, whereas the Riemann invariant ahead of the shock can be written as

$$R_-^{ahead} = -\frac{2c_1}{\gamma - 1},$$

while the Riemann invariant behind the shock is

$$R_{-}^{behind} = u - \frac{2c_2}{\gamma - 1},$$

hence,

$$\frac{R_{-}^{behind}}{c_1} = \frac{u}{c_1} - \frac{2}{\gamma - 1}\frac{c_2}{c_1}.$$

Substituting for c_2/c_1 according to Eq. (3.61) on the right-hand side of the latter equation we have

$$\frac{R_{-}^{behind}}{c_1} = \frac{u}{c_1} - \frac{2}{\gamma - 1}\left[1 + \left(\frac{\gamma - 1}{2}\right)\frac{u}{c_1} + \frac{(\gamma - 1)(\gamma + 1)^2}{64}\frac{u^3}{c_1^3}\right]$$

$$= -\frac{2}{\gamma - 1} - \frac{(\gamma + 1)^2}{32}\frac{u^3}{c_1^3}$$

and, therefore,

$$\frac{R_{-}^{behind} - R_{-}^{ahead}}{c_1} = -\frac{(\gamma + 1)^2}{32}\frac{u^3}{c_1^3}. \qquad (3.63)$$

Hence, we deduce that the change in the Riemann invariant across the shock only appears in terms of third or higher order in the shock strength.

3.13   Thickness of the Shock Wave Region

This text utilizes the von Neumann-Richtmyer method of dealing with shocks by introducing an artificially large viscosity into the equations of motion. This large viscosity, as we shall see in the following chapter, smears out the shock transition region so that the shock acquires a thickness somewhat greater than the spacing of the grid size used in the numerical computations. In reality, shock regions have a much smaller thickness; in fact, a thickness so small that we may consider the parameters on either side of the shock to undergo discontinuous jumps in their values.

When viscosity and thermal conduction are included in the equations, these effects remove the tendency for the solutions to become discontinuous so that the flow parameters vary very rapidly and continuously across the shock region. The thickness of this region has received the attention of many investigators, including Landau and Lifshitz [8], Curle and Davis [9] and von Mises [12], to mention but a few. Accordingly, we will now consider the typical thickness of this region where significant changes are taking place but, more importantly, to see if there is any justification for representing the shock as the position where the flow parameters undergo discontinuous jumps.

Besides the normal pressure force that is always directed *into* a fluid element as a result of the surrounding fluid, there exists shear and normal stress components that also act on the fluid element and these are due to velocity gradients in the flow. In the case of one-dimensional motion in, say, the x-direction, any shearing components vanish as the fluid particles adjacent to the four faces of the rectangular fluid element, which are parallel to the x-direction, move at the same velocity as the element itself. Consequently, only the normal stress components on the two faces perpendicular to the x-direction remain. These normal stress components in fluid flow are, in general, very small in comparison to the shear stress components, but they manifest themselves when the velocity gradient, $\partial u/\partial x$, becomes very large, particularly within the shock wave region. Accordingly, the pressure p term that appears in the equations of motion as outlined in Chap. 1 has an additional normal viscous stress component included; as a result, the equations of continuity, momentum and energy can be written in the following form [13],

$$\frac{\partial \rho}{\partial t} + \frac{\partial}{\partial x}(\rho u) = 0$$

$$\rho \frac{\partial u}{\partial t} + \rho u \frac{\partial u}{\partial x} = -\frac{\partial}{\partial x}(p - \sigma_x)$$

$$\frac{\partial}{\partial t}\left[\rho\left(\frac{1}{2}u^2 + e\right)\right] + \frac{\partial}{\partial x}\left[\rho u\left(\frac{1}{2}u^2 + e\right) + u(p - \sigma_x)\right] = 0$$

where the viscous force per unit area is given by the equation [13],

$$\sigma_x = \frac{4\mu}{3}\frac{\partial u}{\partial x}$$

where μ is the coefficient of viscosity and thermal conduction has been neglected in these equations.

Let us consider a steady-state shock travelling in the positive x-direction with speed U_s and we seek a travelling-wave solution [14] where all quantities are functions of $x - U_s t$ which we designate by the symbol, ω, hence, $\omega = x - U_s t$. In terms of this travelling-wave, the continuity equation can be written as

$$-U_s\frac{d\rho}{d\omega} + \frac{d}{d\omega}(\rho u) = 0,$$

and integrating this equation gives

$$-U_s\rho + \rho u = \text{constant}$$

and applying the boundary conditions at $x = +\infty$ ahead of the shock, namely, $u = 0$ and $\rho = \rho_0$, where ρ_0 is the ambient air density, we find that the constant of integration is $-U_s\rho_0$, hence,

$$-U_s\rho + \rho u = -U_s\rho_0$$

and, therefore,

$$\frac{\rho}{\rho_0} = \frac{U_s}{U_s - u}. \qquad (3.64)$$

The momentum equation yields,

$$-\rho U_s \frac{du}{d\omega} + \rho u \frac{du}{d\omega} + \frac{d}{d\omega}(p - \sigma_x) = 0,$$

that is,

$$(-\rho U_s + \rho u)\frac{du}{d\omega} + \frac{d}{d\omega}(p - \sigma_x) = 0.$$

By using the continuity equation this becomes,

$$-U_s\rho_0 \frac{du}{d\omega} + \frac{d}{d\omega}(p - \sigma_x) = 0,$$

and after integrating we obtain,

$$-U_s\rho_0 u + (p - \sigma_x) = \quad \text{constant.}$$

Applying the boundary conditions at $x = +\infty$, namely, $u = 0, p = p_0$ and $\sigma_x = 0$, where p_0 is the ambient air pressure, we find that the constant of integration is p_0, hence,

$$-U_s\rho_0 u + p - \sigma_x = p_0$$

and, therefore,

$$p - \sigma_x = p_0 + U_s\rho_0 u \qquad (3.65)$$

In the same way, the energy equation becomes,

$$-U_s\frac{d}{d\omega}\left[\frac{1}{2}\rho u^2 + \rho e\right] + \frac{d}{d\omega}\left[u\left(\frac{1}{2}\rho u^2 + \rho e\right) + u(p - \sigma_x)\right] = 0.$$

Since the internal energy per unit mass e is given by $e = p/(\gamma - 1)\rho$, we can write this latter energy equation as

$$-U_s \frac{d}{d\omega}\left[\frac{1}{2}\rho u^2 + \frac{p}{\gamma - 1}\right] + \frac{d}{d\omega}\left[u\left(\frac{1}{2}\rho u^2 + \frac{p}{\gamma - 1}\right) + u(p - \sigma_x)\right] = 0$$

and on integrating, we obtain,

$$-U_s\left[\frac{1}{2}\rho u^2 + \frac{p}{\gamma - 1}\right] + \left[u\left(\frac{1}{2}\rho u^2 + \frac{p}{\gamma - 1}\right) + u(p - \sigma_x)\right] = \text{constant.}$$

Applying the boundary conditions at $x = +\infty$, namely, $u = 0$, $p = p_0$, $\rho = \rho_0$ and $\sigma_x = 0$, accordingly, the constant of integration is $-U_s p_0/(\gamma - 1)$. Hence, this energy equation becomes

$$-\frac{1}{2}\rho U_s u^2 - \frac{U_s p}{\gamma - 1} + \frac{U_s p_0}{\gamma - 1} + \frac{1}{2}\rho u^3 + \frac{pu}{\gamma - 1} + u(p - \sigma_x) = 0.$$

Substituting the relationship for $p - \sigma_x$ according to Eq. (3.65) in this latter equation, we find that

$$-\frac{1}{2}\rho U_s u^2 - \frac{p}{\gamma - 1}(U_s - u) + \frac{U_s p_0}{\gamma - 1} + \frac{1}{2}\rho u^3 + u p_0 + U_s \rho_0 u^2 = 0.$$

Let us now eliminate the pressure p in the above equation by using Eq. (3.65), hence, after a few algebraic steps the following equation is obtained;

$$-\frac{1}{2}\rho U_s u^2 - \left(\frac{U_s - u}{\gamma - 1}\right)\frac{4}{3}\mu\frac{du}{d\omega} + \frac{\gamma u p_0}{\gamma - 1} - \frac{U_s^2 \rho_0 u}{\gamma - 1} + \frac{\gamma \rho_0 U_s u^2}{\gamma - 1} + \frac{1}{2}\rho u^3 = 0, \quad (3.66)$$

where the relationship, $\sigma_x = (4\mu/3)(\partial u/\partial x)$, has been substituted. By dividing Eq. (3.66) by ρ_0 and using Eq. (3.64), we find that

$$-\frac{1}{2}\frac{U_s^2 u^2}{(U_s - u)} - \frac{4(U_s - u)\mu}{3(\gamma - 1)\rho_0}\frac{du}{d\omega} + \frac{uc_0^2}{\gamma - 1} - \frac{U_s^2 u}{\gamma - 1} + \frac{\gamma U_s u^2}{\gamma - 1} + \frac{1}{2}\left(\frac{U_s}{U_s - u}\right)u^3 = 0,$$

where we have used the equation for the speed of sound c_0, namely, $c_0^2 = \gamma p_0/\rho_0$. By multiplying the previous equation by $(U_s - u)(\gamma - 1)$ and after grouping various terms, the following equation results,

$$\frac{(U_s - u)^2}{\rho_0}\frac{4}{3}\mu\frac{du}{d\omega} = \frac{1}{2}(\gamma + 1)U_s u^2(U_s - u) + uc_0^2(U_s - u) - U_s^2 u(U_s - u),$$

hence, we can finally write it in the following manner,

$$\frac{(U_s - u)}{\rho_0} \frac{4}{3} \mu \frac{du}{d\omega} = u \left[\frac{1}{2}(\gamma + 1)U_s u + c_0^2 - U_s^2 \right]. \tag{3.67}$$

One can observe that $du/d\omega = 0$ when $u = 0$ and also when

$$\left[\frac{1}{2}(\gamma + 1)U_s u + c_0^2 - U_s^2 \right] = 0.$$

Solving this latter equation for u, one finds that

$$u = \frac{2c_0}{\gamma + 1} \left(\frac{U_s}{c_0} - \frac{c_0}{U_s} \right) = \frac{2c_0}{\gamma + 1} \left(M - \frac{1}{M} \right), \tag{3.68}$$

which is just Eq. (3.39b) where M is the Mach number and u is the particle or material velocity behind the shock. Consequently, $u = 0$ and u given by Eq. (3.68) are the extreme values of the particle velocity downstream and upstream, respectively, of the propagating shock. In order to determine the width of the shock wave region, one can integrate Eq. (3.67) to obtain the velocity variation across the shock. Letting

$$u_1 = \frac{2c_0}{(\gamma + 1)} \left(\frac{U_s}{c_0} - \frac{c_0}{U_s} \right) = \frac{2c_0}{(\gamma + 1)} \left(M - \frac{1}{M} \right)$$

and re-writing Eq. (3.67) in the form,

$$(U_s - u) \frac{4\mu}{3\rho_0} \frac{du}{d\omega} = \left(\frac{\gamma + 1}{2} \right) U_s u(u - u_1). \tag{3.69}$$

Since $U_s > u$ and $0 \leq u \leq u_1$ for u lying between its values at $x = \pm \infty$, we can see that the slope, $(du/d\omega)$, is negative through the shock region.

Integrating this latter equation by using a partial fraction expansion for the quantity, $u(u - u_1)$, one can show that the following equation can be written as,

$$\left(\frac{U_s - u_1}{u_1} \right) \ln \left(\frac{u_1 - u}{U_s} \right) - \frac{U_s}{u_1} \ln \left(\frac{u}{U_s} \right) = \frac{3(\gamma + 1)\rho_0 U_s}{8\mu} \omega + \text{ constant.} \tag{3.70}$$

The constant of integration can be determined by choosing the origin ($\omega = 0$) at $u = u_1/2$, which corresponds to the point of inflection ($d^2u/d\omega^2 = 0$) of the velocity profile [15], hence, the constant of integration is given by

$$\text{constant } = \left(\frac{U_s - u_1}{u_1} \right) \ln \left(\frac{u_1}{2U_s} \right) - \frac{U_s}{u_1} \ln \left(\frac{u_1}{2U_s} \right)$$

and by substituting this value back in Eq. (3.70), we can write the equation in the following form,

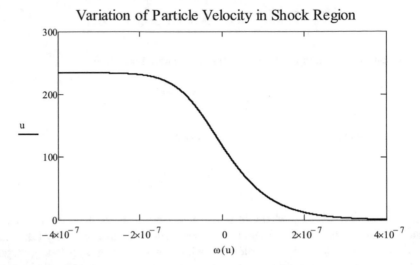

Fig. 3.7 A snapshot of the profile of the particle velocity u (in ms^{-1}) as a function of ω across the shock is shown (see text)

$$\omega(u) = \frac{8\mu}{3(\gamma+1)\rho_0 U_s}\left[\left(\frac{U_s-u_1}{u_1}\right)\ln\left(\frac{2u_1-2u}{u_1}\right) - \frac{U_s}{u_1}\ln\left(\frac{2u}{u_1}\right)\right] \qquad (3.71)$$

Let us consider a propagating shock in air ($\gamma = 1.4$), and let us assume the following parameters for the ambient pressure p_0 and density ρ_0 ahead of the shock wave; $p_0 = 1.01 \times 10^5 Nm^{-2}$, $\rho_0 = 1.25 kgm^{-3}$. With a Mach number of 1.5 and with a coefficient of viscosity of $\mu = 2 \times 10^{-5} Nm^{-2}s$ [15], Fig. 3.7 shows a snapshot of the velocity profile through the shock transition according to Eq. (3.71). The extent of this region ΔW (in metres) is defined here as the width where u has decreased from 90% to 10% of its value at $x = -\infty$. Hence, we find that

$$\Delta W = |\omega(0.9u_1) - \omega(0.1u_1)| = 2.57 \times 10^{-7}m,$$

which is extremely narrow and of the same order of magnitude as the molecular mean-free path. This justifies the general assumption that the changes taking place in the flow parameters on traversing this region can be considered to undergo discontinuous jumps in their values.

The density ratio, ρ/ρ_0, according to Eq. (3.64) is given by

$$\frac{\rho}{\rho_0} = \frac{U_s}{U_s - u},$$

and from Eq. (3.65) in conjunction with Eq. (3.67), we can write the pressure ratio, p/p_0, in the following form,

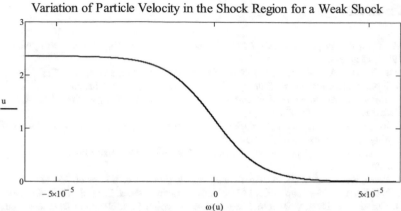

Fig. 3.8 A snapshot of the profile of the particle velocity u (in ms^{-1}) as a function of ω across the shock is shown for a weak shock (see text)

$$\frac{p}{p_0} = 1 + \frac{\gamma u}{c_0^2} \left[\frac{\left(\frac{\gamma-1}{2}\right)uU_s + c_0^2}{U_s - u} \right].$$

Similar profiles of pressure and density ratios can be plotted by using these equations, and the numerical results for the pressure and density behind the shock are found to be in agreement with the values given by Eqs. (3.25) and (3.28).

If we consider the case of a weak shock where the pressure ratio across the shock is only 1% above the ambient pressure, p_0, hence, $p/p_0 = 1.01$. In this case, the particle velocity u_1 satisfies the weak shock approximation according to Eq. (3.58) as $u_1/c_0 \ll 1_<$. This corresponds to a Mach number of 1.004 and a snapshot of the particle velocity profile across the shock is shown in Fig. 3.8.

The width of the shock region in this case is

$$\Delta W = |\omega(0.9u_1) - \omega(0.1u_1)| = 3.3 \times 10^{-5} m,$$

which is minute value, corresponding to a small fraction of a millimetre. Even for this weak shock, despite a considerable increase in width over the previous value, some justification still exists for treating the shock position as a discontinuity.

3.14 Conclusions

This ends our discussion of the various shock wave relationships but we will be returning to them again in the remaining chapters. In particular, the various relationships will be compared with the numerical solutions obtained when dealing with plane shocks as well as in Chaps. 5 and 6, where the limiting form of some of the equations will be required when dealing with very strong shock waves.

References

1. M. A. Saad, *Compressible Fluid Flow*, (Prentice-Hall, Inc., Englewood Cliffs, New Jersey 1985), Chapter 4
2. J. Billingham, A.C. King, *Wave Motion* (Cambridge University Press, New York, 2000)
3. C.E. Needham, *Blast Waves* (Springer, Berlin, Heidelberg, 2010) Chapter 3
4. V. Babu, *Fundamentals of Gas Dynamics*, 2nd edn. (John Wiley & Sons Ltd., Chichester, United Kingdom, 2015) Chapter 3
5. O. Regev, O.M. Umurhan, P.A. Yecko, *Modern Fluid Dynamics for Physics and Astrophysics* (Springer, New York, 2016) Section 6.5.2
6. J.H.S. Lee, *The Gas Dynamics of Explosions* (Cambridge University Press, New York, 2016) Chapter 1
7. R. Blum, Normal Shock Waves in a Compressible Fluid. Am. J. Physiol. **35**, 428 (1967)
8. L.D. Landau, *E. M. Lifshitz* (Fluid Mechanics, Pergamon Press, London, 1966) Chapter 9
9. N. Curle, H. J. Davies, *Modern Fluid Dynamics*, vol.2, (Van Nostrand Reinhold Company, London, 1971), Chapter 3
10. R. C. Binder, *Advanced Fluid Mechanics*, vol. 1, (Prentice-Hall, 1958), Chapter 2
11. J.D. Anderson, *Modern Compressible Flow with Historical Perspective*, 3rd edn. (McGraw-Hill, New York, 2003) Chapter 7
12. R. von Mises, *Mathematical Theory of Compressible Fluid Flow* (Dover Publications, Inc., Mineola, New York, 2004), Chapter 3
13. J.D. Anderson Jr., *Computational Fluid Dynamics: The Basics with Applications* (McGraw Hill, New York, 1995) Chapter 2
14. F. H. Harlow, *LA-2412 Report* (Los Alamos Scientific Laboratory of the University of California, Los Alamos, New Mexico, 1960), Chapter 9
15. R.K. Anand, H.C. Yadav, The effects of viscosity on the structure of shock waves in a non-ideal gas. Acta Phys. Pol. A **129**, 28 (2016)

Chapter 4
Numerical Treatment of Plane Shocks

4.1 Introduction

This chapter describes the numerical procedure that is used to solve a number of fluid flow problems involving plane shock waves. An alternative form of the differential equations of fluid flow in Lagrangian form is derived. A dissipative mechanism in the form of an *artificial viscosity* is introduced into the equations in order to deal with shocks. A set of difference equations corresponding to the differential equations is presented and the numerical solution of these difference equations is obtained. The numerical results are then compared with the theoretical predictions.

4.2 The Need for Numerical Techniques

An analytic solution of the partial differential equations describing the flow of a compressible fluid is, in general, a difficult task due to the nonlinearity of the equations and this difficulty demonstrates the need for more powerful techniques to solve the equations. Accordingly, it is necessary to turn to numerical methods in which the differential equations are replaced by finite difference equations that describe the properties of the fluid at discrete positions and at discrete times. Once the initial and boundary conditions have been specified, one can follow the changes that evolve in the properties of the fluid by solving the set of difference equations at successive time steps (generally known as the *time-marching procedure*). However, a solution is severely hampered by the presence of shocks. In the case of non-viscous flow the shock front is a propagating discontinuity and across which the hydrodynamic functions of pressure, density, temperature and velocity exhibit abrupt changes. The boundary conditions that connect the values on either side of the shock are supplied by the Rankine-Hugoniot equations but their application is complicated as the shock is in motion. In addition, the motion of the shock surface

is not known in advance but, instead, is governed by the differential equations
themselves. In reality, however, when the effects of viscosity and thermal conduction are included in the momentum and energy equations the shock wave is no longer
a discontinuous front, but a very steep continuous transition region whose length is a
few molecular mean free paths as previously noted. Accordingly, the spatial grid size
in the finite difference equations would need to be much smaller than the molecular
mean free path in order to realistically model the shock transition region. This aspect
will be referred to at a later stage.

In order to deal with shocks Von Neumann and Richtmyer [1] introduced a
technique that avoids the need for any boundary conditions referred to above and
treats the calculation as if there were no shocks present at all. Instead, the shocks
appear as near-discontinuities that move through the fluid with essentially the correct
speed and across which the thermodynamic variables approximate the predictions of
the Rankine-Hugoniot relations. The method proposed by Von Neumann and
Richtmyer introduces a dissipative mechanism in the form of an *artificial viscosity*
into the equations so that the shocks are replaced by a thin layer where the pressure,
density, temperature and velocity vary rapidly but continuously and it means that the
shock acquires a thickness somewhat greater than the spacing of the points used in
the numerical procedure. Consequently, the difficult problem of applying boundary
conditions at the shock is avoided. The form of the artificial viscosity introduced by
Von Neumann and Richtmyer for the one-dimensional case is given by the equation,

$$q = -\frac{(\rho_0 \kappa \Delta x)^2}{v} \frac{\partial v}{\partial t} \left| \frac{\partial v}{\partial t} \right|, \tag{4.1}$$

where propagation takes place in the x-direction; v is the specific volume, ρ_0 is the
initial density, Δx is the interval length or spatial grid size used in the numerical
procedure and κ is a number close to unity [1, 2]. If larger values of κ are used it is
found that the changes in the parameters like pressure, specific volume etc. across the
shock are too sluggish while smaller values of κ result in large oscillations in the
parameters behind the shock. This aspect will be investigated in due course by
considering some numerical results having differing values of κ. Since the objective
is to make the shock region as narrow as possible, selecting $\kappa \approx 1.2$ to 1.5 represents
a compromise and results in the shock region having a thickness somewhat larger
than the grid interval Δx in the numerical procedure. The term q is additional to the
pressure forces and is large in regions of shocks (where the gas is compressed) and
negligible outside the shock region. The article by Von Neumann and Richtmyer [1]
provides a complete picture of the technique which includes the mathematical
procedure and the stability of the differential/difference equations involved.

4.3 Lagrangian Equations in Plane Geometry with Artificial Viscosity

Let us now consider an alternative derivation of the equations of fluid flow in Lagrangian form with artificial viscosity included. Here, we will adopt the Lagrangian notation used by Zel'dovich and Raizer [3] and in the article by C. F. Sprague [2].

4.3.1 Continuity Equation

Let us consider a one-dimensional model of the fluid flow in, say, the x-direction. We will concentrate our attention on a small infinitesimal mass of fluid moving with the flow at some instant in time which we take to be at $t = 0$. At this instance let us assume that this small element of fluid is located at x_0, has length dx_0, cross-sectional A and density $\rho(x_0, 0)$. The mass of this fluid element is $\rho(x_0, 0)Adx_0$. As this fluid element moves its density and volume may change but its mass remains constant. After a time t the fluid element is located at $x = x(x_0, t)$ with a length, dx, where $dx = x(x_0 + dx_0, t) - x(x_0, t)$ and with density $\rho(x_0, t)$. Conservation of mass implies that

$$\rho(x_0, 0)Adx_0 = \rho(x_0, t)Adx$$

Hence,[1]

$$\frac{1}{\rho(x_0, t)} = \frac{1}{\rho(x_0, 0)} \frac{\partial x}{\partial x_0}.$$

Writing the left-hand side in terms of the specific volume v rather than the density ρ we have

$$v(x_0, t) = \frac{1}{\rho(x_0, 0)} \frac{\partial x}{\partial x_0},$$

differentiating with respect to time yields the following continuity equation,

---

[1]Partial derivatives are used here to indicate the changes in position and time of *specific particles*; nonetheless, it should be understood that these partial derivatives imply that we are in fact following the path taken by *specific particles* of fluid according to the Lagrangian description.

$$\frac{\partial v}{\partial t} = \frac{1}{\rho(x_0, 0)} \frac{\partial u}{\partial x_0}, \tag{4.2}$$

where the fluid velocity u is

$$u = \frac{\partial x}{\partial t}, \tag{4.3}$$

where $u \equiv u(x_0, t)$. Partial derivatives are used to indicate that we are following specific particles of fluid; for example, the partial derivatives above could also be written in the following manner;

$$\frac{\partial}{\partial t} \equiv \left(\frac{\partial}{\partial t}\right)_{SpecificParticle}$$

In fact, the *substantial* or *material* derivative, D/Dt, referred to in Chap. 1 could also be written as,

$$\frac{D}{Dt} \equiv \left(\frac{\partial}{\partial t}\right)_{SpecificParticle} \tag{4.4}$$

4.3.2   Equation of Motion

The conservation of momentum equation is a mathematical expression of Newton's law of motion and requires that the rate of change of momentum equals the applied external force. Using the information in relation to the derivation of the continuity equation above, we can write Newton's law in the form;

$$\rho(x_0, 0)dx_0 A \frac{\partial u(x_0, t)}{\partial t} = Ap(x_0, t) - Ap(x_0 + dx_0, t),$$

where A is the cross-sectional area over which the force is applied and p is the pressure. Hence, the momentum equation becomes,

$$\rho(x_0, 0) \frac{\partial u}{\partial t} = -\frac{\partial p(x_0, t)}{\partial x_0}, \tag{4.5}$$

which should be compared with the one-dimensional Lagrangian form of the momentum equation in Chap. 1, namely, Eq. (1.47).

When artificial viscosity q is included Eq. (4.5) becomes

$$\rho(x_0, 0) \frac{\partial u}{\partial t} = -\frac{\partial}{\partial x_0}(p + q). \tag{4.6}$$

4.3.3 Equation of Energy Conservation

The energy balance equation (Eq. (1.54) Chap. 1) when written in terms of partial derivatives (to indicate that we are following the motion of a specific particle of fluid) becomes;

$$\frac{\partial e}{\partial t} = -p \frac{\partial v}{\partial t} \tag{4.7}$$

and with artificial viscosity included we have,

$$\frac{\partial e}{\partial t} = -(p + q) \frac{\partial v}{\partial t}. \tag{4.8}$$

Using the equation for the internal energy per unit mass for an ideal gas, namely;

$$e = \frac{pv}{\gamma - 1} \tag{4.9}$$

and differentiating we obtain,

$$\frac{\partial e}{\partial t} = \frac{1}{\gamma - 1} \left(p \frac{\partial v}{\partial t} + v \frac{\partial p}{\partial t} \right). \tag{4.10}$$

Substituting this in Eq. (4.8) yields,

$$\frac{1}{\gamma - 1} \left(p \frac{\partial v}{\partial t} + v \frac{\partial p}{\partial t} \right) = -(p + q) \frac{\partial v}{\partial t},$$

hence,

$$[\gamma p + (\gamma - 1)q] \frac{\partial v}{\partial t} + v \frac{\partial p}{\partial t} = 0 \tag{4.11}$$

It is straightforward to show that Eq. (4.11), becomes

$$\frac{\partial p}{\partial t} = \frac{1}{\rho}[\gamma p + (\gamma - 1)q]\frac{\partial \rho}{\partial t}, \tag{4.12}$$

when it is written in terms of the density ρ rather than in terms of the specific volume. We have now completed the derivation of the relevant differential equations: these equations or, more accurately, their corresponding *difference equations* will be used later on in this chapter to numerically solve some simple problems involving plane shocks.

4.4   Artificial Viscosity

The Von Neumann-Richtmyer method for dealing with shocks avoids the difficult problem of having to apply boundary conditions connecting the parameters on both sides of a discontinuity. Their introduction of an artificial viscosity term into the equations of motion eliminates this discontinuity and smears out the shock transition into a thin region in which the various parameters, such as, pressure, particle velocity, density etc. exhibit rapid but continuous changes. The form of the artificial viscosity term that they introduced is given by Eq. (4.1).

In this section the Von Neumann-Richtmyer method for dealing with a steady-state plane shock in the presence of dissipation in the form of an artificial viscosity will be discussed and we will follow their analysis. The purpose of this analysis is to show that the expression for q, as given by Eq. (4.1), meets certain requirements as set out in their article.

4.4.1   Equations for Plane-Wave Motion with Artificial Viscosity

The fundamental equations for plane-wave motion with artificial viscosity q included have already been established and they are summarized below (with x_0 simply replaced by x):

$$\frac{\partial v}{\partial t} = \frac{1}{\rho_0}\frac{\partial u}{\partial x} \quad \text{(Continuity)}$$

$$\rho_0\frac{\partial u}{\partial t} = -\frac{\partial}{\partial x}(p + q) \quad \text{(Momentum)}$$

$$\frac{\partial e}{\partial t} = -(p+q)\frac{\partial v}{\partial t} \quad \text{(Energy)}$$

$$e = \frac{pv}{\gamma - 1} \quad \text{(Caloric equation of state)}$$

where ρ_0 is the initial density, considered uniform throughout and, as we have already seen, q according to Von Neumann and Richtmyer is given by the equation,

$$q = -\frac{(\rho_0 \kappa \Delta x)^2}{v} \frac{\partial v}{\partial t}\left|\frac{\partial v}{\partial t}\right|.$$

Using the continuity equation, the expression for q can also be written as

$$q = -\frac{(\kappa \Delta x)^2}{v} \frac{\partial u}{\partial x}\left|\frac{\partial u}{\partial x}\right|, \quad (4.13)$$

which takes the form of a nonlinear viscosity. According to Von Neumann and Richtmyer [1], these equations must meet the following requirements:

1. The equations must possess solutions without discontinuities.
2. The width of the shock region must be small and of the same order as the interval Δx used in the numerical computation.
3. The effect of terms containing q must be negligible outside the shock region.
4. The Hugoniot relations must hold across the shock.

4.4.2   A Steady-State Plane Shock with Artificial Viscosity

Von Neumann and Richtmyer considered a long pipe containing a fluid at rest and into which a piston is pushed at constant speed as shown in Fig. 4.1. After a sufficiently long time the shock wave has moved a considerable distance from the piston and propagates down the tube with a constant velocity U_s.

In the absence of artificial viscosity the specific volume v and the velocity u at some instance in time are shown plotted in Fig. 4.1 as solid lines, whereas in the presence of viscosity they are shown as broken lines. Under steady-state conditions the quantities, u, v, p and e depend on x and t through the combination;

$$\omega = x - U_s t \quad (4.14)$$

and Von Neumann and Richtmyer define the mass crossing unit area per unit time as

$$m = \rho_0 U_s. \quad (4.15)$$

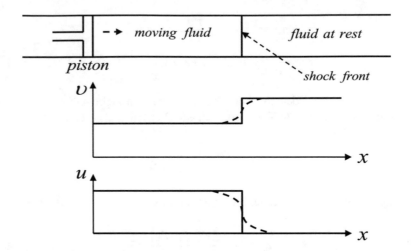

Fig. 4.1 A steady-state plane shock wave moving down a tube

With ω and m defined, we have in relation to the momentum equation;

$$\frac{\partial u}{\partial t} = \frac{du}{d\omega}\frac{\partial \omega}{\partial t} = -U_s\frac{du}{d\omega}$$

and

$$\frac{\partial}{\partial x}(p+q) = \frac{d}{d\omega}(p+q)\frac{\partial \omega}{\partial x} = \frac{d}{d\omega}(p+q),$$

hence, the momentum equation becomes

$$m\frac{du}{d\omega} = \frac{d}{d\omega}(p+q) \tag{4.16}$$

Similarly, the energy equation becomes,

$$\frac{de}{d\omega} + (p+q)\frac{dv}{d\omega} = 0 \tag{4.17}$$

For the continuity equation, we have

$$\frac{\partial v}{\partial t} = \frac{dv}{d\omega}\frac{\partial \omega}{\partial t} = -U_s\frac{dv}{d\omega}$$

and

$$\frac{\partial u}{\partial x} = \frac{du}{d\omega}\frac{\partial \omega}{\partial x} = \frac{du}{d\omega},$$

yielding the following equation for continuity,

$$-m\frac{dv}{d\omega} = \frac{du}{d\omega} \tag{4.18}$$

Equations (4.16) and (4.18) give

$$-m^2\frac{dv}{d\omega} = \frac{d}{d\omega}(p+q) \tag{4.19}$$

Consider now the following expansion;

$$\frac{d}{d\omega}[(p+q)v] = (p+q)\frac{dv}{d\omega} + v\frac{d}{d\omega}(p+q),$$

hence,

$$(p+q)\frac{dv}{d\omega} = \frac{d}{d\omega}[(p+q)v] - v\frac{d}{d\omega}(p+q)$$

and substituting this latter equation in Eq. (4.17) gives,

$$\frac{de}{d\omega} + \frac{d}{d\omega}[(p+q)v] - v\frac{d}{d\omega}(p+q) = 0$$

and by using Eq. (4.19) this becomes,

$$\frac{de}{d\omega} + \frac{d}{d\omega}[(p+q)v] + m^2 v\frac{dv}{d\omega} = 0 \tag{4.20}$$

Integrating Eqs. (4.18), (4.19) and (4.20), yields,

$$mv + u = C_1 \tag{4.21}$$

$$m^2 v + p + q = C_2 \tag{4.22}$$

$$e + (p+q)v + \frac{1}{2}m^2 v^2 = C_3, \tag{4.23}$$

where C_1, C_2 and C_3 are constants of integration. Von Neumann and Richtmyer consider initial (subscript i) and final (subscript f) values to be denoted by:

$$\text{as } \omega \to \infty; \ v \to v_i, \ p \to p_i, \ e \to e_i, \ q \to 0;$$

$$\text{and as } \omega \to -\infty; \ v \to v_f, \ p \to p_f, \ e \to e_f, \ q \to 0.$$

Hence, Eq. (4.22) yields,

$$m^2 v_i + p_i = C_2 \quad \text{and} \quad m^2 v_f + p_f = C_2$$

and, therefore,

$$m^2 v_i + p_i = m^2 v_f + p_f \tag{4.24}$$

and substituting for m gives,

$$\frac{U_s^2}{v_i^2} (v_i - v_f) = p_f - p_i.$$

If we let $v_i = 1/\rho_0$ be the initial density and $v_f = \rho$ the final density with similar expressions for p_i and p_f, then the latter equation becomes,

$$U_s^2 \rho_0^2 \left(\frac{1}{\rho_0} - \frac{1}{\rho} \right) = p - p_0,$$

hence,

$$U_s = \sqrt{\frac{\rho}{\rho_0} \left(\frac{p - p_0}{\rho - \rho_0} \right)}, \tag{4.25}$$

which is Eq. (3.7) of Chap. 3 and is one of the Hugoniot relationships that gives the shock speed in terms of the pressure and density changes across the shock. Similarly, in terms of initial and final values, Eq. (4.23) gives,

$$e_i + p_i v_i + \frac{1}{2} m^2 v_i^2 = C_3$$

and

$$e_f + p_f v_f + \frac{1}{2} m^2 v_f^2 = C_3.$$

It, therefore, follows that,

$$e_f - e_i + p_f v_f - p_i v_i + \frac{1}{2} m^2 \left(v_f^2 - v_i^2 \right) = 0,$$

hence,

$$e_f - e_i = p_i v_i - p_f v_f + \frac{1}{2} m^2 \left(v_i^2 - v_f^2 \right).$$

Substituting Eq. (4.24) in this latter equation, yields,

$$
\begin{aligned}
e_f - e_i &= p_i v_i - p_f v_f + \frac{1}{2} \frac{(p_f - p_i)}{v_i - v_f} \left(v_i^2 - v_f^2 \right) \\
&= p_i v_i - p_f v_f + \frac{1}{2} (p_f - p_i)(v_i + v_f) \\
&= \frac{1}{2} (p_i + p_f)(v_i - v_f),
\end{aligned}
\tag{4.26}
$$

which is the other Hugoniot equation, namely, Eq. (3.9) of Chap. 3; consequently, requirement 4 is satisfied.

4.4.3   Variation in the Specific Volume Across the Shock

In order to determine the shape of the shock Von Neumann and Richtmyer considered solution satisfying the condition, $\partial v / \partial t \leq 0$; this is equivalent to the condition, $dv/d\omega \geq 0$, which corresponds to a shock wave moving to the right. Since we already have $\partial v / \partial t = -U_s (dv/d\omega)$ and $m = \rho_0 U_s$, then the expression for q can be written as

$$qv = (m\kappa\Delta x)^2 \left(\frac{dv}{d\omega} \right)^2 \tag{4.27}$$

Let us now obtain an expression for the left-hand side of this latter equation; using Eqs. (4.22) and (4.23), we have

$$e + \frac{1}{2} m^2 v^2 + v(C_2 - m^2 v) = C_3,$$

that is,

$$e - \frac{1}{2} m^2 v^2 = C_3 - v C_2. \tag{4.28}$$

In the case of a perfect gas we have $e = pv/(\gamma - 1)$ and substituting this in Eq. (4.28), gives

$$pv = \frac{(\gamma - 1)}{2} m^2 v^2 + (\gamma - 1)C_3 - v(\gamma - 1)C_2. \qquad (4.29)$$

Equation (4.22) implies that

$$qv = C_2 v - pv - m^2 v^2$$

and by using Eq. (4.29) in this latter equation, gives

$$qv = C_2 v - \left[\frac{(\gamma - 1)}{2} m^2 v^2 + (\gamma - 1)C_3 - v(\gamma - 1)C_2\right] - m^2 v^2$$

$$= \gamma v C_2 - (\gamma - 1)C_3 - \left(\frac{\gamma + 1}{2}\right) m^2 v^2 \qquad (4.30)$$

after a few algebraic steps. The right-hand side of this latter equation vanishes for $v = v_i$ and for $v = v_f$, hence,

$$\gamma v_i C_2 - (\gamma - 1)C_3 - \left(\frac{\gamma + 1}{2}\right) m^2 v_i^2 = 0$$

$$\gamma v_f C_2 - (\gamma - 1)C_3 - \left(\frac{\gamma + 1}{2}\right) m^2 v_f^2 = 0.$$

By subtracting these latter two equations and solving for C_2, we find that

$$C_2 = \left(\frac{\gamma + 1}{2\gamma}\right) m^2 (v_i + v_f) \qquad (4.31)$$

and substituting this result back to determine C_3, we find that

$$C_3 = \frac{1}{2}\left(\frac{\gamma + 1}{\gamma - 1}\right) m^2 v_i v_f \qquad (4.32)$$

Having determined C_2 and C_3, we can now substitute these back in Eq. (4.30), yielding,

$$qv = \left(\frac{\gamma + 1}{2}\right) m^2 (v - v_f)(v_i - v) \qquad (4.33)$$

after a few lines of algebraic manipulations. Inserting this value for qv in Eq. (4.27), gives the following differential equation,

$$(m\kappa \Delta x)^2 \left(\frac{dv}{d\omega}\right)^2 = \left(\frac{\gamma + 1}{2}\right) m^2 (v - v_f)(v_i - v)$$

and as the m's cancel across, we have the following equation to be solved;

$$(\kappa \Delta x)^2 \left(\frac{dv}{d\omega}\right)^2 = \left(\frac{\gamma+1}{2}\right)(v - v_f)(v_i - v) \tag{4.34}$$

In order to solve Eq. (4.34) Von Neumann and Richtmyer used the following substitutions;

$$\psi = v - \left(\frac{v_i + v_f}{2}\right), \quad \psi_0 = \frac{v_i - v_f}{2} \quad \text{and} \quad \varphi = \frac{\psi}{\psi_0},$$

hence,

$$\kappa \Delta x \frac{d\varphi}{d\omega} = \pm \left(\frac{\gamma+1}{2}\right)^{1/2} (1 - \varphi^2)^{1/2}.$$

Consequently,

$$\omega = \pm \left(\frac{2}{\gamma+1}\right)^{1/2} \kappa \Delta x \int \frac{d\varphi}{(1 - \varphi^2)^{1/2}}$$

and by making the substitution, $\varphi = Sin(x)$, one can integrate the latter equation to obtain,

$$\omega = \pm \left(\frac{2}{\gamma+1}\right)^{1/2} (\kappa \Delta x) arcSin\varphi \tag{4.35}$$
$$= \pm \omega_0 arcSin\varphi,$$

where

$$\omega_0 = \left(\frac{2}{\gamma+1}\right)^{1/2} \kappa \Delta x \tag{4.36}$$

and the constant of integration has been set to zero as it only shifts the ω-axis by that constant amount. Eq. (4.35) gives

$$\varphi = \pm Sin \frac{\omega}{\omega_0} \quad \text{or} \quad \psi = \pm \left(\frac{v_i - v_f}{2}\right) Sin \frac{\omega}{\omega_0},$$

hence,

$$v = \frac{v_i + v_f}{2} \pm \left(\frac{v_i - v_f}{2}\right) Sin \frac{\omega}{\omega_0}. \tag{4.37}$$

Differentiating this latter equation, we obtain,

$$\frac{dv}{d\omega} = \pm \frac{1}{2\omega_0} \left(v_i - v_f\right) Cos \frac{\omega}{\omega_0}. \tag{4.38}$$

Since we are considering the case where the shock wave moves to the right, then $v_i > v_f$ and as a result of the assumption that $(dv/d\omega) \geq 0$ in the shock region, the positive sign in Eq. (4.38) is taken, hence,

$$\frac{dv}{d\omega} = \frac{1}{2\omega_0} \left(v_i - v_f\right) Cos \frac{\omega}{\omega_0} \geq 0,$$

so that ω/ω_0 is confined to the range,

$$-\frac{\pi}{2} \leq \frac{\omega}{\omega_0} \leq \frac{\pi}{2}.$$

Accordingly, Eq. (4.37) can be written as

$$v = \frac{v_i + v_f}{2} + \left(\frac{v_i - v_f}{2}\right) Sin \frac{\omega}{\omega_0}, \tag{4.39}$$

which has a continuous solution within the shock region; $-(\pi/2) \leq (\omega/\omega_0) \leq (\pi/2)$. For $\omega = -(\pi/2)\omega_0$ we have $v = v_f$ and for $\omega = (\pi/2)\omega_0$ we have $v = v_i$. When Eq. (4.39) is pieced together with the particular solutions; $v = v_i$ and $v = v_f$, as stated by Von Neumann and Richtmyer, one obtains the composite continuous solution as shown in Fig. 4.2. The specific volume within the shock region is shown in Fig. 4.2 as a solid line while the particular solutions outside the shock region are shown as broken lines. Hence, requirement 1 is satisfied.

We have already seen that the shock region extends over an interval of width $\pi\omega_0$, where ω_0 is given by Eq. (4.36), hence, the width of the shock region is

$$\pi \left(\frac{2}{\gamma + 1}\right)^{1/2} \kappa\Delta x,$$

which is the order of Δx if κ is close to unity and this satisfies requirement 2.

Outside the shock region $\partial v/\partial t$ is, in general, very small and in the case of a steady-state shock it is zero, so in this region q is negligible in comparison to the pressure p due to the factor $(\Delta x)^2$ in the equation for q; this satisfies requirement 3.

Consequently, Von Neumann and Richtmyer have demonstrated that their expression for q meets all requirements: the equations describing the flow have continuous solutions and these equations can be used for the entire calculation as if no shocks were present at all. Instead, the shocks automatically appear as regions where there are rapid but continuous changes in the velocity, density etc. and have jumps in value that matches the conditions supplied by the Rankine-Hugoniot equations.

Fig. 4.2 Plot of the specific volume as a function of ω showing the shock transition region (see text)

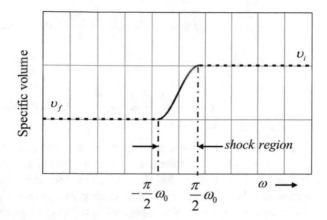

4.5 The Numerical Procedure

A brief outline of the numerical procedure using the *finite difference* representation of the differential equations is presented here. Much more detail on finite difference equations can be found in texts dealing specifically with numerical methods [4, 5] and the interested reader is directed to the literature on this subject and a good starting point is the excellent text by Anderson [4]. Anderson derives a number of different forms of the finite difference expressions and goes on to discuss the order of accuracy of the finite difference quotients obtained. Furthermore, he describes two general approaches; an *explicit* approach and an *implicit* approach as well as the *stability requirement* for the solution of the difference equations involved. Similarly, Ramshaw [6] provides an excellent account of finite-difference approximations and numerical stability requirements.

4.5.1 The Differential Equations for Plane Wave Motion: A Summary

In order to carry out the numerical procedure let us recall the previous equations for plane wave motion; they are re-written here (with x_0 simply replaced by x) and with artificial viscosity included,

$$\frac{\partial v}{\partial t} = \frac{1}{\rho(x,0)} \frac{\partial u}{\partial x}, \quad \text{(Continuity)}$$

$$\rho(x,0) \frac{\partial u}{\partial t} = -\frac{\partial}{\partial x}(p+q), \quad \text{(Momentum)}$$

and

$$[\gamma p + (\gamma - 1)q]\frac{\partial v}{\partial t} + v\frac{\partial p}{\partial t} = 0; \quad \text{(Energy)},$$

where q is given by,

$$q = -\frac{(\kappa\Delta x)^2}{v}\frac{\partial u}{\partial x}\left|\frac{\partial u}{\partial x}\right|. \quad \text{(Artificial viscosity)}$$

The derivatives in the above equations are replaced by a *finite difference* representation of these derivatives in order to obtain a numerical solution. In this discretization process the functions (such as, pressure density etc.) have prescribed values at only a finite number of discrete points in space and time, in contrast to a continuous variation in these functions in the case of an analytical solution.

The highest-order derivatives that appear in the equations above are first-order partial derivatives so that finite difference equations for first-order derivative are required for the numerical procedure. Only a single spatial coordinate in addition to the time t is required for the one-dimensional flow problems presented in this chapter.

4.5.2 Finite Difference Expressions

Since numerical solutions only provide answers at discrete points, a discrete grid is set up as illustrated in Fig. 4.3 which shows a portion of the grid in the x, t-plane. The grid points are denoted by the index j in the x-direction and by the index n for the time. For example, we introduce abbreviations to represent the value of some function f at the point n, j by $f_{n,j}$ where it is understood that $f_{n,j} \equiv f(t_n, x_j)$ which is

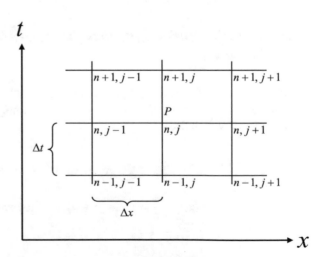

Fig. 4.3 A series of discrete grid points is shown in the x, t plane

the value of the function at time t_n and at position x_j. Uniform grid spacing is assumed with the spacing in the x-direction equal to Δx and the spacing in time equal to Δt as illustrated in Fig. 4.3.

A Taylor series expansion is used at each point in the grid in order to generate finite difference approximations. The Taylor series expansion for $f(x + \Delta x)$ in the case of a continuous function $f(x)$ is

$$f(x + \Delta x) = f(x) + \Delta x \left(\frac{\partial f}{\partial x}\right)_x + \frac{(\Delta x)^2}{2}\left(\frac{\partial^2 f}{\partial x^2}\right)_x + \frac{(\Delta x)^3}{6}\left(\frac{\partial^3 f}{\partial x^3}\right)_x + \cdots\cdots$$

and if $f_{n,j}$ denotes the value of f in the discrete case at node n, j, then $f_{n, j+1}$ at node $n, j + 1$ can be expressed in terms of a Taylor series expansion according to the equation,

$$f_{n,j+1} = f_{n,j} + \Delta x \left(\frac{\partial f}{\partial x}\right)_{n,j} + \frac{(\Delta x)^2}{2}\left(\frac{\partial^2 f}{\partial x^2}\right)_{n,j} + \frac{(\Delta x)^3}{6}\left(\frac{\partial^3 f}{\partial x^3}\right)_{n,j} + \cdots\cdots \quad (4.40)$$

Solving for $(\partial f/\partial x)_{n,j}$ yields

$$\left(\frac{\partial f}{\partial x}\right)_{n,j} = \frac{f_{n,j+1} - f_{n,j}}{\Delta x} - \frac{\Delta x}{2}\left(\frac{\partial^2 f}{\partial x^2}\right)_{n,j} - \frac{(\Delta x)^2}{6}\left(\frac{\partial^3 f}{\partial x^3}\right)_{n,j} - \cdots\cdots$$

The first term on the right-hand side of the latter equation is the finite difference approximation of the partial derivative while the additional terms on the right-hand side are the neglected terms in achieving this approximation. The first neglected term is of the order of Δx, hence, we can write the latter equation as

$$\left(\frac{\partial f}{\partial x}\right)_{n,j} = \frac{f_{n,j+1} - f_{n,j}}{\Delta x} + O(\Delta x)$$

where $O(\Delta x)$ implies "terms of order of Δx". Consequently, we can write

$$\left(\frac{\partial f}{\partial x}\right)_{n,j} = \frac{f_{n,j+1} - f_{n,j}}{\Delta x}$$

where it is understood that the finite difference approximation to the derivative is accurate to first order in Δx. The above finite difference approximation is called the *forward difference* approximation since it includes the node with index $n, j + 1$ in determining the derivative of the function f at the grid point n, j.

Suppose, on the other hand, we carry out the following Taylor expansion,

$$f_{n,j-1} = f_{n,j} - \Delta x \left(\frac{\partial f}{\partial x}\right)_{n,j} + \frac{(\Delta x)^2}{2}\left(\frac{\partial^2 f}{\partial x^2}\right)_{n,j} - \frac{(\Delta x)^3}{6}\left(\frac{\partial^3 f}{\partial x^3}\right)_{n,j} + \ldots\ldots \quad (4.41)$$

hence,

$$\left(\frac{\partial f}{\partial x}\right)_{n,j} = \frac{f_{n,j} - f_{n,j-1}}{\Delta x} + O(\Delta x)$$

and this finite difference representation of the derivative is called a *backward difference* since it uses $f_{n,\,j\,-\,1}$ to the left of the grid point n, j to determine the derivative and it is also accurate to first order in Δx. Suppose we now subtract Eq. (4.40) from Eq. (4.41);

$$f_{n,j+1} - f_{n,j-1} = 2\Delta x \left(\frac{\partial f}{\partial x}\right)_{n,j} + \frac{(\Delta x)^3}{3}\left(\frac{\partial^2 f}{\partial x^2}\right)_{n,j} + \ldots\ldots$$

hence,

$$\left(\frac{\partial f}{\partial x}\right)_{n,j} = \frac{f_{n,j+1} - f_{n,j-1}}{2\Delta x} + O(\Delta x^2).$$

Here we see that the derivative is obtained by using the values of f on either side of the grid point n, j and the finite difference is called the *central difference* and it is accurate to second order in Δx.

Finite difference equation can be produced using any number of points and let us consider an example of a first derivative involving three points according to

$$\left(\frac{\partial f}{\partial x}\right)_{n,j} = \frac{a f_{n,j} + b f_{n,j-1} + c f_{n,j-2}}{\Delta x}. \quad (4.42)$$

Let us now determine the coefficients a, b and c by using the Taylor expansion for $f_{n,\,j\,-\,1}$ and $f_{n,\,j\,-\,2}$ about $f_{n,\,j}$, hence,

$$f_{n,j-1} = f_{n,j} - \Delta x \left(\frac{\partial f}{\partial x}\right)_{n,j} + \frac{\Delta x^2}{2}\left(\frac{\partial^2 f}{\partial x^2}\right)_{n,j} + \ldots\ldots\ldots$$

and

$$f_{n,j-2} = f_{n,j} - 2\Delta x \left(\frac{\partial f}{\partial x}\right)_{n,j} + \frac{(2\Delta x)^2}{2}\left(\frac{\partial^2 f}{\partial x^2}\right)_{n,j} + \dots\dots$$

Therefore, it follows that

$$af_{n,j} + bf_{n,j-1} + cf_{n,j-2} = (a+b+c)f_{n,j} - \Delta x(b+2c)\left(\frac{\partial f}{\partial x}\right)_{n,j}$$

$$+\frac{\Delta x^2}{2}(b+4c)\left(\frac{\partial^2 f}{\partial x^2}\right)_{n,j} + O(\Delta x^3)$$

However, the left-hand side of the latter equation is just equal to $\Delta x \left(\frac{\partial f}{\partial x}\right)_{n,j}$ and, as a result, the terms on the right-hand side becomes,

$$a + b + c = 0$$
$$b + 2c = -1$$
$$b + 4c = 0.$$

The solution of these equations involving the coefficients; a, b and c, yields the following results; $a = 3/2$, $b = -2$ and $c = 1/2$. Hence,

$$\left(\frac{\partial f}{\partial x}\right)_{n,j} = \frac{3f_{n,j} - 4f_{n,j-1} + f_{n,j-2}}{2\Delta x} \tag{4.43}$$

and this difference quotient is accurate to second order in Δx. By using forward grid points rather than backward grid points one can show that an alternative form of the derivative appearing in the latter equation can be written as,

$$\left(\frac{\partial f}{\partial x}\right)_{n,j} = \frac{-3f_{n,j} + 4f_{n,j+1} - f_{n,j+2}}{2\Delta x}. \tag{4.44}$$

Up to now we have concentrated on determining some finite difference formulae in the spatial domain and it is clear that similar formulae pertain to the time domain. For example, the forward difference, backward difference and central difference equations in the time domain are

$$\left(\frac{\partial f}{\partial t}\right)_{n,j} = \frac{f_{n+1,j} - f_{n,j}}{\Delta t},$$

$$\left(\frac{\partial f}{\partial t}\right)_{n,j} = \frac{f_{n,j} - f_{n-1,j}}{\Delta t}$$

and

$$\left(\frac{\partial f}{\partial t}\right)_{n,j} = \frac{f_{n+1,j} - f_{n-1,j}}{2\Delta t},$$

respectively. In the treatment of plane shock waves to be presented, we will use difference equations accurate to first order.

4.5.3   The Discrete Form of the Equations

In the present context, the differential equations for plane wave motion that were summarised in Sect. 4.5.1 are approximated by the following difference equations, where a rectangular network of points denoted by t_n, x_j are used, with *uniform* increments, Δt and Δx, so that $t_n = n\Delta t$ and $x_j = j\Delta x$ and where Δx and Δt are taken as constant values throughout the calculation and where $n = 0, 1, 2. \ldots\ldots\ldots N$ and $j = 0, 1, 2, \ldots\ldots\ldots J$;

$$u_{n+1,j} = u_{n,j} - \frac{\Delta t}{\rho_0 \Delta x}\left(p_{n,j} - p_{n,j-1} + q_{n,j} - q_{n,j-1}\right), \tag{4.45}$$

$$v_{n+1,j-1} = v_{n,j-1} + \frac{\Delta t}{\rho_0 \Delta x}\left(u_{n+1,j} - u_{n+1,j-1}\right), \tag{4.46}$$

$$q_{n+1,j-1} = -\frac{2(\kappa\Delta x)^2}{v_{n,j-1} + v_{n+1,j-1}}\left[\left(\frac{u_{n+1,j} - u_{n+1,j-1}}{\Delta x}\right)\left|\frac{u_{n+1,j} - u_{n+1,j-1}}{\Delta x}\right|\right], \tag{4.47}$$

and

$$p_{n+1,j-1} = \frac{\left(\frac{\gamma+1}{\gamma-1}v_{n,j-1} - v_{n+1,j-1}\right)p_{n,j-1} + 2q_{n+1,j-1}\left(v_{n,j-1} - v_{n+1,j-1}\right)}{\frac{\gamma+1}{\gamma-1}v_{n+1,j-1} - v_{n,j-1}}. \tag{4.48}$$

The latter equation is obtained from Eq. (4.11) by writing it in the following difference form;

$$\left[\gamma p_{n+1,j-1} + (\gamma - 1)q_{n+1,j-1}\right]\left(\frac{v_{n+1,j-1} - v_{n,j-1}}{\Delta t}\right) + v_{n+1,j-1}\left(\frac{p_{n+1,j-1} - p_{n,j-1}}{\Delta t}\right) = 0$$

and solving for $p_{n+1,j-1}$.

Suppose the quantities, $u_{n,j}$, $v_{n,j}$, $q_{n,j}$ and $p_{n,j}$ are known for $j = 0, 1, 2\ldots J$ for some value of n, then $u_{n+1,1}$ can be determined from Eq. (4.45). The quantity $u_{n+1,0}$ at $t_n + \Delta t$ in Eq. (4.46) is generally known from the boundary conditions, for example, in the case of piston motion, so that $v_{n+1,0}$ can be found from Eq. (4.46) and one can proceed to determine the values of $q_{n+1,0}$ and $p_{n+1,0}$ from Eqs. (4.47) and (4.48). This completes the cycle for $j = 1$ and one can repeat the process for $j = 2$, etc. When the process is completed, corresponding to $j = J$, one knows the values of all the quantities at $t_n + \Delta t$ and then it is just a matter of repeating the process to find the quantities at $t_n + 2\Delta t$, etc. Other boundary conditions would be required at a closed end of a tube (corresponding to $j = J$) and, in this case, $u_{n+1, J} = 0$.

4.6 Stability of the Difference Equations

The system of difference equations above is *explicit*; that is, each of the quantities on the left-hand side is determined from quantities on the right-hand side which are known at a short time Δt earlier. However, the *explicit* method comes with a penalty as we are not at liberty to choose Δx and Δt separately; if Δx is chosen then Δt must be less than some prescribed value, otherwise the numerical solution quickly develops very large oscillations with increasing n, and eventually these oscillations can grow exponentially to swamp the entire calculation. This growth is due to *numerical instability* of the difference equations; accordingly, the stability of the numerical procedure is of paramount importance in order to obtain a workable solution. This behaviour has been discussed by Anderson [4] and others [1, 2, 6, 7]. The actual stability requirement is dependent on the form of the difference equations employed in the analysis and it is expressed in the form of an inequality containing Δx and Δt. Depending on the specific forms of the equations used this stability analysis leads, in general, to two separate stability conditions; one called the *von Neumann stability condition* and the other called the *Courant-Friedrichs-Lewy* (CFL) *condition*. A rigorous account of stability analysis is outside the scope of the present text and, accordingly, one is directed to the text by Anderson [4] for an appreciation of the stability requirements that pertain to several specific types of equations.

Von Neumann and Richtmyer [1] examined the stability of the difference equations for the specific form of the artificial viscosity as given by Eq. (4.1). In normal regions, that is, outside the shock region, where the artificial viscosity is negligible, they found that the stability condition is given by

$$\frac{s_0 \Delta t}{\Delta x} \le 1,$$

which is the usual Courant condition and s_0 is specified as the nominal speed of sound. In the shock region they found the condition to be slightly more stringent, but not significantly so, and given by the equation;

$$\frac{s_{0f} \Delta t}{\Delta x} \le \frac{\gamma^{1/2}}{2\kappa},$$

where s_{0f} is specified as the speed of sound in the material behind the shock.

4.7   Grid Spacing

We now come to the question of the choice of grid spacing Δx in order to identify the position of the shocks with some accuracy [8]. Recalling from our previous discussion that the typical shock thickness is the order of the molecular mean-free-path l_{mfp}, and it would appear essential to have the grid spacing $\Delta x \approx l_{mfp}$ to realistically model fluid flow in the presence of shocks. However, having $\Delta x \approx l_{mfp}$ implies that the time-step Δt must be less than l_{mfp}/c in order to satisfy the CFL condition above, where c is the signal propagation speed. Typically, $l_{mfp} \approx 10^{-5} cm$ and this implies that the computing time necessary to follow a specific time interval could become prohibitive. Accordingly, the choice of grid size Δx represents a compromise between accuracy in locating the position of the shock and the total computing time required for any specific application. With artificial viscosity included we will see in due course that the shock acquires a thickness of the order of a few Δx.

4.8   Numerical Examples of Plane Shocks

In the following sections we will investigate the numerical solution to several examples of plane shock waves.

4.8.1   Piston Generated Shock Wave

Let us now consider an example of a perfect gas contained in a semi-infinite cylindrical pipe terminated by a piston as illustrated in Fig. 4.4. The piston is

Fig. 4.4 Schematic diagram showing piston motion at constant velocity in a tube. The position of the piston and the shock front are sketched as a function of time

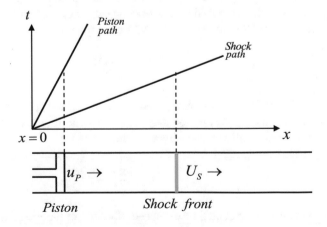

suddenly pushed into the pipe at $t = 0$ at a constant speed u_P. A shock wave is formed immediately as the piston begins to move; initially, the position of the piston and the shock front coincide but at later times the shock front races ahead of the piston. The objective here is to determine the subsequent motion of the shock wave and the associated pressure and density jumps across the shock.

The following values were chosen for the numerical procedure; $\Delta x = 0.1$, $\Delta t = 0.01$, $J = 500$, $N = 1000$, $p(x, 0) = 1$, $v(x, 0) = 1$, $\gamma = 1.4$, piston velocity of $u_p(0, t) = 0.3$ and κ in the artificial viscosity term was taken as 1.2.

Figure 4.5 shows the results of the numerical procedure for the fluid velocity at two different times as a function of position; alternatively, Fig. 4.6 shows the fluid velocity at two different positions as a function of time. Both plots show the position of the shock as a rapid change in fluid velocity. The velocity of the shock from the plots is estimated to be 1.38 ± 0.02 (by finding the peak-value of q; see Sect. 4.8.4), so let us compare this with that predicted by Eq. (3.40), namely;

$$U_s = \frac{(\gamma + 1)}{4} u_P + \sqrt{\frac{1}{16} (\gamma + 1)^2 u_P^2 + c_1^2},$$

where $c_1 = \sqrt{1.4}$ is the speed of sound in the air ahead of the shock. Using the values of u_P and c_1 we obtain $U_S = 1.377$ which is in excellent agreement with the

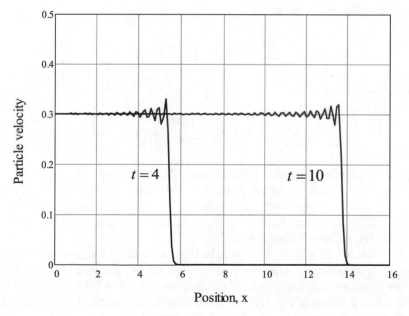

Fig. 4.5 Particle velocity as a function of position at $t = 4$ and at $t = 10$ is shown for the piston moving into the tube at constant velocity. For the numerical procedure the following parameters apply; $\gamma = 1.4$, $\kappa = 1.2$, $\Delta x = 0.1$ and $\Delta t = 0.01$ (see text)

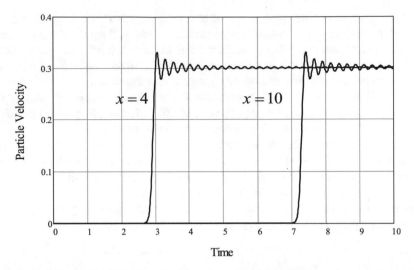

Fig. 4.6 Particle velocity as a function of time for two different positions; $x = 4$ and $x = 10$ is shown for the piston moving into the tube at constant velocity. For the numerical procedure the following parameters apply; $\gamma = 1.4$, $\kappa = 1.2$, $\Delta x = 0.1$ and $\Delta t = 0.01$ (see text)

value above. The pressure and density jumps across the shock are given by Eqs. (3.25) and (3.28), namely,

$$\frac{p_2}{p_1} = 1 + \frac{2\gamma}{\gamma + 1}\left(M_1^2 - 1\right)$$

and

$$\frac{\rho_2}{\rho_1} = \frac{(\gamma + 1)M_1^2}{\left[(\gamma - 1)M_1^2 + 2\right]},$$

respectively, where $M_1 = U_S/c_1$. Accordingly, these equations give $p_2 = 1.413$ and $\rho_2 = 1.278$. The corresponding plots of the pressure and density are shown in Figs. 4.7 and 4.8, respectively, and a numerical estimate of these jumps yields $p_2 = 1.413 \pm 0.001$ and $\rho_2 = 1.278 \pm 0.001$ for the pressure and density in the relatively flat portions behind the shock.

Plots of the fluid velocity are shown in Figs. 4.9 and 4.10 where the step size is reduced by an order of magnitude. One can observe that the shock front displays a sharper transition with much reduced oscillations. An expanded plot in the vicinity of the shock is shown in Fig. 4.11. It can be seen that the width of the shock front is approximately four to five spatial (or time) increments.

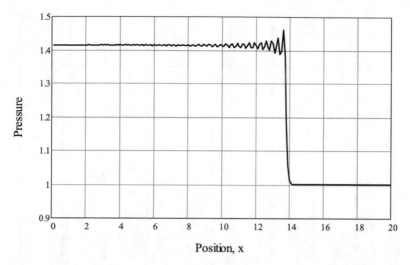

Fig. 4.7 Pressure as a function of position at $t = 10$ is shown for the piston moving into the tube at constant velocity. For the numerical procedure the following parameters apply; $\gamma = 1.4$, $\kappa = 1.2$, $\Delta x = 0.1$ and $\Delta t = 0.01$ (see text)

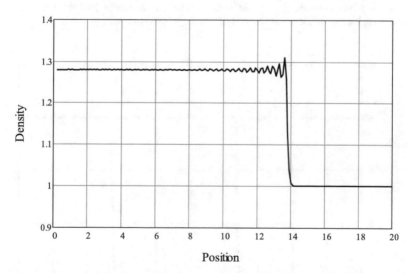

Fig. 4.8 Density as a function of position at $t = 10$ is shown for the piston moving into the tube at constant velocity. For the numerical procedure the following parameters apply; $\gamma = 1.4$, $\kappa = 1.2$, $\Delta x = 0.1$ and $\Delta t = 0.01$ (see text)

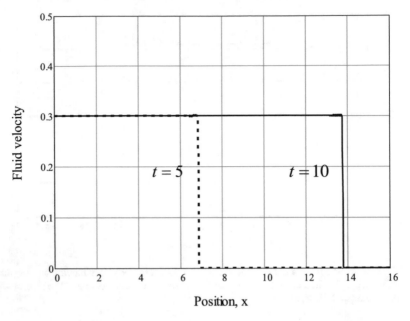

Fig. 4.9 Particle or fluid velocity as a function of position at $t = 5$ and at $t = 10$ is shown for the piston moving into the tube at constant velocity. For the numerical procedure the following parameters apply; $\gamma = 1.4$, $\kappa = 1.5$, $\Delta x = 0.01$ and $\Delta t = 0.005$ (see text)

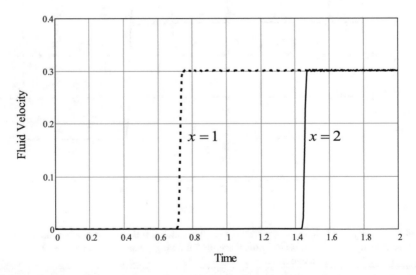

Fig. 4.10 Particle or fluid velocity as a function of time at $x = 1$ and at $x = 2$ is shown for the piston moving into the tube at constant velocity. For the numerical procedure the following parameters apply; $\gamma = 1.4$, $\kappa = 1.5$, $\Delta x = 0.01$ and $\Delta t = 0.005$ (see text)

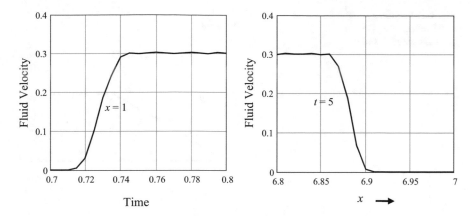

Fig. 4.11 Expanded view of Figs. 4.9 and 4.10 showing the typical shock thickness (see text)

4.8.2   *Linear Ramp*

Let us consider again the piston motion in a tube but, in this instance, we will assume that the piston is uniformly accelerated to a constant speed according to the equations;

$$u(0, t) = at \; ; \;\; 1 \geq t \geq 0$$
$$u(0, t) = 1; \;\; t > 1$$

as illustrated in Fig. 4.12. In addition to the piston motion, the following values were chosen for the numerical procedure: $\Delta x = 0.3$, $\Delta t = 0.05$, $\gamma = 1.4$ and $\kappa = 1.2$. Let us assume that the initial pressure and specific volume in the tube have the following arbitrary values; $p(x, 0) = 1$ and $v(x, 0) = 1$.

The particle velocity as a function of position is shown in Figs. 4.13 and 4.14 at different times. The broken line in these plots indicates the time taken for the shock wave to form according to Eq. (2.26) where we observe that the forward front begins to takes on a vertical profile. In fact, Fig. 4.13 should be taken in conjunction with Fig. 2.8 to compare the similarity between the numerical and analytical results. Using Eq. (2.26), namely,

$$t_{shock} = \frac{2c_0}{(\gamma + 1)a}$$

and noting that $c_0 = \sqrt{\gamma}$ as both the initial pressure and specific volume are unity, we find that the time taken for the shock to form is $t_{shock} = 98.6$ (Arb. units). For times less that this value one can easily verify from the plots that the forward front of the

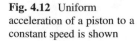

Fig. 4.12 Uniform
acceleration of a piston to a
constant speed is shown

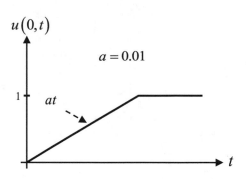

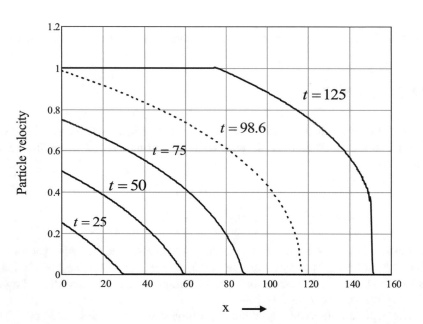

Fig. 4.13 Particle velocity is shown as a function of position at the times indicated for the
uniformly accelerated piston. The following parameters apply: $\gamma = 1.4$, $\kappa = 1.2$, $\Delta x = 0.3$ and
$\Delta t = 0.05$ (see text)

wave moves with just the sonic velocity $c_0 = 1.18$ (Arb. units). For example, let us
take the plot corresponding to $t = 50$ in Fig. 4.13, we can observe that the forward
front has moved a distance equal to approximately 59 arbitrary units, giving a
velocity of 1.18 which is in excellent agreement with the value of c_0. At much
later times ($t > t_{shock}$) as shown in Fig. 4.14 one can see the familiar vertical profile as
the shock wave advances down the tube.

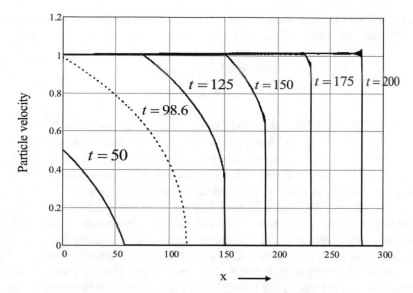

Fig. 4.14 Particle velocity is shown as a function of position at the times indicated for the uniformly accelerated piston. The following parameters apply: $\gamma = 1.4$, $\kappa = 1.2$, $\Delta x = 0.3$ and $\Delta t = 0.05$ (see text)

4.8.3 Piston Motion According to the Law $u = at^n$; $n > 0$.

When a piston moves according to the law $u = at^n$ we found in Chap. 2 that the time and place of formation of the shock wave are given by

$$t_{shock} = \left(\frac{2c_0}{a}\right)^{\frac{1}{n}} \frac{1}{\gamma + 1} \left[\gamma\left(\frac{n+1}{n-1}\right) + 1\right]^{\frac{n-1}{n}}.$$

and

$$x_{shock} = 2c_0 \left(\frac{2c_0}{a}\right)^{\frac{1}{n}} \left[\frac{\gamma}{\gamma + 1} + \frac{n-1}{n+1}\right] \frac{1}{(n-1)^{\frac{n-1}{n}}[\gamma - 1 + n(\gamma + 1)]^{\frac{1}{n}}},$$

respectively. In the case of the linear ramp ($n = 1$) where the piston moves with uniform accelerated velocity we found that the shock wave forms at the forward front of the wave whereas in this case the shock wave forms at some intermediate point. So let us now investigate this using the numerical procedure and in order to do so we will consider the case with $n = 2$. Let us assume the piston moves according to;

$$u(t) = at^2; \quad 10 \geq t \geq 0$$
$$= 1; \quad t > 10,$$

where $a = 0.01$. As in the case of the linear ramp, we will assume that $\gamma = 1.4$ and $c_0 = \sqrt{\gamma}$ as the pressure and density in the undisturbed medium have unity values. Using these parameters in the above equations one finds that $t_{shock} = 14.62$ and $x_{shock} = 14.63$ (both in arbitrary units). For the numerical procedure the following spatial and time increments were assumed; $\Delta x = 0.3$ and $\Delta t = 0.05$. Figure 4.15 shows the particle velocity as a function of position at five different times. One can verify from the plots that the forward front of the wave advances at the sonic speed c_0 while the velocity profile begins to exhibit an almost vertical profile with increasing time and that the onset of this vertical profile occurs at an intermediate point and not at the forward front of the wave. The broken line in Fig. 4.15 corresponds to $t = 14.6$ and represents the approximate time taken for the shock to form according to the above equation for t_{shock} while its position is in conformity with the equation above for x_{shock}. Figure 4.16 shows a plot of the artificial viscosity q as a function of position at times similar to those chosen for Fig. 4.15 (the artificial viscosity is negligible at $t = 7.5$ and at $t = 10$ in comparison to the viscosity at larger values of t and, accordingly, they are not visible in the scale adopted for this figure). Since the artificial viscosity only becomes dominant at the shock front we not only see a significant increase in the value of q at $t = 14.6$, but also a distinctive narrowing of the q-profile that one expects to observe in the vicinity of the shock front. Accordingly, it can be seen that the numerical results obtained here are in accord with the analytical predictions.

4.8.4 Tube Closed at End: A Reflected Shock

Let us now consider again our original example of piston motion in a tube but, in this instance, we will assume that the tube is closed at some position L on the right-hand side so that the following boundary condition applies, namely, $u(t, L) = 0$. Here we take $L = 80$ units, and with $\Delta x = 0.4$ in this particular case, we will divide the space into 200 evenly spaced points separated Δx units apart. The following parameters are assumed: $\gamma = 1.4$, $\kappa = 1.5$, $\Delta x = 0.4$, $\Delta t = 0.05$, $\rho_0 = 1$. Similar to the example in Sect. 4.8.1, the piston motion is given by $u_p(t, 0) = 0.3$ and the boundary condition at the end of the tube is also included in the numerical procedure.

Plots showing the particle velocity, pressure and density are shown in Figs. 4.17, 4.18 and 4.19 for three different times. In addition, a plot of the artificial viscosity at $t = 50$ is included as shown in Fig. 4.20 which enables one to determine position of the shock front with some accuracy from the maximum value of q. The plots

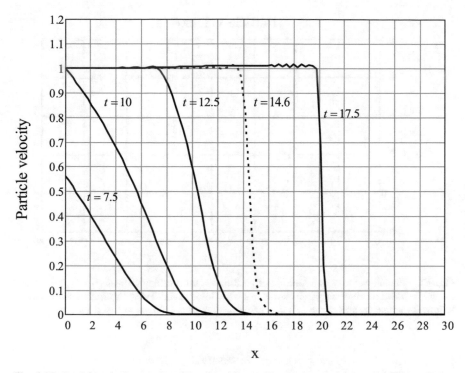

Fig. 4.15 Particle velocity as a function of position is shown when the piston moves according to the law $u = at^n$ (see text)

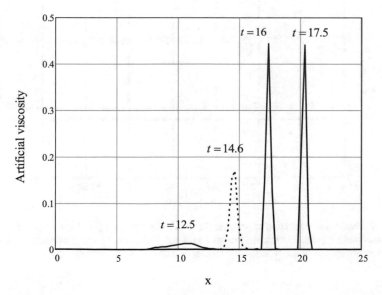

Fig. 4.16 Artificial viscosity q as a function of position is shown at times similar to those chosen for Fig. 4.15 (see text)

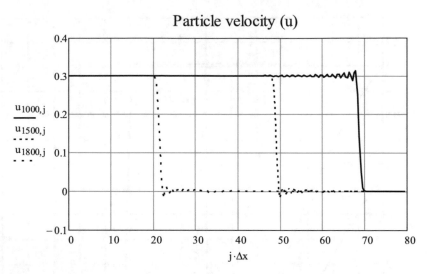

Fig. 4.17 Particle velocity as a function of position for three different times ($t = 50$, $t = 75$ and $t = 90$) is shown for constant piston motion in a tube closed at end. For the numerical procedure the following parameters apply: $\gamma = 1.4$, $\kappa = 1.5$, $\Delta x = 0.4$, $\Delta t = 0.05$, $\rho_0 = 1$ (see text)

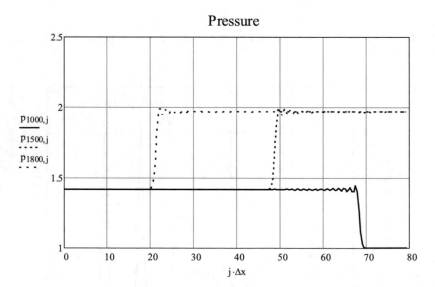

Fig. 4.18 Pressure as a function of position for three different times ($t = 50$, $t = 75$ and $t = 90$) is shown for constant piston motion in a tube closed at end. For the numerical procedure the following parameters apply: $\gamma = 1.4$, $\kappa = 1.5$, $\Delta x = 0.4$, $\Delta t = 0.05$, $\rho_0 = 1$ (see text)

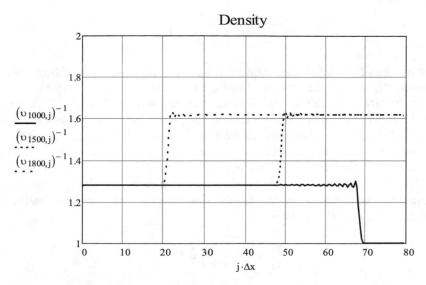

Fig. 4.19 Density as a function of position for three different times ($t = 50$, $t = 75$ and $t = 90$) is shown for constant piston motion in a tube closed at end. For the numerical procedure the following parameters apply: $\gamma = 1.4$, $\kappa = 1.5$, $\Delta x = 0.4$, $\Delta t = 0.05$, $\rho_0 = 1$ (see text)

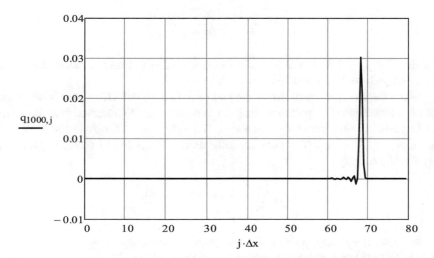

Fig. 4.20 Artificial viscosity as a function of position at $t = 50$ is shown. For the numerical procedure the following parameters apply: $\gamma = 1.4$, $\kappa = 1.5$, $\Delta x = 0.4$, $\Delta t = 0.05$, $\rho_0 = 1$ (see text)

corresponding to $n = 1000$, namely, $u_{1000,\,j}$, $p_{1000,\,j}$ and $1/\upsilon_{1000,\,j}$ shows particle velocity, pressure and density, respectively, as a function of position for the advancing incident shock front towards the closed end. As before, using Eq. (3.40) and writing $U_s = U_i$ for the incident shock, we have,

$$U_i = \frac{(\gamma + 1)}{4} u_p + \sqrt{\frac{1}{16}(\gamma + 1)^2 u_p^2 + c_1^2}$$

we find that $U_i = 1.377$ after noting that $u_p = 0.3$ and $c_1 = \sqrt{1.4}$, giving a Mach number M_1 of 1.164. Let us now compare this value of U_i with the numerical value from the plot shown in Fig. 4.17. In order to do this we must determine the position of the shock front and for this we refer to the plot of the artificial viscosity. Since artificial viscosity only becomes significant at the shock where velocity gradients are large, we can locate its position by determining the location of the maximum value of q. We find from Fig. 4.20 that q_{Max} occurs at $j = 171$, so its position is $171 \times \Delta x = 68.4$, and the time to reach this position is $1000 \times \Delta t = 50$, hence, $(U_i)_N = 68.4/50 = 1.368$ where the subscript N denotes the numerically estimated value. Using Eq. (3.25), namely,

$$\frac{p_2}{p_1} = 1 + \frac{2\gamma}{\gamma + 1}\left(M_1^2 - 1\right)$$

we find that $p_2 = 1.414$ since $p_1 = 1$. From Fig. 4.18 we estimate that $(p_2)_N = 1.413 \pm 0.001$ in the flat portion of the pressure profile. By using Eq. (3.17a) for the density ratio

$$\frac{\rho_2}{\rho_1} = \frac{(\gamma - 1) + (\gamma + 1)(p_2/p_1)}{(\gamma + 1) + (\gamma - 1)(p_2/p_1)}$$

we find that $\rho_2 = 1.279$, which should be compared with the numerically estimated value of $(\rho_2)_N = 1.278 \pm 0.001$ shown in Fig. 4.19.

When the incident shock strikes the end of the tube a reflected shock is produced and the air velocity behind the shock goes to zero as can be observed from the plots of the particle velocity corresponding to $t = 75$ and $t = 90$ as shown in Fig. 4.17, that is, $u_{1500,\,j}$ and $u_{1800,\,j}$. The pressure ratio across the reflected shock is given by Eq. (3.47), namely,

$$\frac{p_3}{p_2} = \frac{(3\gamma - 1)p_2 - (\gamma - 1)p_1}{(\gamma - 1)p_2 + (\gamma + 1)p_1}.$$

Inserting the values, $p_2 = 1.414$ and $p_1 = 1$, we find that $p_3 = 1.967$, which should be compared with the numerically estimated value of $(p_3)_N = 1.964 \pm 0.002$ shown Fig. 4.18. The corresponding density ratio according to Eq. (3.48) is

$$\frac{\rho_3}{\rho_2} = \frac{(\gamma - 1) + (\gamma + 1)(p_3/p_2)}{(\gamma + 1) + (\gamma - 1)(p_3/p_2)},$$

giving $\rho_3 = 1.617$, which should be compared to the numerically estimated value of $(\rho_3)_N = 1.616 \pm 0.001$ shown in Fig. 4.19.

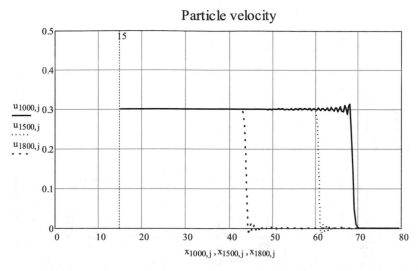

Fig. 4.21 Particle velocity as a function of the position of the fluid elements at the times indicated for constant piston motion in a tube closed at the end. For the numerical procedure the following parameters apply; $\gamma = 1.4$, $\kappa = 1.5$, $\Delta x = 0.4$ and $\Delta t = 0.05$ (see text)

Figures 4.17, 4.18, 4.19 and 4.20 gives the value of the quantities plotted as a function of the initial position of the fluid elements, however, we need these quantities as a function of the position of the fluid elements at the times indicated in order to determine, for example, the velocity of the reflected shock. Accordingly, the discrete form of Eq. (4.3) is used in conjunction to those already presented in Sect. 4.5.3 and this gives the position of the fluid elements as $x_{1000,\,j}$, $x_{1500,\,j}$ and $x_{1800,\,j}$ at the times indicated where j identifies that fluid element whose initial position was $j\Delta x$. When plotted, we obtain the following results for the particle velocity, pressure and density as shown in Figs. 4.21, 4.22 and 4.23. We observe from these plots that the marker at $x = 15$ shows the position of the piston at $t = 1000\Delta t$: this position is consistent with what we expect as the piston moves at a constant velocity of 0.3 (Arb. units), so that its new position at this time is $x = 0.3 \times 1000\Delta t = 15$ from its initial position at $x = 0$.

We can use these numerical results to make an estimate of the velocity of the reflected shock wave and compare it with the theoretical predictions according to the equations presented in Sect. 3.11. Specifically, Eq. (3.54) gives the Mach number M_r of the reflected shock in terms of the Mach number M_i of the incident shock according to

$$\frac{M_r}{M_r^2 - 1} = \frac{M_i}{M_i^2 - 1} \sqrt{1 + \frac{2(\gamma - 1)\left(M_i^2 - 1\right)\left(\gamma + \frac{1}{M_i^2}\right)}{(\gamma + 1)^2}}.$$

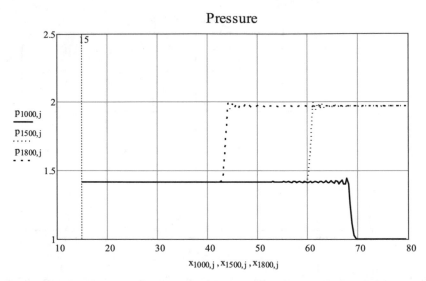

Fig. 4.22 Pressure as a function of the position of the fluid elements at the times indicated for constant piston motion in a tube closed at the end. For the numerical procedure the following parameters apply; $\gamma = 1.4$, $\kappa = 1.5$, $\Delta x = 0.4$ and $\Delta t = 0.05$ (see text)

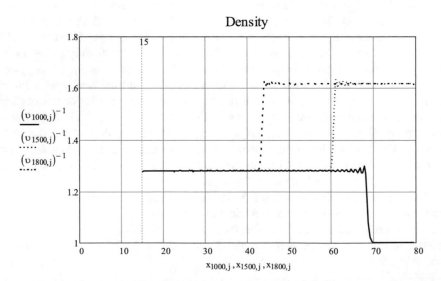

Fig. 4.23 Density as a function of the position of the fluid elements at the times indicated for constant piston motion in a tube closed at the end. For the numerical procedure the following parameters apply; $\gamma = 1.4$, $\kappa = 1.5$, $\Delta x = 0.4$ and $\Delta t = 0.05$ (see text)

We have already established that $M_i = 1.164$ and $\gamma = 1.4$, and as all quantities are known on the right-hand side of the above equation we obtain, after substitution, the following quadratic equation for M_r,

$$3.448M_r^2 - M_r - 3.448 = 0.$$

Solving this quadratic, we find (ignoring the negative root) that $M_r = 1.155$ which is the Mach number of the reflected shock relative to the air into which it is moving. We have already shown from Eqs. (3.25) and (3.28) that $p_2 = 1.414$ and $\rho_2 = 1.279$, so that the sonic velocity in the air that has been traversed by the incident shock is given by

$$c_2 = \sqrt{\frac{\gamma p_2}{\rho_2}},$$

hence,

$$c_2 = 1.244.$$

From Sect. 3.11 we have the relationship for the velocity of the reflected shock, U_r, in terms of the Mach number, M_r, the piston velocity u_p and sound speed c_2 according to the equation,

$$M_r = \frac{U_r + u_p}{c_2}.$$

Substituting the known quantities in this latter equation, we find that

$$U_r = 1.137.$$

Referring now to the numerical results and let us take, for example, the pressure plot shown in Fig. 4.22. We find that the position of the reflected shock corresponding to the plots of $p_{1500, \, j}$ and $p_{1800, \, j}$ occur at $x_{1500, \, 122} = 60.65$ and at $x_{1800, \, 53} = 43.57$, respectively. The specific values for j were determined by locating the position where the artificial viscosity q attains its maximum value in each case. Since the time difference between these maxima is $(1800 - 1500)\Delta t = 15$, the velocity of the reflected shock wave is given by

$$U_r = \frac{60.65 - 43.57}{15}$$
$$= 1.138,$$

which is in very good agreement with the value above.

It should be noted that any fluid element located down stream of the incident shock wave remains at its initial position until impacted by the shock. One can see this clearly from a typical numerical output shown in Fig. 4.24, where we consider a particular mass element of the fluid designated by $j = 140$, so that its initial position is $j\Delta x = 56$. Once the shock arrives and it does so at $t = 41$ in this particular case, the fluid element begins to move and we can estimate its speed of movement from the slope of the solid line in the plot of $x_{n,\,140}$ versus $n\Delta t$ shown in Fig. 4.24, and (based on the position of the markers supplied in this figure) we find that

$$Slope = \frac{65.2 - 56}{71.3 - 41} = 0.303$$

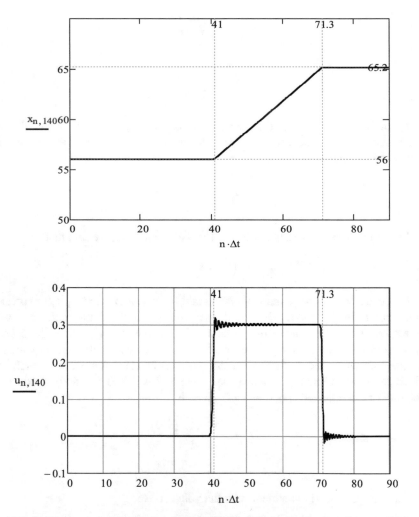

Fig. 4.24 Position and particle velocity of a typical fluid element as a function of time (solid lines) prior to and after been impacted by the incident and reflected shock waves (see text)

which, as expected, is in accordance with the speed of the piston motion. The velocity of the incident shock wave is estimated to be

$$U_i = \frac{140\Delta x}{41} = 1.366,$$

which is in good agreement with the value already determined from Figs. 4.17, 4.18, 4.19 and 4.20. One can also see that this fluid element's velocity is suddenly reduced to zero when impacted by the reflected shock, and this is in accordance with the remarks made at the beginning of Sect. 3.11. Further confirmation of these assertions is presented in the numerical plot of $u_{n,\,140}$ versus $n\Delta t$ which is also shown in Fig. 4.24.

4.8.5 The Numerical Value of κ for the Artificial Viscosity

In the previous numerical examples we have taken κ to have a value close to unity in the artificial viscosity term. An expanded view of Fig. 4.17 at $t = 50$ in the vicinity of the shock is shown in Fig. 4.25 with the addition of other plots having different values of κ; the broken line is a repeat of the plot in Fig. 4.17 with $\kappa = 1.5$. One observes that the smaller value of κ gives rise to large oscillations in the velocity behind the shock front while simultaneously tending to reduce the width of the shock transition region. On the other hand, the larger value of κ, corresponding to $\kappa = 3$, exhibits much reduced oscillations behind the shock but at the expense of making the change across the shock region too sluggish and smeared out over a larger number of grid intervals. Adopting values of κ in the range, 1.2–1.5, represents a compromise between these two extremes.

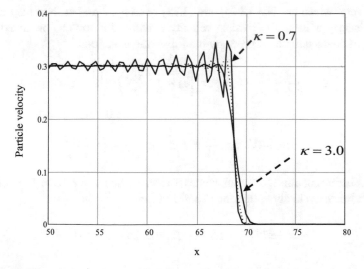

Fig. 4.25 Particle velocity is shown with different values of κ (see text)

4.8.6   Piston Withdrawal Generating an Expansion Wave

In Sect. 2.7.2 we discussed the centered expansion wave that was produced by the sudden withdrawal of a piston in a tube at a constant velocity $-u_0$. Here we will consider this particular problem numerically and investigate the changes that take place to particle velocity, pressure and density within the tube and compare the results with those predicted using the method of characteristics.

Initially, the piston is located at $x = 0$ and it is withdrawn at constant speed $u_0 = -0.3$ (Arb. units). The following increments; $\Delta x = 0.1$ and $\Delta t = 0.01$, were assumed for the numerical procedure and the other parameters chosen are; $\gamma = 1.4$, $\kappa = 1.2$ and $\rho_0 = 1$. Initially, the air is at rest in the tube and the pressure and density are assumed to have unity values, which implies that the ambient sonic velocity is given by $c_0 = \sqrt{\gamma}$.

In this example we consider a specific time following the piston's withdrawal and this time is taken to be $t = 5$ (Arb. units), consequently, the position of the piston at this time is $-0.3 \times 5 = -1.5$ (Arb. units).

The equations for the particle velocity and the sound speed according to the isentropic relations within the expansion fan are

$$u = \frac{2}{\gamma + 1}\left(\frac{x}{t} - c_0\right)$$

and

$$c = \left(\frac{\gamma - 1}{\gamma + 1}\right)\frac{x}{t} + \frac{2c_0}{\gamma + 1},$$

respectively, which are Eqs. (2.74) and (2.75) of Chap. 2 and each of the quantities varies linearly with x. The particle velocity u within the tube as predicted by the method of characteristics is given by the following equations,

$$u = \frac{2}{\gamma + 1}\left(\frac{x}{t} - c_0\right) \quad \text{for} \quad \left[c_0 - \left(\frac{\gamma + 1}{2}\right)u_0\right]t < x < c_0 t$$

$$= 0 \quad \text{for} \quad x > c_0 t$$

$$= -0.3 \quad \text{for} \quad -u_0 t < x < \left[c_0 - \left(\frac{\gamma + 1}{2}\right)u_0\right]t.$$

By substituting the numerical values for γ, c_0 and t we have the following equation for the velocity within the expansion fan,

$$u = 0.833\left(\frac{x}{5} - 1.183\right) \quad \text{for} \quad 4.116 < x < 5.916$$

and outside the expansion fan we have

$$u = 0 \ \text{ for } \ x > 5.916$$
$$= -0.3 \ \text{ for } \ -1.5 < x < 4.116$$

and these theoretical predictions for the particle velocity are shown plotted in Fig. 4.26.

The particle velocity following the numerical integration of the difference equations (that is, Eqs. 4.45, 4.46, 4.47 and 4.48) is also shown plotted in Fig. 4.27 and one can observe good agreement with the analytical solution using the method of characteristics.

The corresponding equations for the pressure and density within the expansion fan are

$$\frac{p}{p_0} = \left[\left(\frac{\gamma - 1}{\gamma + 1} \right) \frac{x}{c_0 t} + \frac{2}{\gamma + 1} \right]^{\frac{2\gamma}{\gamma - 1}} \tag{4.49}$$

and

$$\frac{\rho}{\rho_0} = \left[\left(\frac{\gamma - 1}{\gamma + 1} \right) \frac{x}{c_0 t} + \frac{2}{\gamma + 1} \right]^{\frac{2}{\gamma - 1}} \tag{4.50}$$

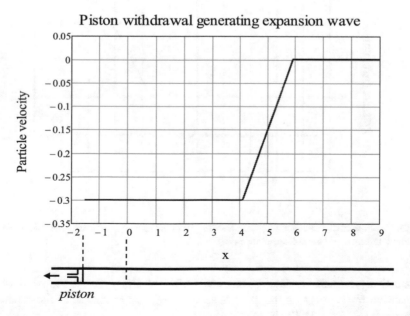

Fig. 4.26 Particle velocity within the tube is shown as a function of position at $t = 5$ for piston withdrawal at constant speed. The piston's position is also shown (see text)

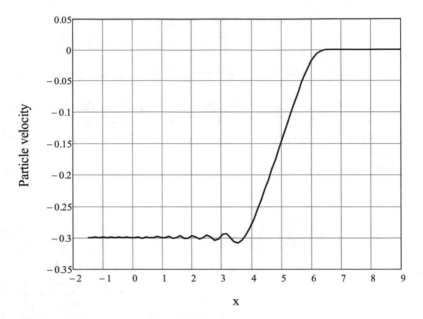

Fig. 4.27 Numerical result showing the particle velocity within the tube as a function of position at $t = 5$ (Arb. units) for piston withdrawal at constant speed (see text)

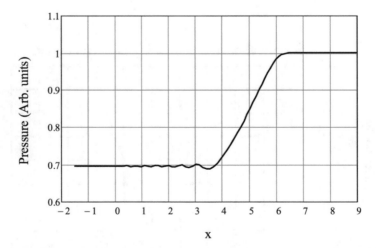

Fig. 4.28 Numerical result showing the pressure within the tube at $t = 5$ (Arb. units) following the piston withdrawal at constant speed (see text)

respectively, which are Eqs. (2.83) and (2.84) of Chap. 2. At the *head* of the expansion fan $x = c_0 t$, so that $p = p_0$ and $\rho = \rho_0$, and at the *tail* of the expansion fan $x = [c_0 - ((\gamma + 1)/2)u_0]t$, so that

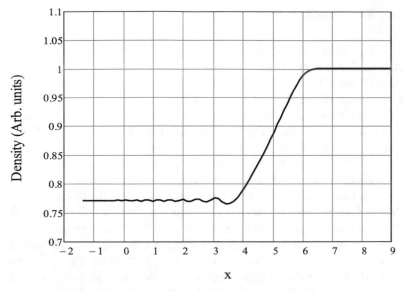

Fig. 4.29 Numerical result showing the density within the tube at $t = 5$ (Arb. units) following the piston withdrawal at constant speed (see text)

$$\frac{p_{Tail}}{p_0} = \left[1 - \left(\frac{\gamma - 1}{2}\right)\frac{u_0}{c_0}\right]^{\frac{2\gamma}{\gamma - 1}} \tag{4.51}$$

and

$$\frac{\rho_{Tail}}{\rho_0} = \left[1 - \left(\frac{\gamma - 1}{2}\right)\frac{u_0}{c_0}\right]^{\frac{2}{\gamma - 1}}. \tag{4.52}$$

As arbitrary values of $p_0 = 1$ and $\rho_0 = 1$ were used for the numerical procedure, we find from Eqs. (4.51) and (4.52) that the pressure p and density ρ at the *tail* of the expansion fan are 0.694 and 0.771, respectively.

The numerical results showing the pressure and density can be seen in Figs. 4.28 and 4.29. In the relatively flat region of each plot, corresponding to the tail of the expansion fan, we find that $(p)_N = 0.694 \pm 0.001$ and $(\rho)_N = 0.771 \pm 0.001$, where the subscript N denotes the numerically estimated values. These results are in excellent agreement with the theoretical values.

Since the piston's withdrawal speed is relatively slow the variation of pressure and density within the expansion wave appear linear from the plots. However, the actual variations follow a power law according to Eqs. (4.49) and (4.50) and this would become more evident at much greater withdrawal speeds (see Appendix A).

4.8.7 The Shock Tube

The previous examples lead us naturally to apply the numerical procedure to the shock tube [9–11] which is an important device for the study of shock waves and shock wave interactions in the laboratory.

An exact analytical solution to the shock tube problem can be obtained using the *method of characteristics* [12]. As a result, the shock tube problem has become an important test for the accuracy of computational fluid dynamic (CFD) codes. Accordingly, the numerical solution can be compared with the analytical solution which allows one to ascertain how accurately the code resolves shock discontinuities and reproduces the correct density, pressure and velocity profiles.

A shock tube consists of a long tube that is divided into two chambers by a diaphragm. One side contains gas at high pressure, p_4, and the other side contains gas at a lower pressure, p_1. Both chambers can contain different gases and, hence, have different gas constants, R, and different specific heat ratios, γ, although in the example to be discussed here we will assume that both chambers contain air with $\gamma = 1.4$. At $t = 0$ the diaphragm is ruptured and the high-pressure gas, called the *driver gas*, rushes into the low-pressure chamber called the *driven section*. The interface or *contact surface* between the two gases that were initially separated by the diaphragm, behaves like a piston which drives a *shock wave* into the low-pressure gas to increase the pressure. During the time interval that the shock wave moves down the tube, an *expansion wave* propagates into the high-pressure chamber to reduce the pressure. Figure 4.30 illustrates the flow in the different regions following the rupture of the diaphragm. The gases in regions 2 and 3 are at the same pressure and they move with the same velocity [9], whilst the strength of the shock wave, p_2/p_1, and the velocities of the gases are dependent on the initial pressure ratio, p_4/p_1, across the diaphragm: the required expressions are derived below.

By using Eqs. (3.27) and (3.41) we have the following relationships for the velocity U_s of the shock wave and the particle velocity u_p behind the shock in terms of the pressure ratio across the shock;

$$U_S = c_1 \left[\frac{(\gamma - 1) + (\gamma + 1)\frac{p_2}{p_1}}{2\gamma} \right]^{1/2} \tag{4.53}$$

and

$$u_2 = \frac{c_1}{\gamma} \left(\frac{p_2}{p_1} - 1 \right) \left(\frac{\frac{2\gamma}{\gamma+1}}{\frac{\gamma-1}{\gamma+1} + \frac{p_2}{p_1}} \right)^{1/2}, \tag{4.54}$$

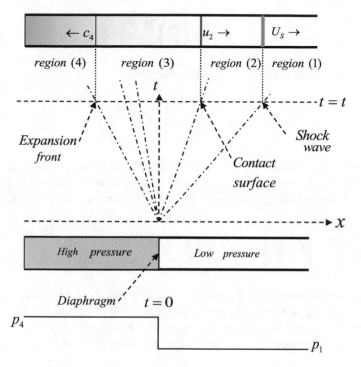

Fig. 4.30 Shock tube showing the air motion shortly after the diaphragm is ruptured

where u_2 has been substituted for u_p in the latter equation. Let us now apply Eq. (2.17) across the expansion wave:

$$du = -\frac{2}{\gamma - 1} dc$$

so that

$$\int_{u_3}^{u_4=0} du = -\frac{2}{\gamma - 1} \int_{c_3}^{c_4} dc,$$

hence

$$u_3 = \frac{2c_4}{\gamma - 1} \left(1 - \frac{c_3}{c_4} \right), \tag{4.55}$$

where c_3 and c_4 are the local sound speeds in regions 3 and 4. However, the flow is isentropic between regions 3 and 4 and, as a result, we have

$$\frac{c_3}{c_4} = \left(\frac{p_3}{p_4}\right)^{\frac{\gamma-1}{2\gamma}},$$

and when this is substituted in Eq. (4.55) we obtain

$$u_3 = \frac{2c_4}{\gamma - 1}\left[1 - \left(\frac{p_3}{p_4}\right)^{\frac{\gamma-1}{2\gamma}}\right]. \tag{4.56}$$

The pressure and fluid velocity on either side of the contact surface are the same [9], so that, $p_2 = p_3$ and $u_2 = u_3$, hence, eliminating u_2 and u_3 from Eqs. (4.54) and (4.56) and substituting p_2 for p_3 gives the following shock tube equation,

$$\frac{p_4}{p_1} = \frac{p_2}{p_1}\left[1 - \frac{(\gamma - 1)(c_1/c_4)(p_2/p_1 - 1)}{\sqrt{2\gamma[2\gamma + (\gamma + 1)(p_2/p_1 - 1)]}}\right]^{-\frac{2\gamma}{\gamma-1}}, \tag{4.57}$$

which gives the shock strength, p_2/p_1, implicitly as a function of the diaphragm pressure ratio, p_4/p_1. Once p_2/p_1 has been determined we can use Eq. (4.53) to find the shock speed U_S and u_2 and u_3 are determined from Eq. (4.54). The density ratio ρ_2/ρ_1 and temperature ratio T_2/T_1 across the shock are obtained from equations,

$$\frac{\rho_2}{\rho_1} = \frac{(\gamma - 1) + (\gamma + 1)(p_2/p_1)}{(\gamma + 1) + (\gamma - 1)(p_2/p_1)}, \tag{4.58}$$

and

$$\frac{T_2}{T_1} = \frac{p_2}{p_1}\left[\frac{(\gamma + 1) + (\gamma - 1)(p_2/p_1)}{(\gamma - 1) + (\gamma + 1)(p_2/p_1)}\right]. \tag{4.59}$$

We can also determine the pressure ratio, p_3/p_4, by noting that

$$\frac{p_3}{p_4} = \frac{p_3}{p_1}\frac{p_1}{p_4} = \left(\frac{p_2}{p_1}\right)/\left(\frac{p_4}{p_1}\right).$$

Similarly, it is straightforward to show that the density ratio, ρ_3/ρ_4 is given by the following equation,

$$\frac{\rho_3}{\rho_4} = \left(\frac{p_2/p_1}{p_4/p_1}\right)^{1/\gamma} \tag{4.60}$$

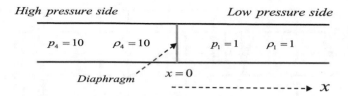

Fig. 4.31 Diagram illustrating the initial conditions for the shock tube

(a) Initial conditions

In this particular example we will assume that the initial pressure p_4 in the driver section is 10 (Arb. units) while the pressure p_1 in the driven section is 1 (Arb. unit): equal temperatures are assumed in either chamber, so that the initial density in the driver section is 10 (Arb. units) while the initial density in the driven section is 1 (Arb. unit) as illustrated in Fig. 4.31. The air in both sections is initially at rest. Assuming the following parameters apply for the numerical procedure; $\gamma = 1.4$, $\kappa = 1.2$, $\Delta x = 0.1$ and $\Delta t = 0.01$.

The fundamental shock tube equation, namely, Eq. (4.57) contains the sonic velocities c_1 and c_4 and, in general, as $c = \sqrt{\gamma p/\rho}$, so in this particular case $c_1 = c_4 = \sqrt{1.4} = 1.183$.

(b) Comparison of the theoretical and numerical results

Using Eq. (4.57) with $p_4/p_1 = 10$ we find that $p_2/p_1 = 2.848$ and substituting this in Eq. (4.53) and Eq. (4.54) we obtain $U_s = 1.902$ and $u_2 = 0.971$. The density ratio ρ_2/ρ_1 across the shock according to Eq. (4.58) is found to be 2.044. Similarly, using Eq. (4.60), we find that the density ratio ρ_3/ρ_4 is 0.408, so that $\rho_3 = 4.08$.

Let us now compare the values above with the results of the numerical procedure as shown in Figs. 4.32, 4.33 and 4.34. From the plots of pressure and density we estimate that $(p_2)_N = 2.85 \pm 0.03$, $(\rho_2)_N = 2.04 \pm 0.02$ and $(\rho_3)_N = 4.06 \pm 0.04$ in the relatively flat sections of the plots, where the subscript N denotes the numerically estimated value. The velocity plot gives $(u_2)_N = 0.97 \pm 0.01$, while we estimate the position of the shock front to be $x = 7.6$ in a time interval of 4 units, giving a shock velocity of 1.9. It can be seen that these values are in good agreement with the theoretical values above. (See Appendix B where some of these quantities are determined over a much longer time interval).

4.8.8 The Effect of Amplitude on Wave Propagation

In this section we will investigate numerically some aspects in relation to amplitude effects that were discussed in Chaps. 1 and 2. We will consider (a), wave profile distortion, (b), piston motion with small velocity and (c), incremental piston motion.

(a) Wave Profile Distortion

The discussion presented in Sect. 1.8, Chap. 1 referred to small-amplitude disturbances that propagate at the speed of sound and all parts of a wave profile

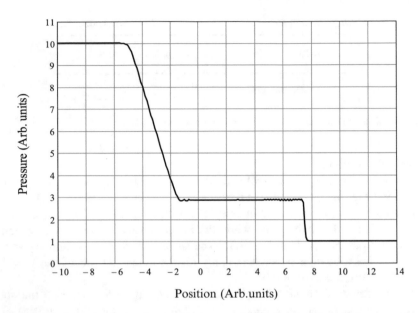

Fig. 4.32 Pressure in the shock tube as a function of position at $t = 4$ (Arb. units) is shown. For the numerical procedure the following parameters apply; $\gamma = 1.4$, $\kappa = 1.2$, $\Delta x = 0.1$ and $\Delta t = 0.01$ (see text)

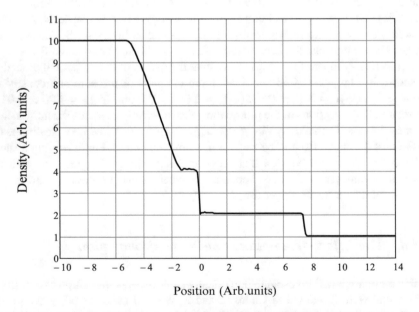

Fig. 4.33 Density in the shock tube as a function of position at $t = 4$ (Arb. units) is shown. For the numerical procedure the following parameters apply; $\gamma = 1.4$, $\kappa = 1.2$, $\Delta x = 0.1$ and $\Delta t = 0.01$ (see text)

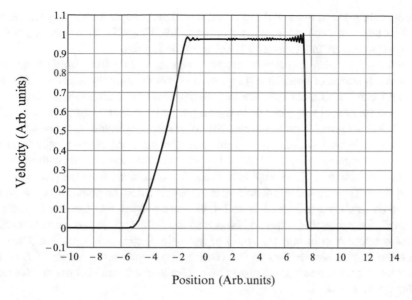

Fig. 4.34 Particle velocity in the shock tube as a function of position at $t = 4$ (Arb. units) is shown. For the numerical procedure the following parameters apply; $\gamma = 1.4$, $\kappa = 1.2$, $\Delta x = 0.1$ and $\Delta t = 0.01$ (see text)

propagate at this speed, consequently, the profile retains its shape with the passage of time. However, when the amplitude of the disturbance increases, as observed in Chap. 2, this behaviour no longer applies as different parts of the profile propagate at different speeds and the profile changes its shape and becomes distorted as it propagates. We will now investigate the effect of amplitude distortion numerically by considering the propagation of a single sinusoidal pulse down a tube that is generated by appropriate piston motion and we will follow the progress of the disturbance as it moves away from the source. In this context, it is useful to recall our discussion of acoustic wave distortion that was presented at the end of Sect. 1.8. Initially, we will assume that the disturbance has small amplitude and we will then proceed to increase the amplitude and observe the changes taking place in the profile. Let us assume that the velocity of the piston can be represented by the equation,

$$u(0,t) = u_m Sin(a\pi t) \ \text{ for } (1/a) \geq t \geq 0$$

$$= 0 \ \text{ for } t > 1/a,$$

where u_m is the maximum amplitude of the velocity. Assuming unity values for the ambient pressure and density within the tube and by taking $\gamma = 1.4$ we find that the velocity of sound is $\sqrt{\gamma} = 1.18$. In order to carry out the numerical procedure the following parameters are assumed; $\kappa = 1.2$, $a = 0.04$, $\Delta x = 0.3$ (Arb. units) and $\Delta t = 0.05$ (Arb. units). Initially, we will assume that $u_m = 0.01$ (Arb. units) which is approximately 1% of the sound speed but, nonetheless, represents a very loud sound wave in terms of acoustic intensity (see Sect. 1.9, Chap. 1); thereafter, u_m will be increased in increments to about 10% of the sound speed.

Figure 4.35 shows the particle velocity when $u_m = 0.01$ (Arb. units) at different times as a function of position and Fig. 4.36 shows the particle velocity at different positions as a function of time. One can observe from these plots that no discernable change in sinusoidal shape occurs in the original profile as it propagates (at least for those values of position and time chosen). When the amplitude is increased to 0.03 (Arb. units), however, one observes that the forward front of the profile becomes steeper as illustrated in Fig. 4.37 and the pulse encountered at a fixed position show a steeper transition to its final amplitude at increased distance from the piston as seen in Fig. 4.38. The onset of this distortion is more evident in the plots corresponding to $u_m = 0.05$ as shown in Figs. 4.39 and 4.40. Here, we see that the profile becomes distorted at much earlier times and begins to show the typical saw-tooth behaviour that was referred to in our discussion of acoustic distortion in Sect. 1.8, Chap. 1. Once the amplitude increases to 0.1 (Arb. units), representing about 10% of the sound velocity, the onset of the distortion occurs at much earlier times as illustrated in Figs. 4.41 and 4.42.

(b) Piston Motion with Small Velocity

In terms of amplitude effects, it is also interesting to return to the example discussed in Sect. 4.8.4 where the piston is suddenly pushed into a tube at constant speed and the tube is closed at one end. In that example we assumed the piston's

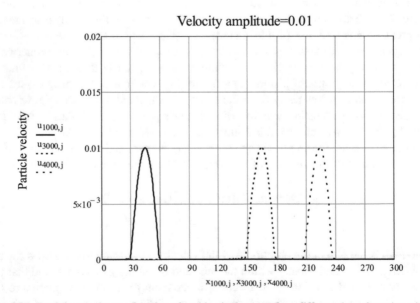

Fig. 4.35 Particle velocity as a function of position is shown at three different times for a sinusoidal pulse propagating down a tube. For the numerical procedure the following parameters apply; $\kappa = 1.2$, $a = 0.04$, $\Delta x = 0.3$ and $\Delta t = 0.05$ (see text)

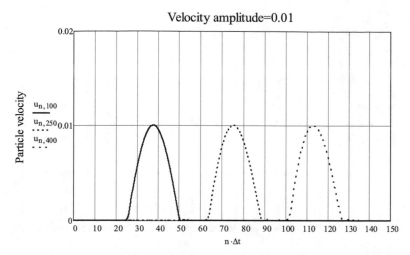

Fig. 4.36 Particle velocity as a function of time is shown at three different positions for a sinusoidal pulse propagating down a tube. For the numerical procedure the following parameters apply; $\kappa = 1.2$, $a = 0.04$, $\Delta x = 0.3$ and $\Delta t = 0.05$ (see text)

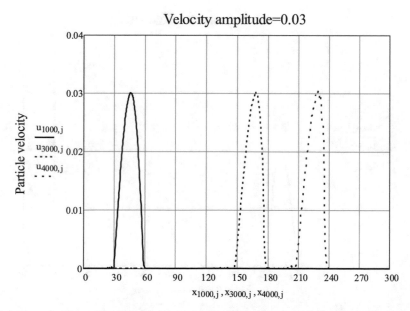

Fig. 4.37 Particle velocity as a function of position is shown at three different times for a sinusoidal pulse propagating down a tube. For the numerical procedure the following parameters apply; $\kappa = 1.2$, $a = 0.04$, $\Delta x = 0.3$ and $\Delta t = 0.05$ (see text)

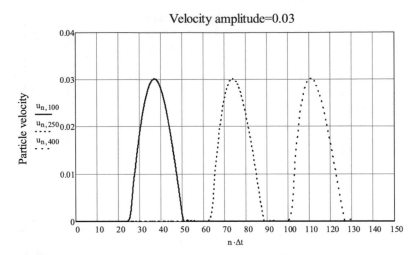

Fig. 4.38 Particle velocity as a function of time is shown at three different positions for a sinusoidal pulse propagating down a tube. For the numerical procedure the following parameters apply; $\kappa = 1.2$, $a = 0.04$, $\Delta x = 0.3$ and $\Delta t = 0.05$ (see text)

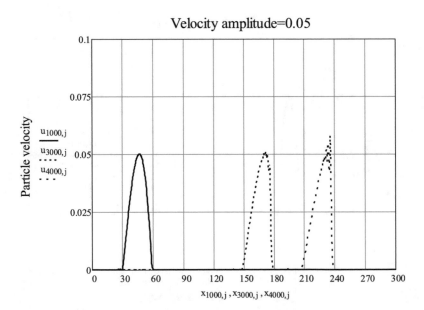

Fig. 4.39 Particle velocity as a function of position is shown at three different times for a sinusoidal pulse propagating down a tube. For the numerical procedure the following parameters apply; $\kappa = 1.2$, $a = 0.04$, $\Delta x = 0.3$ and $\Delta t = 0.05$ (see text)

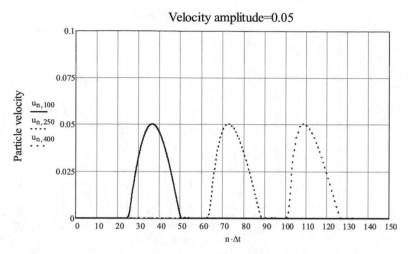

Fig. 4.40 Particle velocity as a function of time is shown at three different positions for a sinusoidal pulse propagating down a tube. For the numerical procedure the following parameters apply; $\kappa = 1.2$, $a = 0.04$, $\Delta x = 0.3$ and $\Delta t = 0.05$ (see text)

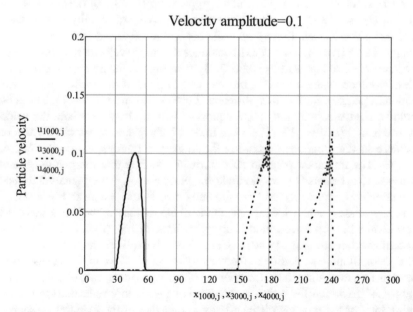

Fig. 4.41 Particle velocity as a function of position is shown at three different times for a sinusoidal pulse propagating down a tube. For the numerical procedure the following parameters apply; $\kappa = 1.2$, $a = 0.04$, $\Delta x = 0.3$ and $\Delta t = 0.05$ (see text)

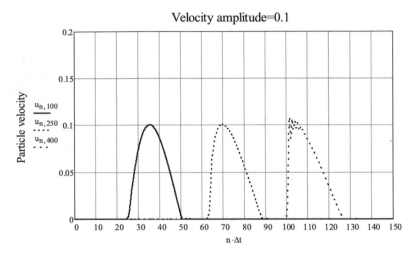

Fig. 4.42 Particle velocity as a function of time is shown at three different positions for a sinusoidal pulse propagating down a tube. For the numerical procedure the following parameters apply; $\kappa = 1.2$, $a = 0.04$, $\Delta x = 0.3$ and $\Delta t = 0.05$ (see text)

speed was given by $u(t, 0) = 0.3$ which is approximately 30% of the speed of sound (equal to $\sqrt{\gamma} = 1.18$ as the ambient pressure and density have unity values) and we found that a shock wave formed immediately and raced ahead of the piston with a velocity of 1.38 (Arb. units). Let us now reduce the piston velocity to about 0.03% of the sound speed so that $u(t, 0) = 0.0003$. By carrying out the numerical calculations with this reduced value of $u(t, 0)$, one can obtain plots of the particle velocity as a function of position for the three different times as shown in Fig. 4.43. These plots should be compared with the velocity plots shown in Fig. 4.17 where the velocity was taken as $u(t, 0) = 0.3$. The magnitude of the piston's velocity is the only difference in the programs generating the numerical outputs shown in Figs. 4.21 and 4.43. The important point to note from Fig. 4.43 is that the piston-generated disturbance (represented by, for example, the plot for $u_{1000, j}$) propagates at the speed of sound which can be verified from the numerical values shown in Fig. 4.43.

We have already noted that the magnitude of the velocity of the shock wave in the plot shown in Fig. 4.17 was found to be approximately 1.38 (Arb. units). This value was confirmed by using Eq. (3.40) in Chap. 3 which explicitly includes the velocity of the piston motion in addition to the velocity of sound. We could of course use the same equation to confirm the velocity of the disturbance in the case of the plot shown in Fig. 4.43 in which the piston's velocity is considerably reduced. On the other hand, let us for the moment disregard any knowledge of the material presented in Chap. 3 or, for that matter, any knowledge in relation to the speed of propagation of small-amplitude disturbances that was discussed in Section 1.8 where the speed of sound c_0 appears explicitly. Hence, by disregarding this knowledge, it is interesting to note that the equations summarized in Sect. 4.5.1, or their equivalent difference equations which are used in the numerical calculations, makes no explicit reference

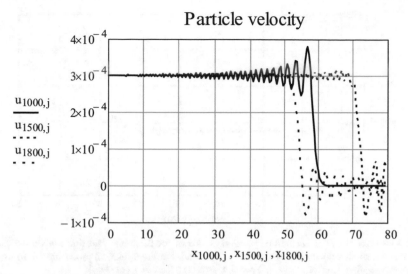

Fig. 4.43 Particle velocity as a function of position is shown for three different times ($t = 50$, $t = 75$ and $t = 90$) when the piston moves with small velocity in a tube closed at one end. For the numerical procedure the following parameters apply: $\gamma = 1.4$, $\kappa = 1.5$, $\Delta x = 0.4$, $\Delta t = 0.05$ and $\rho_0 = 1$ (see text)

to a fundamental propagation speed of c_0 but, nonetheless, by examining the results of the numerical output this propagation speed appears automatically as a fundamental disturbance speed as demonstrated in Fig. 4.43.

Associated pressure plots for this reduced value $u(t, 0)$ of are also shown in Fig. 4.44, and in the case of the plot of $p_{1000, j}$, the magnitude of the pressure perturbation, Δp, is estimated to be 0.00035 ± 0.00001 above the ambient value of unity in the relatively flat portion of the pressure profile. Based on our discussion in Chap. 1, this perturbation is given by Eq. (1.82), namely,

$$\Delta u = \frac{c_0 \Delta p}{\gamma p_0},$$

and by using this equation, one finds that

$$\Delta p = \frac{\gamma p_0 \Delta u}{c_0} = \frac{1.4 \times 1 \times 0.0003}{\sqrt{1.4}} = 0.000353,$$

so that the estimated value is in good agreement with the theoretically predicted value as given by Eq. (1.82).

The important point to note from this discussion is that any initial discontinuity that is generated in the motion of the fluid by virtue of the sudden piston motion will continue to propagate as a discontinuity but, more importantly, its speed of propagation is just equal to the speed of sound when the amplitude of the disturbance is small.

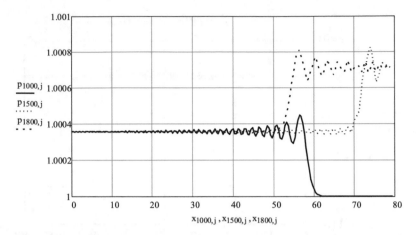

Fig. 4.44 Pressure as a function of position is shown for three different times when the piston moves with small velocity in a tube closed at one end. For the numerical procedure the following parameters apply: $\gamma = 1.4$, $\kappa = 1.5$, $\Delta x = 0.4$, $\Delta t = 0.05$, $\rho_0 = 1$ (see text)

We have already noted in Chap. 1 that small perturbations about ambient values generate disturbances that propagate at the acoustic velocity and, in these circumstances, the propagation of the disturbance satisfies the following wave equation;

$$\frac{\partial^2 \Delta u(x,t)}{\partial x^2} = \frac{1}{c_0^2} \frac{\partial^2 \Delta u(x,t)}{\partial t^2},$$

which is Eq. (1.76) of Chap. 1. Let us set out to solve this equation for the example we are considering here in which the piston is suddenly pushed into a tube at a constant velocity of 0.0003 and the tube is closed at $x = 80$ so that a reflected wave ensues. Using the *separation of variables method* one can write,

$$\Delta u(x,t) = \sum_{n=0}^{\infty} a_n Cos\left[(2n+1)\frac{\pi x}{2L}\right] Sin\left[(2n+1)\frac{\pi c_0 t}{2L}\right], \qquad (4.61)$$

which satisfies the initial conditions in the tube, namely, $u(x,0) = 0$, and the boundary condition, $u(L,t) = 0$ at the end ($L = 80$) of the tube. Specifying the piston velocity as

$$\Delta u(0,t) = 0.0003 \text{ for } t > 0$$

we obtain

$$\sum_{n=0}^{\infty} a_n Sin\left[(2n+1)\frac{\pi c_0 t}{2L}\right] = 0.0003.$$

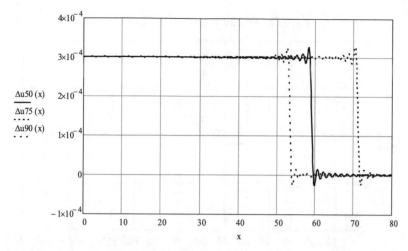

Fig. 4.45 Particle velocity as a function of position arising from the solution of the wave equation is shown for three different times ($t = 50$, $t = 75$ and $t = 90$) when the piston moves with small velocity in a tube closed at one end (see text)

Multiplying this latter equation by $Sin\left[(2m+1)\frac{\pi c_0 t}{2L}\right]$ and performing the integration in the usual manner we obtain the following equation for $\Delta u(x,t)$;

$$\Delta u(x,t) = 0.0003 \sum_{n=0}^{\infty} \left[\frac{(4/\pi)}{2n+1}\right] Cos\left[(2n+1)\frac{\pi x}{2L}\right] Sin\left[(2n+1)\frac{\pi c_0 t}{2L}\right]. \quad (4.62)$$

By summing over 100 Fourier components we obtain the following plots as shown in Fig. 4.45. The quantity $\Delta u50(x)$ in Fig. 4.45 is equal to $\Delta u(x, 50)$ and similar remarks apply for the other quantities plotted. It can be confirmed from these plots that the disturbance propagates at the acoustic velocity as expected and the broken lines in Fig. 4.45 gives the particle velocity due to the reflection from the closed end of the tube.

(c) Incremental Piston Motion

In relation to the formation of shock waves discussed in Sect. 2.4, Chap. 2, we considered how the disturbance-speed depends on the amplitude by considering the propagation of a series of small-amplitude disturbances that are produced by a piston which undergoes a succession of small velocity increments. The discussion in that section drew attention to the fact that the velocity of a particular disturbance travels faster than its predecessors so that later disturbances will eventually catch up with those previously generated and lead to the formation of a shock front.

We will now investigate this aspect numerically and, to do so, we will assume that the piston's velocity within the tube consists of ten small velocity increments with each of magnitude 0.03 and eventually leading to a final velocity of 0.3, similar to

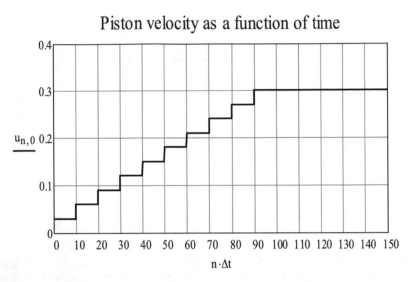

Fig. 4.46 The piston's incremental velocity is shown as a function of time (see text)

that depicted in Fig. 2.3. For the numerical procedure we will use the following parameters; $\gamma = 1.4$, $\kappa = 1.2$, $\Delta x = 1.0$, $\Delta t = 0.1$ and unity values for the ambient pressure and specific volume within the tube are assumed. Each small velocity increment is assigned a duration of $100\Delta t$. The result of the numerical procedure gives the incremental velocity of the piston $u_{n,\,0}$ as a function of time; this is shown in Fig. 4.46.

As a result of the numerical integration of the difference equations one can generate plots of the pressure as a function of the position within the tube and these are shown in Fig. 4.47 at four different times. These numerically generated plots should be compared with those sketched in Fig. 2.5, Chap. 2 to illustrate the "catching-up" process envisaged with the eventual formation of a shock front. One can observe from Fig. 4.47 that the later velocity increments advance faster than their predecessors as can clearly be seen from this figure, and this advance becomes much more evident by comparing the plots at times $t = 200$ and $t = 300$ when the distinction between the individual velocity increments becomes less apparent at $t = 300$ due to the "catching-up" process. In addition, some numerically-generated plots of the pressure as a function of time at three different positions within the tube are shown in Fig. 4.48.

4.8.9   Short Duration Piston Motion: Shock Decay

In the previous sections involving piston motion we found that a shock wave forms immediately when the piston is suddenly pushed at a constant speed and this uniformly propagating shock is maintained provided the piston motion continues.

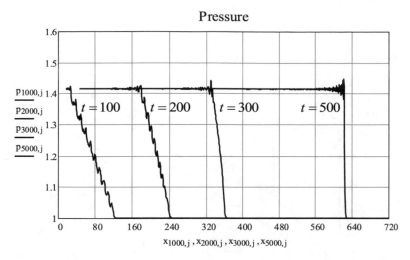

Fig. 4.47 Pressure as a function of position is shown at four different times for the incremental piston motion by numerically integrating the difference equations (see text)

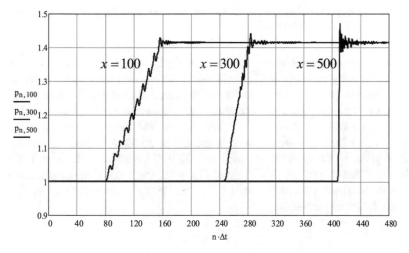

Fig. 4.48 Pressure at three different positions is shown as a function of time for the incremental piston motion by numerically integrating the difference equations (see text)

Alternatively, we found that a shock forms after some finite time when the piston undergoes uniform acceleration to some finite velocity.

In this section we consider the case where the piston is pushed for only a finite duration and thereafter the piston motion is arrested. The fluid adjacent to the piston that was initially set in motion momentarily continues to move and a rarefaction wave is formed that will eventually overtake the shock and reduce its intensity. This overtaking takes place since the shock wave moves with subsonic velocity relative to the air behind it, whereas the head of the rarefaction wave moves at the sonic

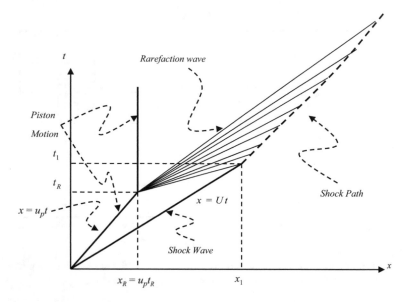

Fig. 4.49 Sketch in the x-t plane showing the rarefaction wave catching up with the shock wave and diminishing its strength (see text)

velocity relative to the air ahead of it. Here, we will consider the decay of the shock wave numerically by using artificial viscosity. Before we do so, let us visualize this catching-up process by referring to the x-t diagram shown in Fig. 4.49. In addition to the shock and piston paths, a rarefaction wave is shown commencing at $t = t_R$ when the piston's motion is arrested. The head of the rarefaction wave has slope $dx/dt = u_p + c$, where c is the speed of sound in the medium behind the shock. This speed is given by the equation, $c = c_0 + (\gamma - 1)(u_p/2)$, where c_0 is the speed of sound in the undisturbed medium. Referring to Fig. 4.49, the head of the rarefaction wave intersects the shock path at the point (x_1, t_1) and these coordinates can be obtained from the following equations,

$$x_1 = Ut_1$$

and

$$x_1 = u_p t_R + \left(u_p + c \right)\left(t_1 - t_R \right).$$

Solving for x_1 and t_1, we find that

$$t_1 = \frac{ct_R}{\left(u_p + c \right) - U} \tag{4.63}$$

and

$$x_1 = \frac{Uct_R}{(u_p + c) - U}.$$ (4.64)

We will be returning to these equations shortly when some numerical results are presented and discussed. In relation to the numerical procedure, it is assumed that the piston is instantaneously accelerated in a tube to a constant velocity of 0.3 (Arb. units) for a finite duration t_R which is taken as some multiple of the time interval Δt; thereafter, the piston motion is arrested. For the numerical calculations we assume that $t_R = 100\Delta t$ and that the other following parameters apply; $\gamma = 1.4$, $\kappa = 1.5$, $\Delta x = 0.4$ and $\Delta t = 0.1$. The undisturbed pressure and specific volume within the tube are assumed to be $p_1 = 1$ and $v_1 = 1$, respectively, so that the ambient speed of sound is $\sqrt{\gamma} = 1.183$.

Figure 4.50 illustrates the results of the numerical calculations with $u_p = 0.3$ (Arb. units) and shows the variation of the particle velocity and pressure in the pulse at different times. The particle velocity in the pulse in which $u = 0$ moves with a velocity equal to the velocity of sound in the undisturbed medium as can be verified from the plot shown in Fig. 4.50 and the velocity exhibits a linear variation with position until it attains its maximum value immediately behind the shock front. For comparison, Fig. 4.51 shows the particle velocity and the pressure plots for a higher piston velocity of $u_p = 0.8$ and in this case we observe a very rapid initial decrease in the shock strength.

For the smaller piston velocity of 0.3 shown in Fig. 4.50 one observes a more progressive decrease in the magnitude of the shock strength and a corresponding incremental increase in the width Δw of the shock wave zone. Within the shock wave zone the particle velocity exhibit a linear variation in these parameters, as observed between the tail of the rarefaction wave and the shock front, as sketched in Fig. 4.52.

Let us now see if these numerical outputs are in accord with the catching-up process predicted according to Fig. 4.49 where we found that the head of the rarefaction wave intersects the shock at the point (x_1, t_1). Using the values adopted for the numerical procedure, namely, $\gamma = 1.4$, $\Delta x = 0.4$, $\Delta t = 0.1$, $u_p = 0.3$ and $c_0 = \sqrt{\gamma} = 1.183$, we find that $U = 1.377$, $c = 1.244$ and $t_R = 100\Delta t = 10$. Inserting these values into Eqs. (4.63) and (4.64) we find that $t_1 = 74.5$ and $x_1 = 102.57$. Hence, the predicted intersection of the shock path with the head of the rarefaction wave occurs at $t_1 = 74.5$. In terms of the numerical calculations, this intersection should correspond to the numerical output where the time-stepping index, n, is set at $n = 745$ since $\Delta t = 0.1$. Hence, Fig. 4.53 shows a plot of the particle velocity, $u_{745, j}$, as a function of the particle position at this instant (solid line in Fig. 4.53). One observes that the particle velocity immediately behind the shock front is close to the value of 0.3 (within the limitations of the numerical calculations using artificial viscosity) and the marker in the plot, corresponding to the position $x_1 = 102.57$, has been inserted. This marker is seen to coincide closely with the position of the shock front according to the numerical calculations. In fact, a plot of the artificial viscosity, $q_{745, j}$, gives a maximum value at $x = 102.3$, which is in good agreement with the theoretical value of 102.57. The other plots shown in the same figure correspond to

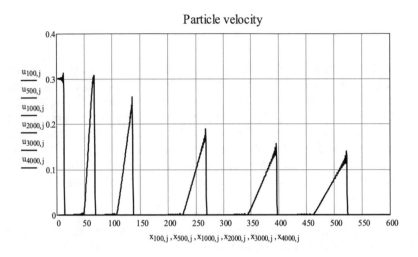

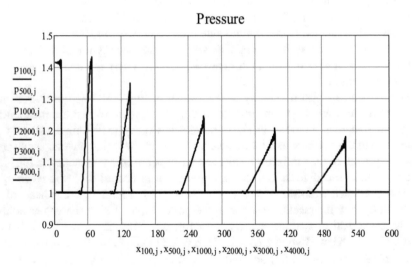

Fig. 4.50 Particle velocity and pressure as a function of position for various times corresponding to short impulsive piston motion. For the numerical procedure the following parameters apply; $u_p = 0.3$, $\gamma = 1.4$, $\kappa = 1.5$, $\Delta x = 0.4$ and $\Delta t = 0.1$ (see text)

somewhat later times where a slight reduction can be noticed in the particle velocity immediately behind the shock, thereby indicating a reduction in its intensity.

The decay of the shock wave has been discussed by several investigators. Chandrasekhar [13], for example, observed that the treatment of shock wave propagation can be simplified if the shock is weak or of moderate strength. Under these circumstances, the increase of entropy of an element of fluid as it crosses the shock front can be ignored and, in addition, the change in the Riemann invariant R_- as the shock front is crossed can be neglected (see our discussion

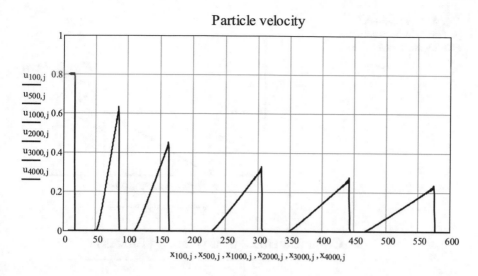

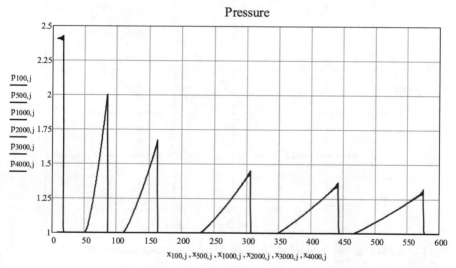

Fig. 4.51 Particle velocity and pressure as a function of position for various times corresponding to short impulsive piston motion. For the numerical procedure the following parameters apply; $u_p = 0.8$, $\gamma = 1.4$, $\kappa = 1.5$, $\Delta x = 0.4$ and $\Delta t = 0.1$ (see text)

presented in Sect. 3.12, Chap. 3). Based on Chandrasekhar observations, Friedrichs [14] has presented a modified approach that permits one to obtain explicit analytical solutions for a number of unsteady gas flow problems that fall within the weak or moderate intensities in which the approximations apply. Kantrowitz [15] has also considered the shock decay process. In the report by Friedrichs [14] an approximate method is presented for the treatment of a decaying

Fig. 4.52 Variation of
particle velocity within the
shock wave zone (see text)

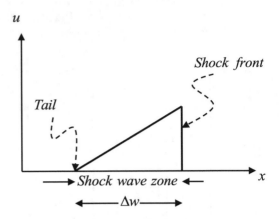

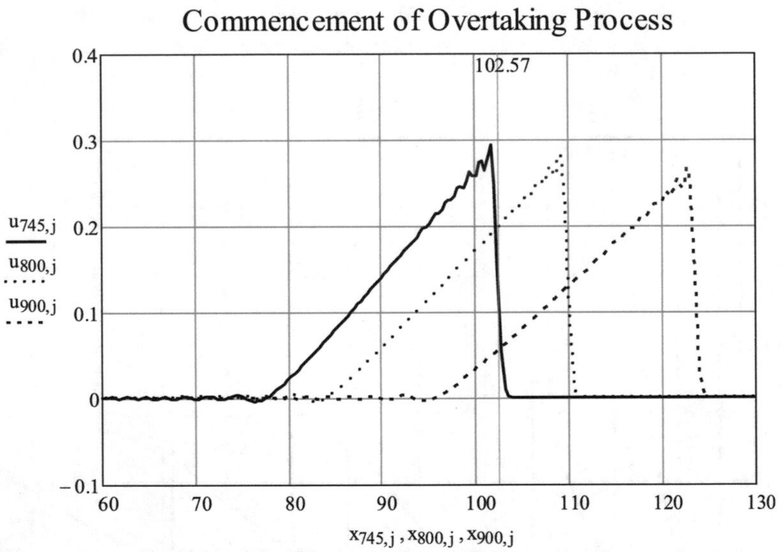

Commencement of Overtaking Process

Fig. 4.53 Plots of particle velocity as a function of position at three different times with $u_p = 0.3$
(see text)

shock wave, that is, a shock wave whose strength is reduced by an overtaking
rarefaction or expansion wave. Friedrichs' treatment of this problem, which will be
briefly discussed below, gives quite accurate results if the shock is weak or of
moderate strength. Friedrichs' method is based on the fact that the flow properties
behind a weak shock and the flow properties behind a simple wave agree up to
terms of second order in the shock strength (see Sect. 3.12, Chap. 3). The progress
of a decaying shock wave can be represented by shocks of varying strength and any
individual shock wave can be identified according to its shock speed. In Friedrichs'
method these individual shock waves are replaced by simple waves that have the
same flow properties behind it as the shocks themselves. The flow of these simple

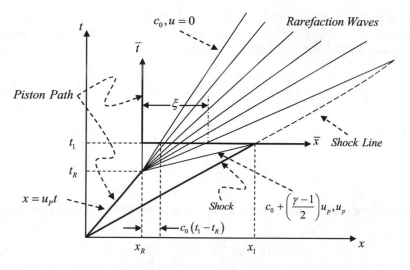

Fig. 4.54 New coordinate system for determining the shock decay process (see text)

waves are calculated independently of the shock waves and the time of intersection of the simple wave with the shock path is then determined. We will follow Friedrichs' analysis as presented below.

The rarefaction wave is a centered simple wave (see Fig. 4.54) and each characteristic is a straight line passing through the point (x_R, t_R) and along which u and c are constant. The point of intersection (x_1, t_1) of the shock with the head or front of the expansion wave becomes the origin of a new coordinate system $(\bar{x}, \bar{t})$ where $\bar{x} = x - x_R$ and $\bar{t} = t - t_1$. It can be seen from Fig. 4.54 that each of the straight line characteristics intersect the $\bar{x}$-axis at a point on this axis which is denoted by the symbol ξ and a typical characteristic is given by the equation,

$$\bar{x} = (u + c)\bar{t} + \xi \tag{4.65}$$

Each value of ξ denotes a specific straight line characteristic as shown in Fig. 4.54 and, hence, u and c are functions of ξ, so that the previous equation for a typical characteristic can be written as

$$\bar{x}(\xi) = [u(\xi) + c(\xi)]\bar{t}(\xi) + \xi \tag{4.66}$$

The intercept, ξ, on the $\bar{x}$ axis for the various rarefaction waves corresponds to the following range;

$$c_0(t_1 - t_R) < \xi \le \left[c_0 + \left(\frac{\gamma + 1}{2}\right)u_p\right](t_1 - t_R)$$

as can be identified from inspection of Fig. 4.54. In order to determine the shock trajectory, $U \equiv U(\bar{x}(\xi), \bar{t}(\xi))$, we follow Friedrichs' analysis and differentiate Eq. (4.66) with respect to ξ, hence,

$$\frac{d\bar{x}}{d\xi} = \bar{t}\left[\frac{du}{d\xi} + \frac{dc}{d\xi}\right] + (u + c)\frac{d\bar{t}}{d\xi} + 1. \tag{4.67}$$

However,

$$\frac{d\bar{x}}{d\xi} = \frac{d\bar{x}}{d\bar{t}}\frac{d\bar{t}}{d\xi} = U\frac{d\bar{t}}{d\xi},$$

where $U = d\bar{x}/d\bar{t}$ represents the velocity of the shock and Eq. (4.67) becomes,

$$[U - (u + c)]\frac{d\bar{t}}{d\xi} - \bar{t}\left(\frac{du}{d\xi} + \frac{dc}{d\xi}\right) = 1. \tag{4.68}$$

This differential equation is to be solved for $\bar{t}(\xi)$ to give the motion of the decaying shock. As the different characteristics have different values of ξ, then $\bar{t}(\xi)$ gives the time that the different characteristics intersect the shock. The shock velocity in the case of the weak shock approximation is given by Eq. (3.57), namely,

$$U = c_0\left[1 + \left(\frac{\gamma + 1}{4}\right)\frac{u}{c_0} + \frac{1}{2}\left(\frac{\gamma + 1}{4}\right)^2\frac{u^2}{c_0^2}\right], \tag{4.69}$$

with c_1 replaced by c_0 in this application. The slope of a typical characteristic is given by the equation (see Fig. 4.54),

$$u(\xi) + c(\xi) = \frac{\xi}{t_1 - t_R} \tag{4.70}$$

and by using the fact that the Riemann invariant R_- is constant (see Sect. 3.12, Chap. 3), we have

$$u(\xi) - \frac{2c(\xi)}{\gamma - 1} = -\frac{2c_0}{\gamma - 1}. \tag{4.71}$$

Solving Eqs. (4.70) and (4.71) for $u(\xi)$ we find that

$$u(\xi) = \frac{2}{\gamma + 1}\left(\frac{\xi}{t_1 - t_R} - c_0\right)$$

Friedrichs [14] defines $\sigma = \left(\frac{\gamma+1}{2}\right)\frac{u}{c_0}$, hence, with this definition we have

$$\sigma(\xi) = \left(\frac{\gamma+1}{2}\right)\frac{u(\xi)}{c_0} = \frac{\xi}{c_0(t_1 - t_R)} - 1 \tag{4.72}$$

and with ξ in the range, $c_0(t_1 - t_R) < \xi \leq \left[c_0 + \left(\frac{\gamma+1}{2}\right)u_p\right](t_1 - t_R)$, we have σ in the range,

$$0 < \sigma \leq \sigma_1,$$

where $\sigma_1 = \left(\frac{\gamma+1}{2}\right)\frac{u_p}{c_0}$. Using Eq. (4.69) for the shock velocity, we can write it in terms of σ, hence, it becomes,

$$U = c_0\left[1 + \frac{1}{2}\sigma + \frac{1}{8}\sigma^2\right],$$

similarly, we have

$$u + c = c_0 + \left(\frac{\gamma+1}{2}\right)u = c_0(1 + \sigma)$$

and, therefore,

$$U - (u + c) = \frac{c_0\sigma}{8}(\sigma - 4).$$

Substituting these latter relationships for U and $(u + c)$ in Eq. (4.68) and using Eq. (4.70) we find that

$$\frac{c_0\sigma}{8}(\sigma - 4)\frac{d\bar{t}}{d\xi} - \frac{\bar{t}}{t_1 - t_R} - 1 = 0,$$

hence,

$$\frac{c_0\sigma}{8}(\sigma - 4)\frac{d\bar{t}}{d\sigma}\frac{d\sigma}{d\xi} - \frac{\bar{t}}{t_1 - t_R} - 1 = 0 \tag{4.73}$$

and after using Eq. (4.72) this latter equation reduces to

$$\sigma(\sigma - 4)\frac{d\bar{t}}{d\sigma} = 8\bar{t} + 8(t_1 - t_R). \tag{4.74}$$

Integrating, we obtain,

$$\frac{1}{8}\int_0^{\bar{t}}\frac{d\bar{t}}{\bar{t} + (t_1 - t_R)} = \int_{\sigma_1}^{\sigma}\frac{d\sigma}{\sigma(\sigma - 4)}.$$

Expressing the right-hand side of this integral in terms of partial fractions according to the equation,

$$\frac{1}{\sigma(\sigma - 4)} = \frac{A}{\sigma} + \frac{B}{\sigma - 4},$$

we find that $A = -(1/4)$ and $B = (1/4)$, hence,

$$\int_0^{\bar{t}} \frac{d\bar{t}}{\bar{t} + (t_1 - t_R)} = 2 \left[\int_{\sigma_1}^{\sigma} \frac{d\sigma}{(\sigma - 4)} - \int_{\sigma_1}^{\sigma} \frac{d\sigma}{\sigma} \right] \tag{4.75}$$

and by carrying out the integration on both side of the latter equation we find that

$$\ln \left[\bar{t} + (t_1 - t_R) \right] - \ln (t_1 - t_R) = 2 \left[\ln (\sigma - 4) - \ln (\sigma_1 - 4) - \ln \sigma + \ln \sigma_1 \right],$$

hence,

$$\ln \left[\frac{\bar{t} + (t_1 - t_R)}{(t_1 - t_R)} \right] = \ln \left[\left(\frac{\sigma - 4}{\sigma} \right) \left(\frac{\sigma_1}{\sigma_1 - 4} \right) \right]^2$$

and, therefore,

$$\bar{t} + (t_1 - t_R) = (t_1 - t_R) \left[\left(\frac{\sigma - 4}{\sigma} \right) \left(\frac{\sigma_1}{\sigma_1 - 4} \right) \right]^2.$$

However, we have already noted that $\bar{t} = t - t_1$, hence, we finally obtain the equation,

$$t = t_R + (t_1 - t_R) \left(\frac{\sigma - 4}{\sigma} \right)^2 \left(\frac{\sigma_1}{\sigma_1 - 4} \right)^2; \quad \sigma_1 \geq \sigma > 0. \tag{4.76}$$

Substituting our previous relationships for $(u + c)$ and ξ in terms of σ in the equation, $\bar{x} = [u(\xi) + c(\xi)]\bar{t} + \xi$, we have

$$\bar{x} = c_0(1 + \sigma)(t - t_1) + c_0(1 + \sigma)(t_1 - t_R).$$

Noting that $\bar{x} = x - x_R$, we arrive at the following equation for x,

$$x = x_R + c_0(1 + \sigma)(t - t_R). \tag{4.77}$$

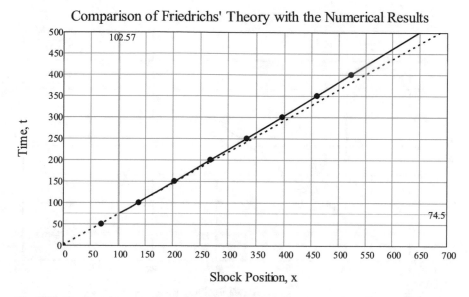

Fig. 4.55 Shock position as a function of time for the decaying shock wave with $u_p = 0.3$ (see text)

Equations (4.76) and (4.77) are the parametric equations for the shock line and we will use these equations for comparison with the numerical results.

In Fig. 4.55 we show a comparison between Friedrichs' theory for the decaying shock wave (represented by the solid line) with the numerical results (indicated by •) obtained from Fig. 4.50 for the position of the shock front at specific times. These positions are determined by locating where the artificial viscosity q attains its maximum value. The broken line in the same figure corresponds to constant piston motion with a velocity of 0.3 (Arb. units). It can be observed that the numerical results are in excellent agreement with Friedrichs' analysis as the ratio, u_p/c_0, coincides with the weak shock approximation.

Let us now use the numerical results presented in Fig. 4.51 in order to compare the outcome of these results with Friedrichs' theory. Here, the piston is similarly pushed for a finite duration but in this case the piston's velocity is 0.8 (Arb. units) which is considerably higher than the velocity of 0.3 that applies to Fig. 4.50. This higher velocity was anticipated to be well outside the weak shock regime and poor agreement with theory was expected; nonetheless, Fig. 4.56 shows that the theory as presented by Friedrichs shows remarkably good agreement with the numerical results. The markers on this plot represent the point (x_1, t_1) where the initial interaction of the shock with the head of the rarefaction wave occurs.

Perhaps, it's not surprising that the agreement is so good when one refers to a table of numerical values presented by Chandrasekhar [13]. In relation to this table Chandrasekhar has demonstrated that in the case of shocks of moderate strength, that

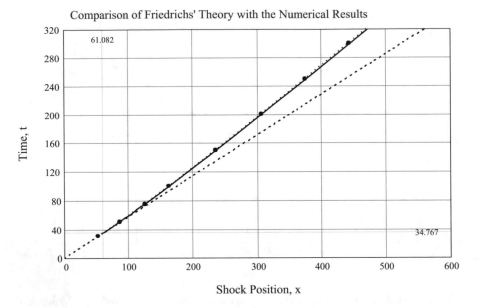

Fig. 4.56 Shock position as a function of time for the decaying shock wave with $u_p = 0.8$ is shown. The solid line is based on Friedrichs' theory and the points (indicated by $\cdot \cdot \bullet \cdot \cdot$) are the numerical results obtained from Fig. 4.51. The broken line represents the shock path for constant piston motion with $u_p = 0.8$ (see text)

is, shocks with a Mach number less than 1.5 or pressure ratios on either side of the shock front that are less than 2.5, then the increase in the entropy of a fluid element as it crosses the shock can be ignored and the change in the Riemann invariant can also be neglected. The Mach number and pressure ratio for the numerical results shown in Fig. 4.56 are, 1.485 and 2.405, respectively.

Based on the finding of Chandrasekhar [13], Friedrichs [14] and Kantrowitz [15], it is found that the shock strength, $\Delta p = (p_2 - p_1)/p_1$, and the width of the shock wave zone Δw vary as $1/\sqrt{t}$ and $\sqrt{t}$, respectively, for large values of the time t. In order to explore this aspect further, let us follow Friedrichs' analysis and write Eq. (4.76) in the following manner,

$$\left(\frac{4-\sigma}{\sigma}\right)^2 \left(\frac{\sigma_1}{4-\sigma_1}\right)^2 = \frac{t-t_R}{t_1-t_R},$$

hence,

$$\frac{\sigma}{4-\sigma} = \frac{\sigma_1}{4-\sigma_1} \sqrt{\frac{t_1-t_R}{t-t_R}}.$$

Adding 1 to both sides of this latter equation and inverting, we have

$$\frac{4-\sigma}{4} = \left[1 + \left(\frac{\sigma_1}{4-\sigma_1}\right)\sqrt{\frac{t_1-t_R}{t-t_R}}\right]^{-1},$$

hence,

$$\sigma = 4 - 4\left[1 + \left(\frac{\sigma_1}{4-\sigma_1}\right)\sqrt{\frac{t_1-t_R}{t-t_R}}\right]^{-1}.$$

In general, the quantity, $\sigma_1/(4-\sigma_1) \ll 1$ in the case of weak shocks and, therefore, by using the following binomial expansion, $(1+x)^{-1} = 1 - x + x^2 - x^3 + ..$ to second order, it follows that

$$\sigma = 4\left(\frac{\sigma_1}{4-\sigma_1}\right)\sqrt{\frac{t_1-t_R}{t-t_R}} - 4\left(\frac{\sigma_1}{4-\sigma_1}\right)^2\frac{t_1-t_R}{t-t_R} + \ldots \qquad (4.78)$$

The width Δw of the shock wave zone according to Eq. (4.77) is given by

$$\Delta w = c_0\sigma(t - t_R)$$

and by substituting for σ from Eq. (4.78), we find that

$$\Delta w = 4c_0\left(\frac{\sigma_1}{4-\sigma_1}\right)\sqrt{(t_1-t_R)(t-t_R)} - 4c_0\left(\frac{\sigma_1}{4-\sigma_1}\right)^2(t_1-t_R) + \ldots \qquad (4.79)$$

and therefore the width of the shock wave zone increases according to the equation,

$$\Delta w \propto \sqrt{t - t_R}. \qquad (4.80)$$

In relation to the shock strength, $(p_2 - p_1)/p_1$, let us use Eq. (3.58) of Chap. 3 in the case of a weak shock. By retaining terms up to second order in this equation we have

$$\frac{p_2}{p_1} = 1 + \gamma\frac{u}{c_0} + \frac{\gamma(\gamma+1)}{4}\frac{u^2}{c_0^2},$$

and, hence, the shock strength according to this latter equation can be written in the following form,

$$\frac{p_2-p_1}{p_1} = \frac{2\gamma}{\gamma+1}\left(\left[\left(\frac{\gamma+1}{2}\right)\frac{u}{c_0}\right]\right) + \frac{2\gamma}{\gamma+1}\frac{1}{2}\left[\left(\frac{\gamma+1}{2}\right)^2\left(\frac{u}{c_0}\right)^2\right]$$

and as σ it defined according to the equation,

$$\sigma = \left(\frac{\gamma + 1}{2}\right) \frac{u}{c_0}$$

the expression for the shock strength becomes

$$\frac{p_2 - p_1}{p_1} = \frac{2\gamma}{\gamma + 1} \left(\sigma + \frac{\sigma^2}{2}\right).$$

By using the value of σ in Eq. (4.78) we find, after retaining terms of the order of $\sigma_1^2/(4 - \sigma_1)^2$, that

$$\frac{p_2 - p_1}{p_1} = \frac{2\gamma}{\gamma + 1} \left[4\left(\frac{\sigma_1}{4 - \sigma_1}\right)\sqrt{\frac{t_1 - t_R}{t - t_R}} + 4\left(\frac{\sigma_1}{4 - \sigma_1}\right)^2 \frac{t_1 - t_R}{t - t_R}\right], \qquad (4.81)$$

and for large values of t we see that the shock strength decreases as

$$\left(\frac{p_2 - p_1}{p_1}\right) \propto \frac{1}{\sqrt{t - t_R}} \qquad (4.82)$$

By using the numerical results shown in Fig. 4.50 one can generate the plots shown in Figs. 4.57 and 4.58 for Δp and Δw as functions of time. It can be observed from these plots that the numerical results are in good agreement with the analytical predictions.

Some other examples of wave motion that are worth mentioning arise from different types of short duration piston motion. The first involves the sudden motion of the piston into a tube, then decelerating it and returning it to its initial position as sketched in Fig. 4.59(a). The resulting numerical output for the pressure pulse as a function of position is shown in Fig. 4.59(b) at three different times and this is an example of a so-called "N-wave" [14]. One can see that the wave region or wave zone Δw is bounded by two shocks; a head shock and a tail shock. The width Δw is further sketched in Fig. 4.59(c) and from some tabulated numerical values shown in Fig. 4.59(d) it is found that Δw increases according to the relation [14],

$$\Delta w \propto \sqrt{t - t_R}$$

as shown in Fig. 4.59(e) where $t_R = 100\Delta t$.

Another type of "N-wave" can be produced by firstly retracting the piston, arresting it and then moving it forward to its initial position as sketched in Fig. 4.60(a). The resulting pressure pulse arising from the numerical calculations is shown plotted in Fig. 4.60(b) for three different times. One can confirm that the width of the wave zone Δw as sketched in Fig. 4.60(c) remains constant with time, while the shock strength, $\Delta p = (p_2 - p_1)/p_1$, for some representative values of the

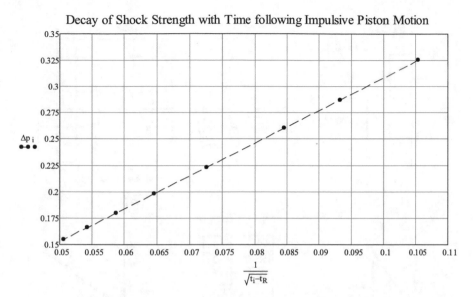

Fig. 4.57 A plot of the decaying shock strength Δp at different time intervals (denoted by the points •) according to the numerical calculations (see text)

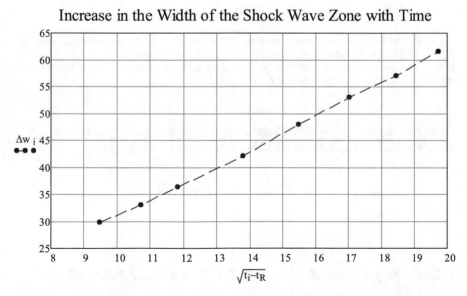

Fig. 4.58 A plot of the width of the shock wave zone Δw at different time intervals (denoted by the points •) according to the numerical calculations (see text)

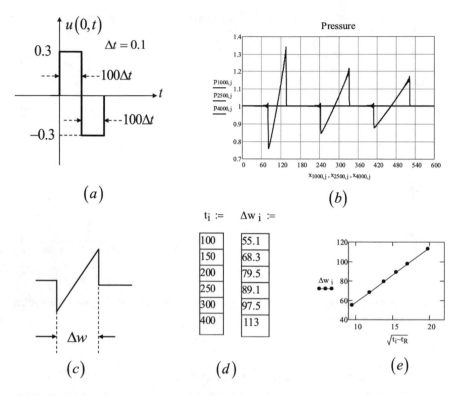

Fig. 4.59 The "N-wave" generated pressure pulse is shown in (**b**) arising from the piston motion as sketched in (**a**). The table (**d**) shows the width of the wave zone Δw_i for some representative values of the time t_i and this is shown plotted in (**e**). For the numerical procedure the following parameters apply; $\gamma = 1.4$, $\kappa = 1.5$, $\Delta x = 0.4$ and $\Delta t = 0.1$ (see text)

time t are tabulated in Fig. 4.60(d) and are shown plotted in Fig. 4.60(e) and, in this case, the shock strength varies with time according to the relation,

$$\Delta p \propto \frac{1}{t - t_R}, \quad \text{where } t_R = 10.$$

4.8.10 Some Numerical Results for Shock Wave Interactions

So far we have considered shock wave generation by several types of piston motion and we have also investigated shock wave reflection from an end wall. In this section we will consider some problems involving shock wave interactions numerically: the first to be considered is the reflection of a shock wave at the boundary of two gases

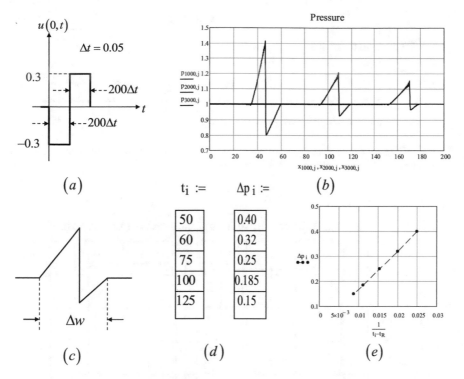

Fig. 4.60 This "N-wave" generated pressure-pulse is shown plotted in (**b**) arising from the piston motion as sketched in (**a**). The table (**d**) shows the shock strength Δp_i for some representative values of the time t_i and this is shown plotted in (**e**). For the numerical procedure the following parameters apply; $\gamma = 1.4$, $\kappa = 1.5$, $\Delta x = 0.2$ and $\Delta t = 0.05$ (see text)

having different specific heat ratios and densities; secondly, we will consider the overtaking of one shock by another and, finally, we will present some numerical results obtained for the head-on collision of two shock waves.

(a) Shock reflection from a plane boundary between two gases

The incident shock wave is assumed to propagate in gas 2 as shown in Fig. 4.61 as a result of the sudden motion of the piston to a constant velocity of 0.3 (Arb. units). From our previous examples with this piston velocity we have already established that the shock speed is 1.377 in gas 2 and the pressure behind the shock is 1.413. The condition for the shock wave to be reflected from the boundary is given by Landau and Lifshitz [16] as

$$\frac{v_{01}}{(\gamma_1 + 1)\frac{p}{p_1} + (\gamma_1 - 1)} < \frac{v_{02}}{(\gamma_2 + 1)\frac{p}{p_1} + (\gamma_2 - 1)},$$

where p is the pressure behind the shock wave which is equal to 1.413 in this particular case.

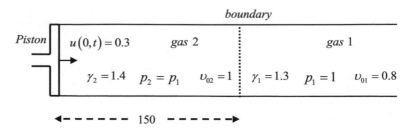

Fig. 4.61 Diagram illustrating piston motion for shock reflection between two different gases at the same pressure but having different specific volumes

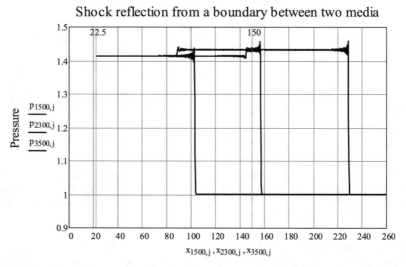

Fig. 4.62 Pressure plots at three different times for the incident and reflected shocks as a function of the position of the fluid elements at the times indicated. For the numerical procedure the following parameters apply; $\kappa = 1.5$, $\Delta x = 0.3$ and $\Delta t = 0.05$ (see text)

Figure 4.62 shows the results of the numerical output with the pressure plotted at three different times as a function of the position of the fluid elements at these times. The marker at 22.5 gives the piston's position at $t = 1500\Delta t$ while the marker at 150 represents the boundary between the two gases. One can observe that a reflected shock results since the numerical values adopted in this case satisfy the above inequality, on the other hand, it is easy to confirm numerically that no reflected shock is obtained if the condition is violated.

(b) Overtaking Shock Waves

Let us now consider the case of two plane shock waves following each other and the overtaking of one shock by the other. The two shocks are generated by assuming the piston has the following velocity;

$$u_{n,0} = 0.8 \text{ if } 0 \le n \le 500$$
$$= 0 \text{ if } 500 < n \le 1250$$
$$= 1 \text{ if } n > 1250$$

where n is the time-stepping index, hence, the numerical outputs for the piston's velocity and position are shown plotted in Fig. 4.63 where $\Delta t = 0.04$ is the time-stepping increment. Pressure plots of the two forward moving shock waves as a function of position are shown plotted at four different times in Fig. 4.64 and the markers in these plots show the position of the piston at these times.

After the second shock overtakes the first shock, a transmitted shock and a weak reflected rarefaction wave result [17] as shown in Fig. 4.65 and the region between them is divided by a contact surface. An expanded view of the numerical output for this region (bounded by the marker at $x = 255$ and forward front of the transmitted shock) is shown in Fig. 4.66. One observes that this contact surface (indicated by the marker at $x = 377$) divides this region into two zones where the density and the temperature (which is proportional to the product of the pressure and specific volume) differ within the region while the pressure and particle velocity are the same on both sides.

(c) Colliding Shock waves

The final example to be considered involves the collision of two shock waves where one of the colliding shocks is produced by short-duration piston motion and the shock's subsequent reflection from an end wall. This returned shock then meets a second outgoing shock wave which is produced by the further motion of the piston. In the example to be considered here we will assume that the piston's velocity is given by,

$$u_{n,0} = 0.4 \text{ if } 0 \le n \le 500$$
$$= 0 \text{ if } 500 < n \le 3000$$
$$= 0.6 \text{ if } n > 3000$$

Piston velocity as a function of time

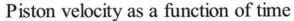

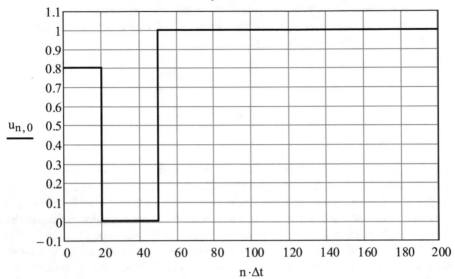

Piston position as a function of time

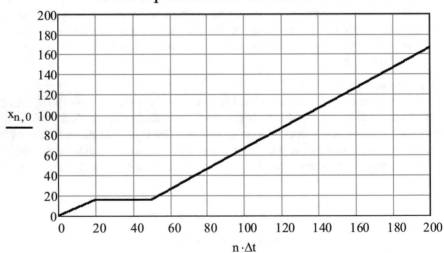

Fig. 4.63 Piston's velocity and position as a function of time are shown for the generation of two shock waves. For the numerical procedure the following parameters apply; $\gamma = 1.4$, $\kappa = 1.5$, $\Delta x = 0.4$ and $\Delta t = 0.04$ (see text)

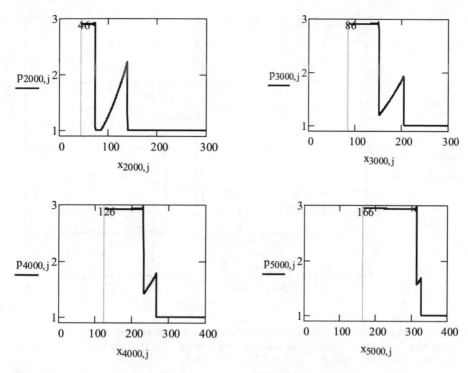

Fig. 4.64 Pressure as a function of position at four different times showing one shock catching up with the other shock. For the numerical procedure the following parameters apply; $\gamma = 1.4$, $\kappa = 1.5$, $\Delta x = 0.4$ and $\Delta t = 0.04$ (see text)

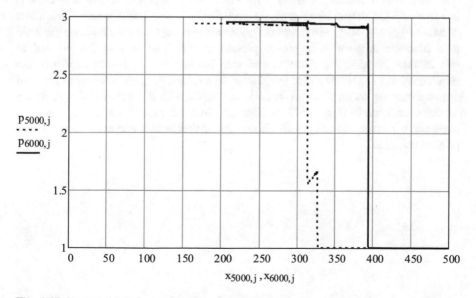

Fig. 4.65 Pressure plotted as a function of position for the two shocks prior to and after the overtaking process. For the numerical procedure the following parameters apply; $\gamma = 1.4$, $\kappa = 1.5$, $\Delta x = 0.4$ and $\Delta t = 0.04$ (see text)

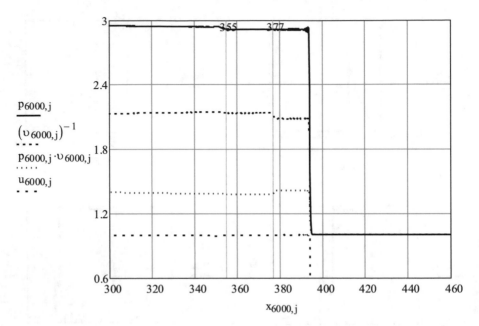

Fig. 4.66 Expanded view of the numerical output at $t = 6000\Delta t$ showing the pressure, density, temperature and particle velocity as a function of position. For the numerical procedure the following parameters apply; $\gamma = 1.4$, $\kappa = 1.5$, $\Delta x = 0.4$ and $\Delta t = 0.04$ (see text)

where n is the time-stepping index and the numerical output for the piston's velocity and position as a function of time $t\ (=n\Delta t)$ is shown in Fig. 4.67 where $\Delta t = 0.04$. It is assumed that the ambient pressure and density have unity values. The short duration $(0 \leq n \leq 500)$ piston motion generates the outgoing shock whose pressure as a function of position is shown plotted in Fig. 4.68 at $t = 2000\Delta t$ and at $t = 2600\Delta t$. Its reflection from an end wall located at $x = 160$ is shown in the same figure at $t = 3100\Delta t$ and at $t = 3300\Delta t$. For $t > 3000\Delta t$ one observes the second outward moving shock wave on its collision course with the reflected shock. Based on our discussion in Sect. 4.8.9 we note that this reflected shock propagates with decreasing intensity and Fig. 4.69 shows the further advancement of these shocks prior to collision.

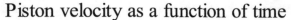

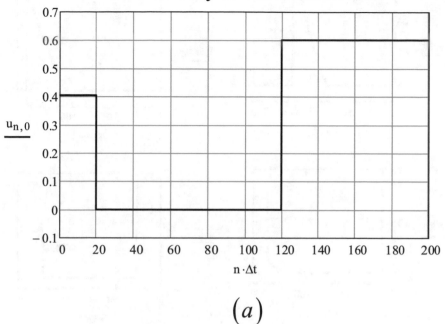

$$(a)$$

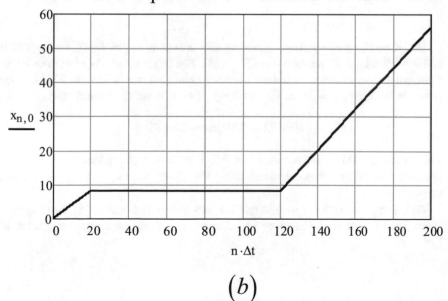

$$(b)$$

Fig. 4.67 Piston velocity (**a**) and piston position (**b**) as a function of time for generating the colliding shock waves. For the numerical procedure the following parameters apply; $\gamma = 1.4$, $\kappa = 1.5$, $\Delta x = 0.4$ and $\Delta t = 0.04$ (see text)

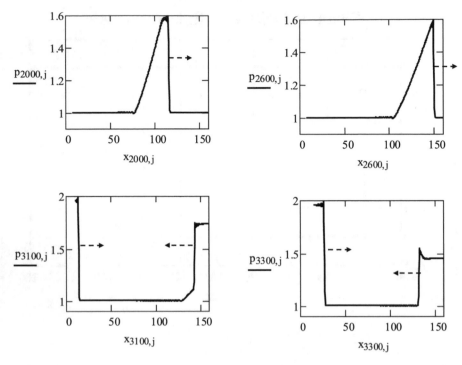

Fig. 4.68 Pressure profiles for the generation of colliding shock waves. For the numerical procedure the following parameters apply; $\gamma = 1.4$, $\kappa = 1.5$, $\Delta x = 0.4$ and $\Delta t = 0.04$ (see text)

Based on the numerical results the shocks collide at approximately $t = 4172\Delta t$ and at position $x \cong 82.8$ as shown in Fig. 4.70. The larger shock, moving from left to right, commences its motion when the time-stepping index n is equal to 3001 and the piston is located at $x = 8$ (see Fig. 4.67(b)). If U is its shock velocity then,

$$U(4172 - 3001)\Delta t + 8 = 82.8$$

and by solving this latter equation we find that $U = 1.597$ which is in excellent agreement with the value calculated using Eq. (3.40) with $u_p = 0.6$, $\gamma = 1.4$ and $c_1 = \sqrt{\gamma}$.

Following the collision we see that the two incident shocks reflect back as new shock waves and continue to move further away from each other as shown in Fig. 4.70.

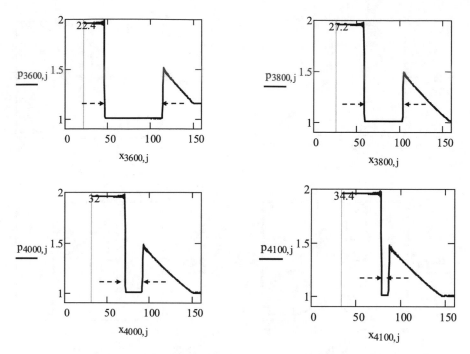

Fig. 4.69 Pressure profiles for the advancing shock fronts prior to colliding. The markers (21.4, 27.2, 32 and 34.4) show the position of the piston at the times indicated. For the numerical procedure the following parameters apply; $\gamma = 1.4$, $\kappa = 1.5$, $\Delta x = 0.4$ and $\Delta t = 0.04$ (see text)

4.9   Conclusions

This concludes our look at some of the numerical results involving plane shock waves and their comparison with theoretical predictions. In the next chapter we move on to consider very strong spherical shocks from an analytical point of view, before proceeding to consider numerical solutions to spherical shock waves in Chap. 6.

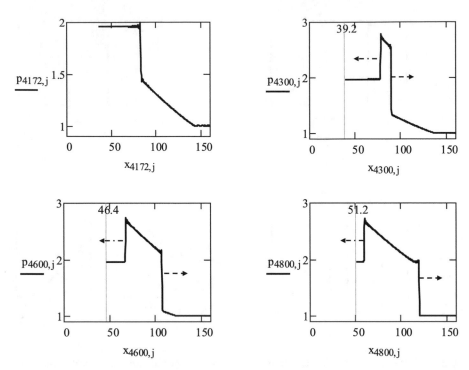

Fig. 4.70 Showing pressure profile at the instant of collision ($t = 4172\Delta t$) and the profiles as the reflected shocks move apart. The markers (39.2, 46.4 and 51.2) show the position of the piston at the times indicated. For the numerical procedure the following parameters apply; $\gamma = 1.4$, $\kappa = 1.5$, $\Delta x = 0.4$ and $\Delta t = 0.04$ (see text)

References

1. J. Von Neumann, R.D. Richtmyer, A method for the numerical calculation of hydrodynamic shocks. J. Appl. Phys. **21**, 232 (1950)
2. C.F. Sprague III, *The Numerical Treatment of Simple Hydrodynamic Shocks Using the Von Neumann-Richtmyer Method, LA-1912 Report* (Los Alamos Scientific Laboratory of the University of California, Los Alamos, 1955)
3. Ya.B. Zel'dovich, Yu.P. Raizer, *Physics of Shock Waves and High-Temperature Hydrodynamic Phenomena*, (Dover Publications, Inc., Mineola, 2002), Chapter 1
4. J.D. Anderson Jr., *Computational Fluid Dynamics: The Basics with Applications* (McGraw-Hill, New York, 1995)
5. T.J. Chung, *Computational Fluid Dynamics*, 2nd edn. (Cambridge University Press, New York, 2010), Chapter 3
6. J.D. Ramshaw, *Elements of Computational Fluid Dynamics* (Imperial College Press, London, 2011)
7. K.W. Morton, D.F. Mayers, *Numerical Solution of Partial Differential Equations* (Cambridge University Press, New York, 1994)
8. D. Mihalas, B. Weibel-Mihalas, *Foundations of Radiation Hydrodynamics* (Dover Publications Inc., New York, 1999), p. 273

9. J.D. Anderson, *Modern Compressible Flow with Historical Perspective*, 3rd edn. (McGraw-Hill, 2003), Chapter 7
10. H.W. Liepmann, A. Roshko, *Elements of Gas Dynamics*, (Dover Pub. Inc., Mineola, 1956), Chapter 3
11. N. Curle, H.J. Davies, *Modern Fluid Dynamics*, vol. 2, (Van Nostrand Reinhold Company, London, 1971), Section 3.4
12. G.B. Whitham, *Linear and Nonlinear Waves*, (John Wiley & Sons, Inc., New York, 1999), Chapter 6
13. S. Chandrasekhar, *On the Decay of Plane Shock Waves, Report No. 423* (Ballistics Research Laboratory, Aberdeen Proving Ground, Maryland, 1943)
14. K.O. Friedrichs, *Formation and Decay of Shock Waves, Report: NYU IMM-158 Institute for Mathematics and Mechanics* (New York University, New York, 1947)
15. A. Kantrowitz, One-Dimensional Treatment of Non-steady Gas Dynamics, Section C, in *Fundamentals of Gas Dynamics*, ed. by H. W. Emmons, vol. 2, (Princeton University Press, Princeton, 1958)
16. L.D. Landau, E.M. Lifshitz, *Fluid Mechanics*, (Pergamon Press, London, 1966), Chapter 10
17. R. Courant, K.O. Friedrichs, *Supersonic Flow and Shock Waves: A Manual on the Mathematical Theory of Non-Linear Wave Motion* (New York, Courant Institute of Mathematical Sciences, New York University, 1944), p. 115

Chapter 5
Spherical Shock Waves: The Self-similar Solution

5.1 Introduction

Let us now turn our attention in the remaining chapters to some examples of air motion involving spherical shock waves. This particular chapter deals exclusively with the effects on the surrounding atmosphere following the sudden release of a considerable quantity of energy coming essentially from a point source, which was one of the first attempts to ascertain the propagation of the effects of nuclear weapons into the surrounding atmosphere. The analysis of this particular problem is known as the *self-similar solution*.

5.2 Shock Wave from an Intense Explosion

Nuclear weapons generate very strong shock waves and cause unimaginable destruction over a very wide area. The energy generated is generally measured in tons of TNT (trinitrotoluene) which is a high-explosive chemical material. The energy unit for nuclear explosions is measured in terms of the energy released by the chemical reaction in one ton of TNT which corresponds to the release of 4.2×10^9 Joules of energy. The fission bombs dropped on the cities of Hiroshima and Nagasaki during the end of the Second World War generated explosions with energy in the region of 20 kilotons of TNT, while thermonuclear weapons using the fusion of light elements have their energies measured in megatons (10^6 tons) of TNT. One of the first estimates of the radius of destruction of a nuclear weapon (that was not yet developed!) was given in a series of lectures by Robert Serber [1] in April 1943 at Los Alamos, New Mexico. Serber indicates that if E_0 is the total energy released in the explosion, then the maximum pressure p in the wave-front varies according to the relation, $p \approx E_0/r^3$ rather than the usual $1/r^2$ dependence because the width of the strongly compressed region increases in proportion to r (see Sect. 5.4). An estimate

© The Author(s), under exclusive license to Springer Nature Switzerland AG 2021
S. Prunty, *Introduction to Simple Shock Waves in Air*, Shock Wave and High
Pressure Phenomena, https://doi.org/10.1007/978-3-030-63606-7_5

of the destructive radius expected is dependent on the amplitude of maximum pressure and it follows from the above equation that the radius of destruction produced by an explosion varies as $\sqrt[3]{E_0}$. It was known that a bomb containing 0.25 ton of TNT produces a destructive radius of 150 feet, consequently, the destructive radius for a bomb containing 20,000 tons of TNT would be the order of 1.2 miles.

Once the chain reaction starts in the nuclear material it takes less than one microsecond to release the entire energy of the weapon [2]. The material of the bomb does not move very much in this time so that a very small volume of space is heated to very high temperatures ($\sim10^7$ degrees). These high temperatures imply that the bomb material exerts a very high pressure on the surrounding air which begins to expand with high velocity; a velocity well in excess of the velocity of sound. The outward moving spherical surface of compressed air has at its front a pressure wave known as a *shock front* while the gas flow behind the front is known as a *blast wave*. As the outward expanding shock front moves further from the centre its intensity becomes spread out over a larger surface area; accordingly, its intensity diminishes and it eventually settles down to a normal sound wave at very great distances from the explosion.

5.3   The Point Source Solution

With the development of nuclear weapons there was increased interest in predicting the effects on the surrounding atmosphere of very strong explosions. The determination of the history of the blast wave produced is a very complicated mathematical problem: it requires the solution of a set of coupled partial differential equations with a moving boundary; the boundary itself is headed by a shock front whose trajectory is governed by the equations themselves. One analytical approach was based on the *similarity solution* [3] which was very successful in describing the early stages of a very intense explosion. This solution method relied on the fact that space and time are simply related to each other once the ambient air pressure was negligible in comparison to the very strong pressures generated by the explosion. This meant that the partial differential equations could be reduced to ordinary differential equations which were easier to solve. G. I. Taylor [4] in England, Von Neumann [5, 6] in the United States of America and Sedov [7] in Russia investigated independently *similarity solutions* to the strong shock-point source explosion in air [8]. We will spend some time describing Taylor's analysis in the following sections.

Sir Geoffrey Taylor's paper [4], although published in 1950, was written in 1941 on behalf of the Civil Defence Research Committee of the Ministry of Home Security in Britain following a report, which has become known as the Frisch-Peierls Memorandum, on the feasibility of a "super-bomb" with the sudden release of very large amount of energy from the fission of uranium. The objective of Taylor's investigation was to form an idea of the hydrodynamic effects that might be expected if such a bomb exploded. Taylor's analysis was largely based on dimensional arguments; he found that the radius R of the shock front was proportional to $t^{2/5}$ where t is the time following detonation. Using the equations of hydrodynamics in Eulerian form together with the Rankine-Hugoniot equations he went on to derive a set of coupled ordinary differential equations for the pressure, gas velocity and density behind the shock front. He solved these equations numerically and then derived analytical approximations to the numerical solution which turned out to be remarkably accurate.

5.4 Taylor's Analysis of Very Intense Shocks

In the following sections we will follow Taylor's analysis of the *self-similar solution* and adopt Taylor's notation (however, here we use E_0 for the energy of the explosion rather than E and c_0 rather than a for the sound speed in air). In the case of a nuclear explosion we can regard the energy being liberated almost instantaneously and coming from essentially a point source after the shock wave has traversed a distance that is large in comparison to the dimensions of the nuclear device. The pressure generated by the resulting disturbance in the early stages of the explosion amounts to several tens of thousands of atmospheres and is so large that the normal atmospheric pressure p_0 is negligible in comparison. Consequently, the only dimensional parameters appearing as inputs are the energy of the explosion E_0 and the undisturbed density ρ_0 of the air. Based on the similarity solution, Taylor, Sedov and von Neumann showed independently that any dimensionless function of r and t, such as, pressure, velocity or density depend only on the dimensionless combination $r(\rho_0/E_0 t^2)^{1/5}$ or some function of it. In fact, the disturbance generated in the air is headed by a shock front at $r = R(t)$, where t is the time since the explosion. $R(t)$ can be inferred from dimensional arguments and, as the only other dimensional parameters are E_0 and ρ_0, we can write,

$$R \propto E_0^\alpha t^\beta \rho_0^\delta,$$

and expressing these quantities in terms of their dimensions we have

$$L \propto M^\alpha L^{2\alpha} T^{-2\alpha} T^\beta M^\delta T^{-3\delta}.$$

Equating dimensions on both sides gives

$$\alpha + \delta = 0, \quad 2\alpha - 3\delta = 1 \quad \text{and} \quad -2\alpha + \beta = 0,$$

hence,

$$\alpha = 1/5, \quad \beta = 2/5 \quad \text{and} \quad \alpha = -1/5.$$

Consequently, the radius of the shock front has the following dependence;

$$R(t) \propto \left(\frac{E_0}{\rho_0}\right)^{\frac{1}{5}} t^{\frac{2}{5}}.$$

Letting

$$R(t) = \eta_0 \left(\frac{E_0}{\rho_0}\right)^{\frac{1}{5}} t^{\frac{2}{5}}, \tag{5.1}$$

where η_0 is a dimensionless constant. One can see that the dependence as expressed in this latter equation is in accordance with the dimensionless combination referred to above. The velocity of the shock front is

$$U(t) = \frac{dR}{dt} = \frac{2}{5}\eta_0 \left(\frac{E_0}{\rho_0}\right)^{\frac{1}{5}} t^{-\frac{3}{5}} = \frac{2}{5}\eta_0 \left(\frac{E_0}{\rho_0}\right)^{\frac{1}{5}} \frac{t^{\frac{2}{5}}}{t} = \frac{2}{5}\frac{R}{t},$$

or alternatively,

$$U(t) = \frac{2}{5}\eta_0^{\frac{5}{2}} \left(\frac{E_0}{\rho_0}\right)^{\frac{1}{2}} R^{-\frac{3}{2}} \equiv AR^{-\frac{3}{2}}. \tag{5.2}$$

where A is defined according to the equation,

$$A = \frac{2}{5}\eta_0^{\frac{5}{2}} \left(\frac{E_0}{\rho_0}\right)^{\frac{1}{2}}. \tag{5.3}$$

Writing the dimensionless combination $r(\rho_0/E_0 t^2)^{1/5}$ as

$$\eta = \left(\frac{\rho_0}{E_0 t^2}\right)^{\frac{1}{5}} r$$

and noting that $r = R$ at the shock front, we have

$$\eta_0 = \left(\frac{\rho_0}{E_0 t^2}\right)^{\frac{1}{5}} R,$$

hence,

$$\eta = \eta_0 \frac{r}{R}.$$

Here, we will take $\eta = r/R$ as the dimensionless similarity parameter; η_0 will be determined in due course. The pressure, p_s, the gas velocity, u_s and the density, ρ_s *just behind* the shock front satisfy the following Rankine-Hugoniot equations (see Chap. 3; Eqs. (3.26a), (3.38a) and (3.17a));

$$\frac{U^2}{c_0^2} = \frac{1}{2\gamma}\left[(\gamma - 1) + (\gamma + 1)\frac{p_s}{p_0}\right], \tag{5.4a}$$

$$u = \frac{2\left(\frac{p_s}{p_0} - 1\right)U}{(\gamma - 1) + (\gamma + 1)\frac{p_s}{p_0}}, \tag{5.4b}$$

and

$$\frac{\rho_s}{\rho_0} = \frac{(\gamma - 1) + (\gamma + 1)(p_s/p_0)}{(\gamma + 1) + (\gamma - 1)(p_s/p_0)}, \tag{5.4c}$$

(Taylor uses the symbol y_1 for the pressure ratio, p_s/p_0)

In the case of very strong shocks, namely, $(p_s/p_0) \gg 1$, we have previously noted in Chap. 3 that the latter equations become,

$$\frac{U^2}{c_0^2} = \frac{\gamma + 1}{2\gamma}\frac{p_s}{p_0} \tag{5.5a}$$

$$\frac{u_s}{U} = \frac{2}{\gamma + 1} \tag{5.5b}$$

$$\frac{\rho_s}{\rho_0} = \frac{\gamma + 1}{\gamma - 1} \tag{5.5c}$$

where $U = dR/dt$ is the velocity of the shock front and $c_0 = \sqrt{\gamma p_0/\rho_0}$ is the speed of sound in the undisturbed atmosphere. Hence, the pressure, particle velocity and density *immediately* behind the shock front are

$$p_s = \frac{2\gamma p_0}{(\gamma + 1)c_0^2}U^2 = \frac{2\rho_0}{(\gamma + 1)}U^2,$$

$$u_s = \frac{2}{\gamma + 1}U,$$

and

$$\rho_s = \rho_0\frac{\gamma + 1}{\gamma - 1},$$

respectively. Substituting Eq. (5.2) in these latter equations yields;

$$p_s = \frac{2\rho_0}{(\gamma + 1)} A^2 R^{-3}, \tag{5.6a}$$

$$u_s = \frac{2}{\gamma + 1} AR^{-\frac{3}{2}} \tag{5.6b}$$

and

$$\rho_s = \rho_0 \frac{\gamma + 1}{\gamma - 1}. \tag{5.6c}$$

The self-similarity of the flow implies that the pressure, gas velocity and density can be expressed using the dimensionless similarity variable η and we write,

$$p = \frac{\rho_0 A^2}{\gamma} R^{-3} f(\eta) \tag{5.7a}$$

$$u = AR^{-3/2} \phi(\eta) \tag{5.7b}$$

$$\rho = \rho_0 \psi(\eta), \tag{5.7c}$$

where the quantities, $f(\eta)$, $\phi(\eta)$ and $\psi(\eta)$ are functions only of the dimensionless variable η, (the notation for these distributions follows that in the paper by Taylor). At the shock front where $\eta = 1$ it is clear by comparing Eqs. (5.6a), (5.6b), (5.6c), (5.7a), (5.7b) and (5.7c) that

$$f(1) = \frac{2\gamma}{\gamma + 1}: \quad \text{(pressure)}$$

$$\phi(1) = \frac{2}{\gamma + 1}: \quad \text{(velocity)} \tag{5.8}$$

$$\psi(1) = \frac{\gamma + 1}{\gamma - 1}: \quad \text{(density)}$$

Let us now be substituted Eqs. (5.7a), (5.7b) and (5.7c) in the momentum, continuity and energy equations as demonstrated below.

5.4.1 Momentum Equation

The equation of motion is (see Eq. (1.65), Chap. 1)

$$\frac{\partial u}{\partial t} + u \frac{\partial u}{\partial r} = -\frac{1}{\rho} \frac{\partial p}{\partial r},$$

hence, using Eqs. (5.7a), (5.7b) and (5.7c) above we have

$$\frac{\partial u}{\partial t} = A\frac{\partial}{\partial t}\left(R^{-\frac{3}{2}}\phi(\eta)\right) = -\frac{3}{2}A\phi R^{-\frac{5}{2}}\frac{dR}{dt} + AR^{-\frac{3}{2}}\frac{d\phi}{d\eta}\frac{\partial\eta}{\partial t},$$

now

$$\frac{\partial\eta}{\partial t} = \frac{\partial}{\partial t}\left(rR^{-1}\right) = -rR^{-2}\frac{dR}{dt} = -\frac{\eta}{R}\frac{dR}{dt}.$$

Collecting terms, we have

$$\frac{\partial u}{\partial t} = -\frac{3}{2}A\phi R^{-\frac{5}{2}}\frac{dR}{dt} - AR^{-\frac{5}{2}}\eta\phi'\frac{dR}{dt}$$

where ϕ' denotes differentiation with respect to η. However, $dR/dt = AR^{-3/2}$, hence,

$$\frac{\partial u}{\partial t} = -A^2 R^{-4}\left(\frac{3}{2}\phi + \eta\phi'\right).$$

Similarly,

$$\frac{\partial u}{\partial r} = A\frac{\partial}{\partial r}\left(R^{-\frac{3}{2}}\phi(\eta)\right) = AR^{-\frac{3}{2}}\frac{d\phi}{d\eta}\frac{\partial\eta}{\partial r} = AR^{-\frac{5}{2}}\phi'$$

so that

$$u\frac{\partial u}{\partial r} = A^2 R^{-4}\phi\phi'.$$

The right-hand side of the equation of motion becomes

$$-\frac{1}{\rho}\frac{\partial p}{\partial r} = -\frac{1}{\rho_0\psi}\frac{\partial}{\partial r}\left(\frac{A^2\rho_0}{\gamma}R^{-3}f(\eta)\right) = -\frac{A^2 R^{-3}}{\gamma\psi}\frac{\partial}{\partial r}f(\eta) = -\frac{A^2 R^{-3}}{\gamma\psi}\frac{df}{d\eta}\frac{\partial\eta}{\partial r}$$

$$= -\frac{A^2 R^{-4}}{\gamma\psi}f',$$

where $f' = df/d\eta$. Collecting all terms, we can write the equation of motion as

$$-A^2 R^{-4}\left(\frac{3}{2}\phi + \eta\phi'\right) + A^2 R^{-4}\phi\phi' = -\frac{A^2 R^{-4}}{\gamma\psi}f',$$

so that

$$\phi'(\eta - \phi) = \frac{f'}{\gamma\psi} - \frac{3}{2}\phi. \tag{5.9}$$

This is the equation of motion in terms of f, ϕ and ψ which are all functions of the dimensionless parameter; $\eta = r/R$.

5.4.2 Continuity Equation

Let us now consider the continuity equation (see Eq. (1.63), Chap. 1);

$$\frac{\partial \rho}{\partial t} + u\frac{\partial \rho}{\partial r} + \rho\left(\frac{\partial u}{\partial r} + \frac{2u}{r}\right) = 0,$$

and using Eqs. (5.7a), (5.7b) and (5.7c) we have

$$\frac{\partial \rho}{\partial t} = \rho_0\frac{\partial \psi}{\partial t} = \rho_0\frac{\partial \psi}{\partial \eta}\frac{\partial \eta}{\partial R}\frac{\partial R}{\partial t}$$

$$= -\rho_0\psi'\eta AR^{-\frac{5}{2}},$$

where ψ' denotes differentiation with respect to η. Similarly,

$$\frac{\partial \rho}{\partial r} = \rho_0\frac{\partial \psi}{\partial \eta}\frac{\partial \eta}{\partial r} = \frac{\rho_0\psi'}{R}, \text{hence,}$$

$$u\frac{\partial \rho}{\partial r} = AR^{-\frac{3}{2}}\rho_0\frac{\phi\psi'}{R} = AR^{-\frac{5}{2}}\rho_0\phi\psi'$$

and as before we have

$$\frac{\partial u}{\partial r} = AR^{-\frac{5}{2}}\phi'.$$

Collecting all of the various terms above and substituting them in the continuity equation, we obtain

$$-\rho_0\psi'\eta AR^{-\frac{5}{2}} + AR^{-\frac{5}{2}}\rho_0\phi\psi' + \rho_0\psi\left(AR^{-\frac{5}{2}}\phi' + 2\frac{AR^{-\frac{5}{2}}\phi}{\eta}\right) = 0,$$

and after cancelling some common terms this latter equation reduces to

$$-\psi'\eta + \phi\psi' + \psi\left(\phi' + 2\frac{\phi}{\eta}\right) = 0$$

so that

$$\frac{\psi'}{\psi} = \frac{\phi' + 2\phi/\eta}{\eta - \phi}.$$ (5.10)

5.4.3   Energy Equation

Let us finally consider the equation of energy conservation in the following form (see Eq. (1.67), Chap. 1);

$$\frac{\partial}{\partial t}(p\rho^{-\gamma}) + u\frac{\partial}{\partial r}(p\rho^{-\gamma}) = 0.$$

By carrying out the differentiation with respect to r and t it is straightforward to show that this latter equation becomes

$$-\frac{\gamma p}{\rho}\frac{\partial \rho}{\partial t} + \frac{\partial p}{\partial t} + u\left(-\frac{\gamma p}{\rho}\frac{\partial \rho}{\partial r} + \frac{\partial p}{\partial r}\right) = 0.$$ (5.11)

We already have expressions for the density variations; $\partial\rho/\partial t$ and $\partial\rho/\partial r$, however, the pressure variations are

$$\begin{aligned}
\frac{\partial p}{\partial t} &= -3\frac{\rho_0 A^2}{\gamma}R^{-4}f\frac{dR}{dt} + \frac{\rho_0 A^2}{\gamma}R^{-3}\frac{\partial f}{\partial \eta}\frac{\partial \eta}{\partial R}\frac{dR}{dt} \\
&= -3\frac{\rho_0 A^2}{\gamma}R^{-4}fAR^{-\frac{3}{2}} - \frac{\rho_0 A^2}{\gamma}R^{-4}f'\eta AR^{-\frac{3}{2}} \\
&= -\frac{\rho_0 A^2}{\gamma}\left[3R^{-4}f + R^{-4}f'\eta\right]AR^{-\frac{3}{2}}
\end{aligned}$$

and

$$\frac{\partial p}{\partial r} = \frac{\partial}{\partial r}\left(\frac{\rho_0 A^2}{\gamma}R^{-3}f\right)$$

hence,

$$\frac{\partial p}{\partial r} = \frac{\rho_0 A^2}{\gamma}R^{-3}\frac{\partial f}{\partial \eta}\frac{\partial \eta}{\partial r} = \frac{\rho_0 A^2}{\gamma}R^{-4}f'.$$

When these expressions are substituted in Eq. (5.11) one finds that many common multiplying factors cancel out and the resulting equation can be written in the form;

$$3f + \eta f' + \gamma \frac{\psi'}{\psi} f(-\eta + \phi) - \phi f' = 0. \tag{5.12}$$

By substituting the right-hand side of Eq. (5.10) for ψ'/ψ in Eq. (5.12) and then using Eq. (5.9) for ϕ' in the resulting equation, we can write f' as

$$f' = f \frac{\left[-3\eta + \phi(3 + \gamma/2) - 2\gamma\phi^2/\eta\right]}{\left[(\eta - \phi)^2 - f/\psi\right]}. \tag{5.13}$$

5.5 Derivatives at the Shock Front

On inspection of Eq. (5.13) above it can be seen that all quantities on the right-hand side are known at the shock front ($\eta = 1$), namely,

$$f(1) = \frac{2\gamma}{\gamma + 1}, \quad \phi(1) = \frac{2}{\gamma + 1} \quad \text{and} \quad \psi(1) = \frac{\gamma + 1}{\gamma - 1},$$

hence, the value of $f'(1)$ can be determined by directly substituting these values; giving,

$$\left(\frac{f'}{f}\right)_{\eta=1} = \frac{-3 + \frac{2}{\gamma + 1}\left(3 + \frac{\gamma}{2}\right) - \frac{8\gamma}{(\gamma + 1)^2}}{\left(1 - \frac{2}{\gamma+1}\right)^2 - \frac{2\gamma(\gamma - 1)}{(\gamma + 1)^2}} \tag{5.14}$$

$$= \frac{2\gamma^2 + 7\gamma - 3}{\gamma^2 - 1}$$

and hence,

$$f'(1) = \frac{2\gamma(2\gamma^2 + 7\gamma - 3)}{(\gamma - 1)(\gamma + 1)^2}. \tag{5.15}$$

Using this value for $f'(1)$ in Eq. (5.9) above we obtain the following expression for $(\phi'/\phi)_{\eta = 1}$;

$$\left(\frac{\phi'}{\phi}\right)_{\eta=1} = \frac{\gamma + 9}{2(\gamma + 1)}, \tag{5.16}$$

so that

$$\phi'(1) = \frac{\gamma + 9}{(\gamma + 1)^2}.$$

(5.17)

Similarly, using Eq. (5.10) for ψ'/ψ it is easy to show that

$$\left(\frac{\psi'}{\psi}\right)_{\eta=1} = \frac{5\gamma + 13}{\gamma^2 - 1},$$

(5.18)

so that

$$\psi'(1) = \frac{5\gamma + 13}{(\gamma - 1)^2}.$$

(5.19)

We will be requiring these relationships later on in order to obtain approximate analytical expressions for f, ϕ and ψ.

5.6   Numerical Integration of the Equations

When Eq. (5.13) is substituted in Eqs. (5.9) and (5.10) it follows that corresponding equations can be written for ϕ' and ψ' which are expressed purely in terms of f, ϕ, ψ and η. Taylor numerically solved these equations for f, ϕ and ψ as a function of η starting at $\eta = 1$ while, surprisingly, Sedov [7] was able to obtain an analytical solution [8]. Here, the Runge-Kutta method [9] is used to solve these coupled equations starting at $\eta = 1$ and the following plots of f, ϕ and ψ as a function of η are obtained. The plots of the distributions are shown in Fig. 5.1 for $\gamma = 1.4$. It is seen from these plots that the values corresponding to $\eta = 1$ agree with the boundary conditions as given by Eq. (5.8); for example, $f(1) = 1.167$, $\phi(1) = 0.833$ and $\psi(1) = 6$ for $\gamma = 1.4$. A characteristic feature of this strong explosion scenario is clearly illustrated in the plot of the density distribution $\psi(\eta)$: the explosion blows most of the air away from the centre and piles it up in a thin layer at the shock front with the density decreasing dramatically from the shock front to the centre. Consequently, almost the entire mass that originally occupied the sphere has been stacked up into a thin shell immediately behind the shock front and the inner portion of the sphere is almost completely evacuated. It can also be seen that the pressure drops significantly from the immediate vicinity of the shock front and then remains essentially constant over the majority of the sphere. One can verify from the plot of $f(\eta)$ that the ratio of this constant central pressure to the pressure at the shock front is approximately equal to 0.366. The velocity plot shows an approximate linear variation with η almost over the entire range of η values.

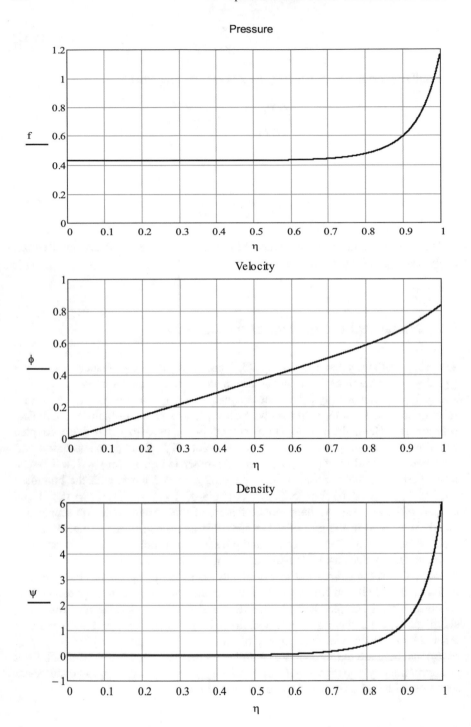

Fig. 5.1 Plots of f, ϕ and ψ for the point source explosion as a function of η for $\gamma = 1.4$ (see text)

5.7   Energy of the Explosion

The energy of the explosion, E_0, comprises two parts; kinetic energy and heat energy, the kinetic energy per unit volume is $\rho u^2/2$ and the heat energy per unit volume is $p/(\gamma - 1)$, so that the energy generated by the explosion is

$$E_0 = \int_0^R \left(\frac{1}{2}\rho u^2 + \frac{p}{\gamma - 1}\right) 4\pi r^2 dr. \tag{5.20}$$

We should in fact subtract from this the pre-shock internal energy of the air engulfed by the shock, namely,

$$\frac{4\pi R^3}{3}\frac{p_0}{\gamma - 1},$$

so that a more accurate expression for the energy of the explosion becomes

$$E_0 = \int_0^R \left(\frac{1}{2}\rho u^2 + \frac{p}{\gamma - 1}\right) 4\pi r^2 dr - \frac{4\pi R^3}{3}\frac{p_0}{\gamma - 1}. \tag{5.21}$$

However, the similarity solution is based on the fact that the external atmospheric pressure p_0 is negligible in comparison to the very high pressures generated by the strong explosion, so that we will use the first expression for E_0, namely, Eq. (5.20). Substituting Eqs. (5.7a), (5.7b) and (5.7c) in Eq. (5.20) yields,

$$E_0 = 4\pi \int_0^R \left(\frac{1}{2}(\rho_0\psi)A^2 R^{-3}\phi^2 + \frac{\rho_0 A^2 R^{-3}f}{\gamma(\gamma - 1)}\right) r^2 dr$$

$$= 4\pi\rho_0 A^2 \int_0^R \left(\frac{1}{2}\psi R^{-3}\phi^2 + \frac{R^{-3}f}{\gamma(\gamma - 1)}\right) r^2 dr.$$

But $r = \eta R$, hence, $dr = Rd\eta$ and therefore, $r^2 dr = R^3\eta^2 d\eta$, consequently,

$$E_0 = 4\pi\rho_0 A^2 \int_0^1 \left(\frac{1}{2}\psi R^{-3}\phi^2 + \frac{R^{-3}f}{\gamma(\gamma - 1)}\right) R^3\eta^2 d\eta,$$

hence,

$$E_0 = \rho_0 A^2 \left[2\pi \int_0^1 \psi\phi^2\eta^2\,d\eta + \frac{4\pi}{\gamma(\gamma-1)} \int_0^1 f\eta^2\,d\eta \right].$$

The quantity in square brackets in this latter equation is just a function of γ, hence we may write it as

$$B(\gamma) = \left[2\pi \int_0^1 \psi\phi^2\eta^2\,d\eta + \frac{4\pi}{\gamma(\gamma-1)} \int_0^1 f\eta^2\,d\eta \right]$$

so that

$$E_0 = \rho_0 A^2 B(\gamma). \tag{5.22}$$

Numerical integration gives the following results for $\gamma = 1.4$;

$$\int_0^1 \psi\phi^2\eta^2\,d\eta = 0.185 \quad \text{and} \int_0^1 f\eta^2\,d\eta = 0.185,$$

hence,

$$B(\gamma) = 5.31, \tag{5.23}$$

and we can identify A according to Eq. (5.22) as,

$$A = \frac{1}{\sqrt{B(\gamma)}} \left(\frac{E_0}{\rho_0} \right)^{\frac{1}{2}}$$

However, Eq. (5.3) gave the following relationship between η_0 and A,

$$A = \frac{2}{5}\eta_0^{\frac{5}{2}} \left(\frac{E_0}{\rho_0} \right)^{\frac{1}{2}},$$

hence,

$$\frac{2}{5}\eta_0^{\frac{5}{2}} = \frac{1}{\sqrt{B(\gamma)}}$$

and therefore, $\eta_0 = 1.033$ when $\gamma = 1.4$. The radius of the shock front according to Eq. (5.1) can now be written as

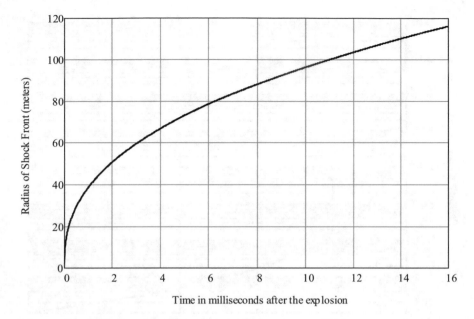

Fig. 5.2   Radius of shock front as a function of time is shown for a point source explosion with energy equivalent to 20 kTons of TNT ($\gamma = 1.4$)

$$R(t) = 1.033 \left(\frac{E_0}{\rho_0}\right)^{\frac{1}{5}} t^{\frac{2}{5}}. \tag{5.24}$$

A typical plot of R versus t for a point source explosion with energy equivalent to 20,000 tons of TNT is shown in Fig. 5.2 (using the fact that the energy released per ton of TNT is 4.2×10^9 Joules).

5.8   The Pressure

The pressure at an arbitrary radial position r is given by Eq. (5.7a), namely,

$$p(r,t) = \frac{\rho_0 A^2}{\gamma} R^{-3} f(\eta),$$

where $\eta = r/R(t)$, hence p becomes

$$p = R^{-3} f \frac{A^2 \rho_0}{\gamma} = R^{-3} f \frac{\rho_0}{\gamma} \frac{E_0}{\rho_0 B(\gamma)} = \left(\frac{1}{\gamma B(\gamma)}\right) R^{-3} E_0 f, \tag{5.25}$$

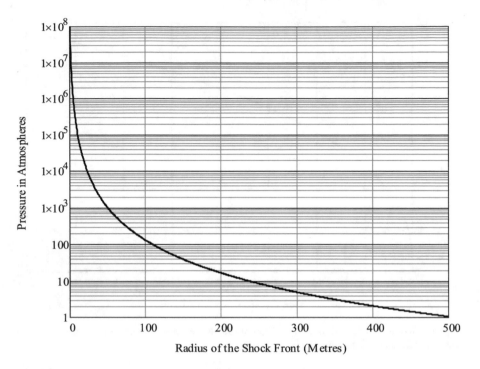

Fig. 5.3 Pressure in atmospheres at the shock front versus shock radius is shown for a point source explosion with energy equivalent to 20 kTons of TNT ($\gamma = 1.4$)

after using Eq. (5.22). The maximum pressure occurs at $r = R$, so that $\eta = 1$ and, hence, $[f]_{\eta\,=\,1} = 1.166$ for $\gamma = 1.4$. Hence,

$$p_{\max.} = 0.157R^{-3}E_0.$$

The pressure in atmospheres at the shock front as a function of shock front radius R for a point source explosion with energy equivalent to 20 kTons of TNT ($\gamma = 1.4$) is shown plotted in Fig. 5.3. Using Einstein's famous mass-energy equation ($\Delta E = c^2 \Delta m$) this implies that approximately one gram of matter has been converted to release this enormous quantity energy equivalent to 8.4×10^{13} Joules! Although the pressure in Fig. 5.3 is plotted down to 1 atmosphere it should be noted that a pressure of this magnitude is well below the strong shock regime that applies here.

Figure 5.4 shows the variation of pressure with distance according to Eq. (5.25) from the centre at the instant when, for example, the shock front has expanded to a radius of 200 m. Beyond the 200 m point the, as yet, undisturbed pressure of 1 atmosphere has been included in the plot.

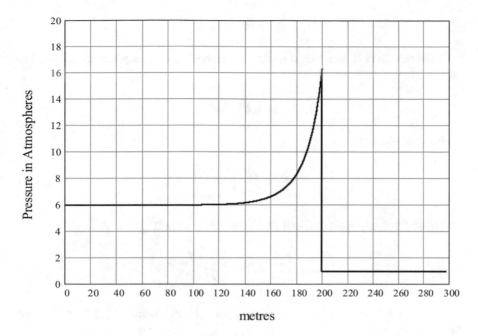

Fig. 5.4 Variation of pressure with distance from the centre at the instant when the shock front has expanded to a radius of 200 *m*. (A point source explosion with energy equivalent to 20 kTons of TNT is assumed and $\gamma = 1.4$)

5.9 The Temperature

The temperature T at any point is related to the pressure and density by the relation,

$$p = \rho RT,$$

hence,

$$\frac{T}{T_0} = \frac{p\rho_0}{p_0\rho} = \frac{p}{p_0\psi} = \left(\frac{1}{\gamma B}\right)\frac{R^{-3}E_0 f}{p_0\psi}. \tag{5.26}$$

5.10 The Pressure-Time Relationship for a Fixed Point

Let us consider a particular radius R_0 and let us assume that the shock front passes over R_0 at time t_0. If p_1 is the pressure as the shock front passes over R_0, then,

$$p_1 \propto R_0^{-3}[f]_{\eta=1}.$$

The pressure at R_0 at a later time t after the shock wave has passed and moved to radius R is

$$p \propto R^{-3}f(\eta).$$

Forming the ratio,

$$\frac{p}{p_1} = \left(\frac{R_0}{R}\right)^3 \frac{f(\eta)}{[f]_{\eta=1}} \tag{5.27}$$

and noting that $\eta = R_0/R$ and $R(t) \propto t^{2/5}$, hence, $\eta = (t_0/t)^{2/5}$ so that

$$\frac{p}{p_1} = \eta^3 \frac{f_{\eta=(t_0/t)^{\frac{2}{5}}}}{[f]_{\eta=1}} = \left(\frac{t_0}{t}\right)^{\frac{6}{5}} \frac{f_{\eta=(t_0/t)^{\frac{2}{5}}}}{[f]_{\eta=1}}. \tag{5.28}$$

Defining a normalized time according to $\tau = t/t_0$, hence, $\tau(\eta) = \eta^{-5/2}$, and therefore,

$$\frac{p}{p_1} = \frac{1}{\tau(\eta)^{\frac{6}{5}}} \frac{f(\eta)}{[f]_{\eta=1}}. \tag{5.29}$$

This pressure ratio p/p_1 is shown plotted in Fig. 5.5 as a function of τ.

For example, Fig. 5.6 shows the pressure (in atmospheres) at 100 meters from a point source explosion with energy equivalent to 20,000 tons of TNT. One can observe that the shock front arrives at approximately 11 *ms* following detonation with a maximum pressure of about 130 atmospheres.

5.11   Taylor's Analytical Approximations for Velocity, Pressure and Density

Following Taylor's analysis [4] and using his notation let us now proceed to obtain approximate analytical expressions for f, ϕ and ψ.

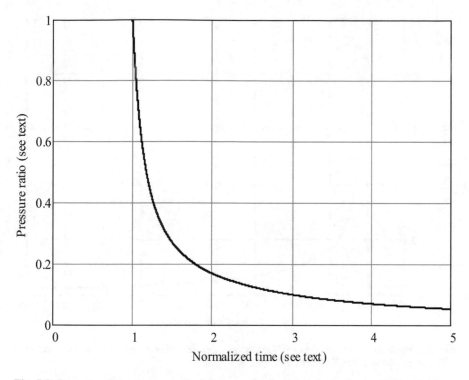

Fig. 5.5 Pressure ratio versus normalized time at a fixed point is shown for $\gamma = 1.4$ (see text)

5.11.1   The Velocity φ

Re-writing Eq. (5.13) again;

$$f' = f \frac{\left[-3\eta + \phi(3 + \gamma/2) - 2\gamma\phi^2/\eta\right]}{\left[(\eta - \phi)^2 - f/\psi\right]},$$

and by writing it in the form, we have

$$\frac{f'}{f}(\eta - \phi)^2 - \frac{f'}{\psi} = -3\eta + \phi(3 + \gamma/2) - 2\gamma\phi^2/\eta. \qquad (5.30)$$

In order to eliminate ψ in the latter equation we use Eq. (5.9), namely,

$$\phi'(\eta - \phi) = \frac{f'}{\gamma\psi} - \frac{3}{2}\phi,$$

which gives,

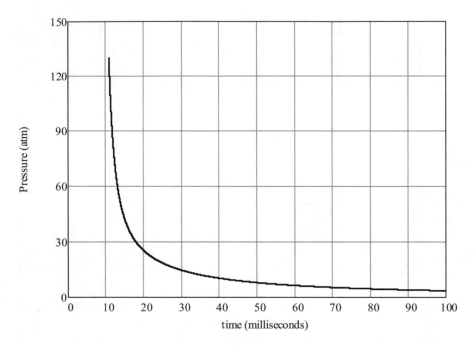

Fig. 5.6 Pressure in atmospheres at 100 meters as a function of time for a point source explosion with energy equivalent to 20,000 tons of TNT ($\gamma = 1.4$)

$$\frac{f'}{\psi} = \frac{3}{2}\gamma\phi + \gamma\phi'(\eta - \phi).$$

Substitution this in Eq. (5.30) yields

$$\frac{f'}{f}(\eta - \phi)^2 = \gamma\phi'(\eta - \phi) + \frac{3}{2}\gamma\phi - 3\eta + \phi\left(3 + \frac{\gamma}{2}\right) - \frac{2\gamma\phi^2}{\eta}$$

and by dividing across by $\eta - \phi$ we obtain

$$\frac{f'}{f}(\eta - \phi) = \gamma\phi' - 3 + \frac{2\gamma\phi}{\eta}. \qquad (5.31)$$

In the interior region f'/f can be neglected as the curve of pressure is relatively flat, consequently,

$$\gamma\phi' - 3 + \frac{2\gamma\phi}{\eta} \cong 0,$$

and the approximate solution of the latter equation for which ϕ vanishes as $\eta \to 0$ is

$$\phi(\eta) = \eta/\gamma, \tag{5.32}$$

as can be seen by direct substitution. This gives a linear relationship and, in fact, the actual variation is quite close to linear in the range: $0 \le \eta \le 0.6$. Taylor suggested the following approximate formula for $\phi(\eta)$ over the complete range; $0 \le \eta \le 1$,

$$\phi(\eta) = \frac{\eta}{\gamma} + \alpha\eta^n. \tag{5.33}$$

As the formula applies at $\eta = 1$, then,

$$\frac{2}{\gamma+1} = \frac{1}{\gamma} + \alpha,$$

hence,

$$\alpha = \frac{\gamma-1}{\gamma(\gamma+1)}. \tag{5.34}$$

Let us now substitute $\phi(\eta) = \eta/\gamma + \alpha\eta^n$ and $\phi' = 1/\gamma + n\alpha\eta^{n-1}$ in Eq. (5.31), hence,

$$\frac{f'}{f}\left(\eta - \frac{\eta}{\gamma} - \alpha\eta^n\right) = 1 + \gamma n\alpha\eta^{n-1} - 3 + \frac{2\gamma}{\eta}\left(\frac{\eta}{\gamma} + \alpha\eta^n\right)$$

$$= \gamma\alpha\eta^{n-1}(n+2).$$

At $\eta = 1$ the latter equation gives

$$(f'/f)_{\eta=1} = n + 2, \tag{5.35}$$

after substituting for α above. However, we already know from Eq. (5.14) that

$$(f'/f)_{\eta=1} = \frac{2\gamma^2 + 7\gamma - 3}{\gamma^2 - 1},$$

hence,

$$n = \frac{7\gamma - 1}{\gamma^2 - 1}. \tag{5.36}$$

By substitution these values for α and n in the equation for $\phi(\eta)$ we obtain the following approximate analytical formula for the velocity;

$$\phi(\eta) = \frac{\eta}{\gamma} + \frac{(\gamma - 1)}{\gamma(\gamma + 1)} \eta^{\frac{7\gamma-1}{\gamma^2-1}}, \tag{5.37}$$

which agrees with the exact values at $\eta = 0$ and at $\eta = 1$. Similarly, with this value of $\phi(\eta)$ substituted in the Eq. (5.10), namely,

$$\frac{\psi'}{\psi} = \frac{\phi' + 2\phi/\eta}{\eta - \phi},$$

it is straightforward to show that $(\psi'/\psi)_{\eta = 1}$ also agrees with the exact value at $\eta = 1$ according to Eq. (5.18).

5.11.2 The Pressure f

Using Eq. (5.31), let us write it in the form,

$$\frac{f'}{f} = \frac{\gamma\phi' - 3 + \frac{2\gamma\phi}{\eta}}{\eta - \phi}. \tag{5.38}$$

We now substitute $\phi(\eta)$ and $\phi'(\eta)$ in this latter equation where $\phi(\eta)$ is given by,

$$\phi(\eta) = \frac{\eta}{\gamma} + \alpha\eta^n.$$

Hence, one obtains,

$$\frac{f'}{f} = \frac{(n + 2)\alpha\gamma^2\eta^{n-2}}{[(\gamma - 1) - \alpha\gamma\eta^{n-1}]}. \tag{5.39}$$

Substituting the values for n and α, we obtain,

$$\frac{df}{f} = \frac{\gamma(2\gamma^2 + 7\gamma - 3)}{(\gamma + 1)} \frac{\eta^{n-2}d\eta}{(\gamma^2 - 1) - (\gamma - 1)\eta^{n-1}},$$

and letting, $x = (\gamma^2 - 1) - (\gamma - 1)\eta^{n-1}$, hence, $dx = -(n - 1)(\gamma - 1)\eta^{n-2}d\eta$. Consequently,

$$\frac{df}{f} = -\frac{(2\gamma^2 + 7\gamma - 3)}{7 - \gamma} \frac{dx}{x}.$$

Integrating, we obtain,

$$\ln f = -\frac{(2\gamma^2 + 7\gamma - 3)}{(7 - \gamma)} \ln\left[(\gamma^2 - 1) - (\gamma - 1)\eta^{n-1}\right] + \text{constant}, \qquad (5.40)$$

and the constant of integration can be obtained since f is known at $\eta = 1$, hence,

$$\ln\left(\frac{2\gamma}{\gamma + 1}\right) = -\frac{(2\gamma^2 + 7\gamma - 3)}{(7 - \gamma)} \ln\left[(\gamma^2 - 1) - (\gamma - 1)\right] + \text{constant},$$

yielding,

$$\text{constant} = \ln\left(\frac{2\gamma}{\gamma + 1}\right) + \frac{(2\gamma^2 + 7\gamma - 3)}{(7 - \gamma)} \ln\left[\gamma^2 - \gamma\right].$$

Substituting this back into Eq. (5.40) and simplifying, we obtain,

$$\ln f = \ln\left(\frac{2\gamma}{\gamma + 1}\right) - \frac{(2\gamma^2 + 7\gamma - 3)}{(7 - \gamma)} \ln\left[\frac{\gamma + 1 - \eta^{n-1}}{\gamma}\right],$$

and finally,

$$f(\eta) = \frac{2\gamma}{\gamma + 1}\left(\frac{\gamma + 1}{\gamma} - \frac{\eta^{\frac{7\gamma-1}{\gamma^2-1}-1}}{\gamma}\right)^{-\frac{2\gamma^2+7\gamma-3}{7-\gamma}} \qquad (5.41)$$

after substituting for n.

5.11.3 The Density ψ

Let us now substitute $\phi(\eta)$ and $\phi'(\eta)$ in Eq. (5.10), where $\phi(\eta)$ is given by,

$$\phi(\eta) = \frac{\eta}{\gamma} + \alpha\eta^n.$$

Hence,

$$\frac{\psi'}{\psi} = \frac{3 + (n + 2)\alpha\gamma\eta^{n-1}}{(\gamma - 1)\eta - \alpha\gamma\eta^n},$$

yielding,

$$\frac{d\psi}{\psi} = \frac{3d\eta}{[(\gamma-1)-\alpha\gamma\eta^{n-1}]\eta} + \frac{(n+2)\alpha\gamma\eta^{n-2}d\eta}{(\gamma-1)-\alpha\gamma\eta^{n-1}}. \tag{5.42}$$

Writing the first term on the right-hand side in Eq. (5.42) as

$$\frac{C_1}{\eta} + \frac{C_2\alpha\gamma\eta^{n-2}}{(\gamma-1)-\alpha\gamma\eta^{n-1}},$$

we find that $C_1 = C_2 = 3/(\gamma - 1)$. Substituting these values back in Eq. (5.42) we obtain

$$\frac{d\psi}{\psi} = \frac{3}{\gamma-1}\frac{d\eta}{\eta} + \frac{\alpha\gamma\left[\frac{3}{\gamma-1}+(n+2)\right]\eta^{n-2}d\eta}{(\gamma-1)-\alpha\gamma\eta^{n-1}}.$$

Substituting for α and n yields,

$$\frac{d\psi}{\psi} = \frac{3}{\gamma-1}\frac{d\eta}{\eta} + \frac{2\gamma(\gamma+5)}{(\gamma+1)^2}\frac{\eta^{n-2}d\eta}{(\gamma-1)-\alpha\gamma\eta^{n-1}}, \tag{5.43}$$

hence,

$$\ln\psi = \frac{3}{\gamma-1}\ln\eta + \frac{2\gamma(\gamma+5)}{(\gamma+1)^2}\int\frac{\eta^{n-2}d\eta}{(\gamma-1)-\alpha\gamma\eta^{n-1}} + \text{constant}. \tag{5.44}$$

In order to evaluate the integral we let $x = (\gamma-1) - \alpha\gamma\eta^{n-1}$, hence, we find that

$$\eta^{n-2}d\eta = -\frac{(\gamma+1)^2}{\gamma(7-\gamma)}dx. \tag{5.45}$$

Substituting Eq. (5.45) in Eq. (5.44) and integrating yields,

$$\ln\psi = \frac{3}{\gamma-1}\ln\eta - \frac{2(\gamma+5)}{(7-\gamma)}\ln\left[(\gamma-1)-\alpha\gamma\eta^{n-1}\right] + \text{constant}.$$

In order to determine the constant of integration we note that $\psi = (\gamma+1)/(\gamma-1)$ when $\eta = 1$, and with these substitutions we find the constant of integration to be

$$\ln\frac{\gamma+1}{\gamma-1} + \frac{2(\gamma+5)}{7-\gamma}\ln\frac{\gamma(\gamma-1)}{(\gamma+1)}.$$

Upon substituting back and simplifying the resulting equation we obtain,

$$\ln \psi = \ln \left(\frac{\gamma+1}{\gamma-1}\right) + \frac{3}{\gamma-1} \ln \eta - \frac{2(\gamma+5)}{(7-\gamma)} \ln \left[\frac{\gamma+1-\eta^{n-1}}{\gamma}\right], \qquad (5.46)$$

which can be finally written as

$$\psi(\eta) = \left(\frac{\gamma+1}{\gamma-1}\right) \eta^{\frac{3}{\gamma-1}} \left[\frac{\gamma+1-\eta^{n-1}}{\gamma}\right]^{-2\left(\frac{\gamma+5}{7-\gamma}\right)}, \qquad (5.47)$$

where $n = (7\gamma - 1)/(\gamma^2 - 1)$.

5.12   The Density for Small Values of η

If η is small let us expand Eq. (5.46) above; leading to;

$$\ln \psi = \ln \left(\frac{\gamma+1}{\gamma-1}\right) + \frac{3}{\gamma-1} \ln \eta - \frac{2(\gamma+5)}{(7-\gamma)} \ln \left[\left(1 - \frac{\eta^{n-1}}{\gamma+1}\right)\left(\frac{\gamma+1}{\gamma}\right)\right]$$

$$= \ln \left(\frac{\gamma+1}{\gamma-1}\right) + \frac{3}{\gamma-1} \ln \eta - \frac{2(\gamma+5)}{(7-\gamma)} \left[\ln \left(\frac{\gamma+1}{\gamma}\right) + \ln \left(1 - \frac{\eta^{n-1}}{\gamma+1}\right)\right].$$

Now $n - 1 = \gamma(7 - \gamma)/(\gamma^2 - 1)$, and with $\gamma = 1.4$ we find that $n - 1 \approx 8$, hence we can neglect the term, $\ln(1 - \eta^{n-1}/(\gamma + 1))$ for small values of η in comparison to the other terms appearing in the latter equation, hence,

$$\ln \psi = \ln \left(\frac{\gamma+1}{\gamma-1}\right) - \frac{2(\gamma+5)}{(7-\gamma)} \ln \left(\frac{\gamma+1}{\gamma}\right) + \frac{3}{\gamma-1} \ln \eta. \qquad (5.48)$$

Taylor [4] defines D according to the equation,

$$\ln D = \ln \left(\frac{\gamma+1}{\gamma-1}\right) - \frac{2(\gamma+5)}{(7-\gamma)} \ln \left(\frac{\gamma+1}{\gamma}\right), \qquad (5.49)$$

and with $\gamma = 1.4$ we find that $D = 1.75$ and, accordingly, we obtain

$$\ln \psi = \ln D + \ln \eta^{\frac{3}{\gamma-1}},$$

so that

$$\psi(\eta) = 1.75 \eta^{7.5}. \qquad (5.50)$$

A comparison of the approximate analytical formulae for the pressure, gas velocity and density with the exact numerical values are shown plotted in Fig. 5.7.

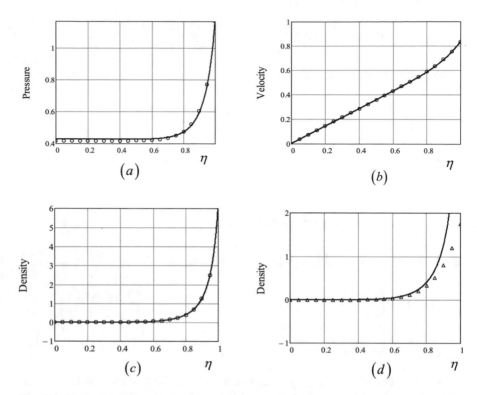

Fig. 5.7 Plots of the pressure, velocity and density as a function of η showing the comparison between the exact values (solid lines) and the approximate analytical values (denoted by $\cdots \oplus \cdots$). In addition, (d) shows a comparison between the exact density ψ and the approximate value of ψ (denoted by $\cdots \Delta \cdots$) for small values of η as given by Eq. (5.50)

Taylor indicates that the density according to Eq. (5.50) decreases rapidly as the centre is approached while noting that the pressure is relatively constant in this region. This would imply that the temperature (based on the equation, $p = \rho RT$) increases proportionally to $\eta^{-7.5}$ and at first sight this might imply a very high concentration of energy near the centre. He goes on to dismiss this assumption by pointing out that the energy per unit volume of a gas is simply $p/(\gamma - 1)$ so that the distribution of energy is uniform in the central region since the pressure is esentially constant (see Fig. 5.1).

5.13   The Temperature in the Central Region

Let us now investigate the temperature left behind by the blast wave in the central region. Using the equation; $p = \rho RT$, the temperature T at any point is related to the pressure p and density ρ by the equation,

$$\frac{p}{p_0} = \frac{\rho T}{\rho_0 T_0},$$

where p_0 and ρ_0 are the nominal atmospheric pressure and density, respectively, hence,

$$\frac{T}{T_0} = \frac{p \rho_0}{p_0 \rho}.$$

However, $p = (1/\gamma B(\gamma)) R^{-3} E_0 f$, and with $\gamma = 1.4$ this yields, $p = 0.134 R^{-3} E_0 f$ so that

$$\frac{T}{T_0} = \frac{0.134 R^{-3} E_0 f}{p_0 \psi}, \tag{5.51}$$

where we have used the fact that $\psi = \rho/\rho_0$. In the central region $f \to 0.426$ and we have seen above that $\psi \to 1.75 \eta^{7.5}$, hence,

$$\frac{T}{T_0} \to \frac{0.134 R^{-3} E_0 (0.426)}{p_0 (1.75)} \eta^{-7.5} = 0.033 \frac{R^{-3} E_0}{p_0} \eta^{-7.5}. \tag{5.52}$$

Taylor considered the case where the shock front has moved out to such an extent that the pressure in the central region is reduced to atmospheric pressure p_0, hence,

$$p_0 = (0.134)(0.426) R^{-3} E_0$$

and, accordingly,

$$\frac{T}{T_0} \to \frac{1}{1.75} \eta^{-7.5}. \tag{5.53}$$

At $\eta = 0.5$, $\eta^{-7.5} = 181$, hence, $T = 103 T_0$ and if $T_0 = 290$ then $T \approx 30,000 K$. As a result, the temperatures left behind by the blast wave are very high but the energy density is low because the density itself is low.

5.14 The Wasted Energy

Taylor [4] also considered the energy left in the atmosphere after the blast wave has propagated away. Following the blast wave the air eventually returns to atmospheric pressure p_0 but it is left at a temperature T_1 which is higher than the original temperature T_0 of the atmosphere. Consequently, the energy required to raise the temperature of the air from T_0 to T_1 is left in the atmosphere which is not available for doing mechanical work on the surrounding atmosphere. This energy is denoted by E_1 and it is wasted as a blast wave producer; let us now determine E_1.

At a particular stage of the disturbance where the pressure is p and the air is at temperature T, it expands adiabatically to atmospheric pressure and the air is left at a temperature $T_1 > T_0$, hence,

$$\frac{T}{T_1} = \left(\frac{p}{p_0}\right)^{\frac{\gamma-1}{\gamma}}, \tag{5.54}$$

where

$$p = \left(\frac{R^{-3}E_0 f}{\gamma B(\gamma)}\right)$$

according to Eq. (5.25). Hence,

$$\frac{T}{T_1} = \left(\frac{R^{-3}E_0 f}{\gamma p_0 B(\gamma)}\right)^{\frac{\gamma-1}{\gamma}}. \tag{5.55}$$

However,

$$p = \rho R T \quad \text{and} \quad p_0 = \rho_0 R T_0,$$

hence,

$$\frac{T}{T_0} = \frac{p \rho_0}{p_0 \rho} = \frac{1}{\psi}\left(\frac{R^{-3}E_0 f}{\gamma p_0 B(\gamma)}\right),$$

so that

$$\frac{T_1}{T_0} = \frac{T_1}{T}\frac{T}{T_0} = \left(\frac{R^{-3}E_0 f}{\gamma p_0 B(\gamma)}\right)^{\frac{1-\gamma}{\gamma}}\frac{1}{\psi}\left(\frac{R^{-3}E_0 f}{\gamma p_0 B(\gamma)}\right) = \frac{f^{\frac{1}{\gamma}}}{\psi}\left(\frac{R^{-3}E_0}{\gamma p_0 B(\gamma)}\right)^{\frac{1}{\gamma}}. \tag{5.56}$$

Now the heat energy per unit mass of air following the passage of the disturbance is

$$h = c_P T_1$$
$$= \frac{\gamma}{\gamma - 1} T_1 R, \tag{5.57}$$

where R is the gas constant. However, $R = p_0/\rho_0 T_0$, so that the latter equation can be written as

$$h = \frac{\gamma}{\gamma - 1} \frac{T_1 p_0}{T_0 \rho_0}.$$

Before the passage of the disturbance the heat energy per unit mass of air is

$$h_0 = \frac{\gamma}{\gamma - 1} \frac{p_0}{\rho_0},$$

so that the increase in heat energy per unit mass is

$$\Delta h = \frac{\gamma}{\gamma - 1} \frac{T_1 p_0}{T_0 \rho_0} - \frac{\gamma}{\gamma - 1} \frac{p_0}{\rho_0} = \frac{\gamma p_0}{(\gamma - 1)\rho_0} \left(\frac{T_1}{T_0} - 1 \right),$$

and the corresponding increase in the heat energy per unit volume is

$$\frac{\Delta h}{V} = \frac{\gamma p_0 \rho}{(\gamma - 1)\rho_0} \left(\frac{T_1}{T_0} - 1 \right) = \frac{\gamma p_0}{(\gamma - 1)} \psi \left(\frac{T_1}{T_0} - 1 \right), \tag{5.58}$$

where $\psi = \rho/\rho_0$ according to Eq. (5.7c). Accordingly, the heat energy wasted when the shock front has expanded to radius R is given by

$$E_1 = 4\pi R^3 \int\limits_0^1 \left[\frac{\gamma p_0 \psi}{\gamma - 1} \left(\frac{T_1}{T_0} - 1 \right) \right] \eta^2 d\eta,$$

and substituting for the ratio T_1/T_0 from Eq. (5.56) in the latter equation gives

$$E_1 = 4\pi R^3 \frac{\gamma p_0}{\gamma - 1} \int\limits_0^1 \left[\left(\frac{R^{-3} E_0}{\gamma p_0 B(\gamma)} \right)^{1/\gamma} f^{1/\gamma} - \psi \right] \eta^2 d\eta. \tag{5.59}$$

However, the pressure at the shock front is given by

$$p_1 = \frac{R^{-3} E_0}{\gamma B(\gamma)} [f]_{\eta=1}, \tag{5.60}$$

and dividing by p_0 on both sides gives,

$$\frac{p_1}{p_0} = \frac{R^{-3} E_0}{\gamma p_0 B(\gamma)} [f]_{\eta=1} \equiv y_1, \tag{5.61}$$

so that y_1 is the pressure in atmospheres at the shock front. Using this in the integral in Eq. (5.59) gives,

$$E_1 = 4\pi R^3 \frac{\gamma p_0}{\gamma - 1} \int_0^1 \left[\left(\frac{y_1}{[f]_{\eta=1}} \right)^{1/\gamma} f^{1/\gamma} - \psi \right] \eta^2 d\eta. \tag{5.62}$$

Dividing across by E_0, the total energy generated in the explosion, we obtain the following equation for the fraction of the energy wasted when the shock front has expanded to a radius where the pressure (in atmospheres) is y_1;

$$\frac{E_1}{E_0} = \frac{4\pi\gamma}{(\gamma - 1)B(\gamma)} \frac{[f]_{\eta=1}}{y_1} \left[\left(\frac{y_1}{[f]_{\eta=1}} \right)^{1/\gamma} \int_0^1 f^{1/\gamma} \eta^2 d\eta - \int_0^1 \psi \eta^2 d\eta \right], \tag{5.63}$$

where Eq. (5.22) has been used for the energy E_0 on the right-hand side of the latter equation. It should be noted here that this formula differs from Taylor's formula by a factor of γ as the energy in the system after the blast wave has passed has been estimated from the enthalpy instead of the internal energy since the final state of the air is obtained by heating it at constant pressure p_0 [10].

With $\gamma = 1.4$ we obtain the following results; $B(\gamma) = 5.317$, $[f]_{\eta = 1} = 1.166$ and

$$\int_0^1 f^{1/\gamma} \eta^2 d\eta = 0.217 \quad \text{and} \quad \int_0^1 \psi \eta^2 d\eta = 1/3 \quad \text{(independent of } \gamma),$$

hence,

$$\frac{E_1}{E_0} = \frac{1}{y_1} \left(1.34 y_1^{1/1.4} - 2.29 \right), \tag{5.64}$$

which is shown plotted in Fig. 5.8.

It can be seen from Fig. 5.8 that E_1/E_0 increases as y_1 decreases and, of course, as y_1 decreases the radius R of the shock front expands to enclose a larger mass of air. The results are shown plotted down to 20 atmospheres as the equations only apply to very strong shocks and, consequently, any conclusions drawn for pressure below about this value would be wholly inaccurate. Talyor discusses this point by comparing the strong shock boundary conditions as given by Eqs. (5.5a), (5.5b) and (5.5c) with the true boundary conditions as given by Eqs. (5.4a), (5.4b) and (5.4c).

Taylor then goes on to discuss the remaining energy, namely, $E_0 - E_1$, in which a part, E_2, is used in doing mechanical work against atmospheric pressure p_0 during the expansion of the heated air. We can determine E_2 by noting that at a particular radius $r = \eta R$ the heated air at pressure p expands against the atmospheric pressure and the work done is given by the equation,

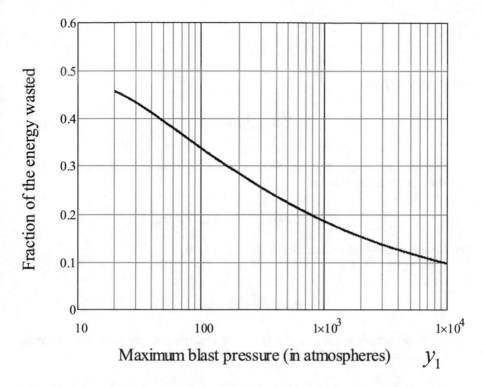

Fig. 5.8 A plot showing the fraction of the energy wasted when the shock front has expanded to a radius where the blast pressure (in atmospheres) is y_1 ($\gamma = 1.4$)

$$W = \int_{V_i}^{V_f} p_0 dV = p_0 \left(V_f - V_i \right)$$

$$= p_0 V_i \left(\frac{V_f}{V_i} - 1 \right),$$

(5.65)

where V_i and V_i are the initial and final volumes, respectively. Assuming the expansion is adiabatic we have

$$\frac{V_f}{V_i} = \left(\frac{p}{p_0} \right)^{1/\gamma},$$

(5.66)

hence, the work done by unit volume of the compressed air is

$$\frac{W}{V_i} = p_0 \left[\left(\frac{p}{p_0} \right)^{1/\gamma} - 1 \right],$$
(5.67)

or alternatively as

$$\frac{W}{V_i} = p_0 \left[\frac{T_1 p}{T p_0} - 1 \right]$$
(5.68)

as given by Taylor [4]. The latter two expressions are identical since in the case of an adiabatic expansion we have

$$T_1/T = (p/p_0)^{(1-\gamma)/\gamma} = (p_0/p)(p/p_0)^{1/\gamma},$$

confirming the equivalence of Eqs. (5.67) and (5.68).

Noting that

$$p/p_0 = \left(y_1 f / [f]_{\eta=1} \right)$$

and with this substitution in Eq. (5.68) we can integrate over the volume to radius R, that is, to $\eta = 1$, thereby obtaining

$$E_2 = W = 4\pi R^3 p_0 \left[\left(\frac{y_1}{[f]_{\eta=1}} \right)^{1/\gamma} \int_0^1 f^{1/\gamma} \eta^2 d\eta - \int_0^1 \eta^2 d\eta \right].$$
(5.69)

Dividing across by E_0 and using Eq. (5.22) for E_0 on the right-hand side gives

$$\frac{E_2}{E_0} = \frac{4\pi}{\gamma B(\gamma)} \left[\left(\frac{y_1}{[f]_{\eta=1}} \right)^{(1-\gamma)/\gamma} \int_0^1 f^{1/\gamma} \eta^2 d\eta - \frac{[f]_{\eta=1}}{3y_1} \right].$$
(5.70)

This ratio is plotted in Fig. 5.9 as a function of y_1.

Taylor then refers to the remaining energy, namely, $E_0 - E_1 - E_2$, which is available for propagation the blast wave; this energy (as a fraction of E_0) is shown plotted in Fig. 5.10.

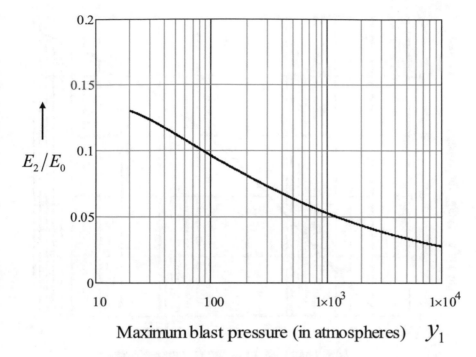

Fig. 5.9 A plot showing the fraction of the energy used in doing work against the atmospheric pressure during the expansion of the heated air when the shock front has expanded to a radius where the blast pressure (in atmospheres) is y_1 ($\gamma = 1.4$)

5.15 Taylor's Second Paper

The relationship between the radius of the shock front and the time measured from the start of the explosion was used by G. I. Taylor [11] to estimate the energy yield of the Trinity test. This relationship, as we have seen, was established by Taylor's in his first paper [4]. However, this estimate is dependent on the value of γ, namely, the ratio of the specific heats of air. Taylor provides two estimates of this energy yield in terms of the energy that would be released by an equivalent chemical explosion using TNT. One estimate, that he considered to be more accurate, gives the energy yield as 16,800 tons of TNT while the second estimate gives 23,700 tons of TNT. He indicates that this latter value may overestimate the energy, but the value has been included to emphasise the typical error that might be expected when the effects of radiation are neglected as well as the variations in the specific heats of air due to the extremely high temperatures associated with nuclear explosions. Let us now consider these estimates as presented by Taylor.

Taylor's estimate of the energy yield was based on photographic records of the Trinity test, which showed the shape of the expanding shock wave as a function of

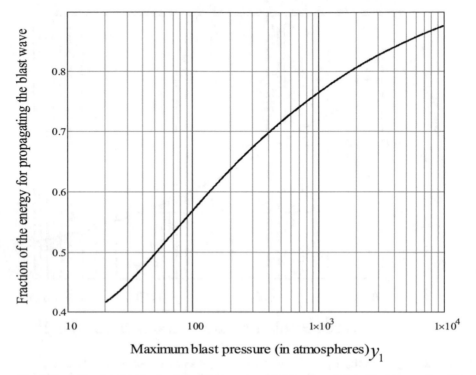

Fig. 5.10 A plot showing the fraction of the energy used for propagating the blast wave when the shock front has expanded to a radius where the blast pressure (in atmospheres) is y_1 ($\gamma = 1.4$)

time following detonation. According to his theoretical predictions in first paper [4] we have seen from Eq. (5.24) that the radius of the shock front has the following time-dependence (with $\gamma = 1.4$);

$$R(t) = 1.033 \left(\frac{E_0}{\rho_0}\right)^{1/5} t^{2/5},$$

hence,

$$(R(t))^{5/2} = (1.033)^{5/2} \left(\frac{E_0}{\rho_0}\right)^{1/2} t.$$

By taking logs to the base 10 of this latter equation, we can write it as,

$$\frac{5}{2} \log_{10} R = \log_{10} t + \log_{10}\left[(1.033)^{5/2} \left(\frac{E_0}{\rho_0}\right)^{1/2}\right].$$

Consequently, a plot of $\log_{10}t$ as abscissa with $(5/2)\log_{10}R$ as ordinate should give a $45°$ straight-line whose intercept determines the energy yield, E_0. Taylor plotted values of R over a range from 20 to 185 meters and spanning a time interval from about 0.24 ms to 62 ms and found that the logarithmic plot coincided almost exactly with his theoretical predictions, these predictions as he stated "made more than four years before the explosion took place". According to Taylor, the straight-line plot that appears in Fig. 1 in his article [11] represents the relation,

$$\frac{5}{2} \log_{10}R - \log_{10}t = 11.915.$$

By taking the density of air as 1.25 kg m^{-3} and with Taylor's first estimate of $E_0 = 16,800$ tons of TNT (hence, $E_0 = 4.2 \times 10^9 \times 16,800$ Joules), we find, after substituting these values for E_0 and ρ_0 in the equation above, that

$$\frac{5}{2} \log_{10}R - \log_{10}t = 6.911.$$

However, it should be noted that Taylor measured the shock front in *cm*, so that the latter equation becomes,

$$\frac{5}{2} \log_{10}\left(\frac{R_{cm}}{100}\right) - \log_{10}t = 6.911,$$

hence,

$$\frac{5}{2} \log_{10}R_{cm} - \log_{10}t = 11.911,$$

which is in very good agreement with the straight-line representing the data of the Trinity test. Taylor points out that this estimate is based on the assumption that the energy available for propagating the blast, and which was not radiated outside the expanding shock front, was 16,800 tons of TNT.

He also considers an alternative possibility by taking into account the temperature behind the shock wave in order to determine a mean value for γ in the range where the product $R^{5/2}t^{-1}$ is constant. From the plot of radius versus time he determines a mean radius of 100 *m* in the range and found that the temperature behind the shock wave at this radius to be about 2800 Kelvin. Using values of c_P that were calculated for nitrogen and oxygen at this temperature, and assuming that the relation $c_P = c_V + R$ still applies at this elevated temperature, he finds that $\gamma \cong 1.3$.

We have seen that the multiplying constant in Eq. (5.24) is 1.033 when $\gamma = 1.4$, hence, to determine the multiplying constant for $\gamma = 1.3$, we refer to the integrals appearing in Sect. 5.7, in this chapter. Evaluating these integrals for $\gamma = 1.3$, we find that

$$\int_0^1 \psi \phi^2 \eta^2 \, d\eta = 0.20955 \quad \text{and} \quad \int_0^1 f \eta^2 \, d\eta = 0.18087$$

and, consequently, we obtain the following values; $B(\gamma) = 7.14$ and $\eta_0 = 0.974$, hence,

$$\frac{5}{2} \log_{10} R = \log_{10} t + \log_{10} \left[(0.974)^{5/2} \left(\frac{E_0}{\rho_0} \right)^{1/2} \right].$$

With $E_0 = 4.2 \times 10^9 \times 23{,}000$ Joules (assuming the energy yield is 23,000 tons of TNT), we find that the latter equation (with R in *cm*) gives,

$$\frac{5}{2} \log_{10} R_{cm} - \log_{10} t = 11.915,$$

which is identical to Taylor's equation for the fit to the numerical data. However, Taylor, on the other hand, finds that the energy yield amounts to 23,700 tons of TNT. The slight discrepancy is due to differences in determining the various quantities used in the analysis; for example, with $\gamma = 1.3$, Taylor gives the integrals as

$$\int_0^1 \psi \phi^2 \eta^2 \, d\eta = 0.221 \quad \text{and} \quad \int_0^1 f \eta^2 \, d\eta = 0.183,$$

while he uses 4.18×10^9 Joules per ton of TNT; clearly, one can see that even these small differences contribute to the error in estimating the yield. Nonetheless, the closeness of the fit to the measured data was a triumph for G. I. Taylor's theoretical predictions and his formula for the radius of the shock front as a function of time is frequently referred to when dimensional methods are discussed.

5.16   Approximate Treatment of Strong Shocks

According to the point source solution of Taylor [4] and von Neumann [5] almost the entire mass of air encompassed by the explosive wave is concentrated in a thin layer behind the shock front while the density of the air is extremely low in the inner region and the pressure is essentially constant in the region of low density. These characteristics can be used to develop simple analytical approximations for the strong shock point-source solution and these are discussed below.

In the following sections we will consider two approximate treatments for strong shocks; one by Chernyi and the other by Bethe. Both use the unique characteristics of the point source solution in order to obtain approximate formulae for the expanding shock wave.

5.16.1   Chernyi's Approximation

Using the characteristics of the point source solution Chernyi [12], has presented a simple analysis of the strong explosion and an account of the method as presented in the text by Zel'dovich and Raizer [13] is outlined below.

Assuming the entire mass m of air is concentrated in a thin layer behind the shock front as illustrated in Fig. 5.11, then one can write,

$$\frac{4\pi}{3}\rho_0 R^3 = 4\pi\rho R^2 \Delta r$$

Using the strong shock condition for the density increase according to the Rankine-Hugoniot equation in the above, namely, $\rho = \rho_0(\gamma + 1)/(\gamma - 1)$, we have

$$\Delta r = \frac{R}{3}\frac{\gamma - 1}{\gamma + 1} \tag{5.71}$$

for the thickness of the layer, and the assumption that all of the air is piled up at the front becomes more accurate as γ approaches unity. As the layer is so thin, we can assume that the air velocity within it is approximately constant and equal to the velocity immediately behind the shock front. If p_i is the pressure on the inner side of the layer (the external atmospheric pressure p_0 is assumed to be zero (as in Taylor's analysis) as $p_i \gg p_0$. Newton's second law of motion for the mass m of air in the layer gives

$$\frac{d}{dt}(mu) = 4\pi R^2 p_i. \tag{5.72}$$

Let us assume that p_i is some fraction, α, of the pressure p_s immediately behind the shock front so that $p_i = \alpha p_s$, where $p_s = 2\rho_0 U_s^2/(\gamma + 1)$ is the strong shock

Fig. 5.11 The concentration of almost the entire mass of air in a sphere of radius R into a thin shell of thickness Δr at the shock front

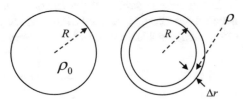

pressure and U_s is the velocity of the shock front as previously noted. The velocity u in Eq. (5.72) is the air velocity *immediately* behind the shock which we have previously seen to be given by the Eq. (3.38b), that is, $u = 2U_s/(\gamma + 1)$, hence, Eq. (5.72) becomes,

$$\frac{d}{dt}\left(\frac{4\pi}{3}\rho_0 R^3 \frac{2}{\gamma+1} U_s\right) = 4\pi R^2 \alpha \frac{2}{\gamma+1}\rho_0 U_s^2, \tag{5.73}$$

after cancelling common terms in the latter equation we can write

$$\frac{1}{3}\frac{d}{dt}\left(R^3 U_s\right) = \alpha R^2 U_s^2,$$

and, furthermore, writing this as

$$\frac{1}{3}\frac{dR}{dt}\frac{d}{dR}\left(R^3 U_s\right) = \alpha R^2 U_s^2$$

so that

$$\frac{1}{3}\frac{d}{dR}\left(R^3 U_s\right) = \alpha R^2 U_s \tag{5.74}$$

since $U_s = dR/dt$. Writing Eq. (5.74) in the form;

$$\frac{1}{3}\frac{d\left(R^3 U_s\right)}{R^3 U_s} = \alpha\frac{dR}{R}$$

and after integrating we obtain

$$\frac{1}{3}\ln R^3 U_s = \alpha\ln R + \text{constant}$$

or

$$U_s = aR^{-3(1-\alpha)}, \tag{5.75}$$

where a is a constant of integration.

Let us now determine the constants, a and α in these equations by taking into account the energy generated in the explosion. The total energy as we have previously seen is made up of kinetic and internal energy according to

$$E_0 = \frac{1}{2}mu^2 + mc_V T. \tag{5.76}$$

The latter term represents the internal energy, so that

$$\text{Internal energy} = mc_V T = mc_V \frac{p_i V}{mR} = c_V \frac{p_i V}{R} = \frac{p_i V}{\gamma - 1},$$

after using the equation of state, $p_i V = mRT$, and noting that $c_P - c_V = R$ and $\gamma = c_P/c_V$. Hence, the internal energy becomes

$$\text{Internal energy} = \frac{4\pi}{3} \frac{R^3}{\gamma - 1} p_i = \frac{4\pi}{3} \frac{R^3}{\gamma - 1} \alpha \frac{2}{\gamma + 1} \rho_0 U_s^2,$$

while the kinetic energy is given by

$$\text{Kinetic energy} = \frac{1}{2} m u^2 = \frac{1}{2} \frac{4\pi}{3} R^3 \rho_0 \left(\frac{2}{\gamma + 1} U_s \right)^2,$$

where the usual strong shock conditions for u and p_i have been substituted. By using these expressions for the kinetic and internal energy in Eq. (5.76) yields,

$$E_0 = \frac{4\pi}{3} R^3 \rho_0 \left[\frac{2 U_s^2}{(\gamma + 1)^2} + \frac{2\alpha U_s^2}{\gamma^2 - 1} \right] \tag{5.77}$$

and let us now substitute Eq. (5.75) in this latter equation so that we can finally obtain

$$E_0 = \frac{4\pi}{3} \rho_0 a^2 \left[\frac{2}{(\gamma + 1)^2} + \frac{2\alpha}{\gamma^2 - 1} \right] R^{3 - 6(1 - \alpha)} \tag{5.78}$$

However, the explosion energy is a constant so that the dependence on R must vanish; this implies that $3 - 6(1 - \alpha) = 0$ so that $\alpha = 1/2$. Substituting this value for α back into Eq. (5.78), we find that

$$E_0 = \frac{4\pi}{3} \rho_0 a^2 \left[\frac{3\gamma - 1}{(\gamma + 1)^2 (\gamma - 1)} \right], \tag{5.79}$$

so that a is given by

$$a = \left[\frac{3}{4\pi} \frac{(\gamma + 1)^2 (\gamma - 1)}{(3\gamma - 1)} \right]^{\frac{1}{2}} \left(\frac{E_0}{\rho_0} \right)^{\frac{1}{2}}. \tag{5.80}$$

Since $\alpha = 1/2$, we have

$$U_s = \frac{dR}{dt} = aR^{-\frac{3}{2}} \qquad (5.81)$$

and after performing the integration we obtain

$$R = \left(\frac{5a}{2}\right)^{\frac{2}{5}} t^{\frac{2}{5}}, \qquad (5.82)$$

and, finally, after substituting for a, we find that the radius of the shock front has the following time-dependence,

$$R(t) = \left[\frac{75}{16\pi} \frac{(\gamma+1)^2(\gamma-1)}{(3\gamma-1)}\right]^{\frac{1}{5}} \left(\frac{E_0}{\rho_0}\right)^{\frac{1}{5}} t^{\frac{2}{5}}. \qquad (5.83)$$

Evaluating the quantity in square brackets for $\gamma = 1.4$ we find that

$$R(t) = 1.01 \left(\frac{E_0}{\rho_0}\right)^{\frac{1}{5}} t^{\frac{2}{5}}, \qquad (5.84)$$

which should be compared with Taylor's numerical analysis [4] which gives the multiplying factor as 1.033. In addition, the central pressure is $p_i = 0.5p_s$ whereas Taylor's analysis gives $p_i = 0.366p_s$.

5.16.2 Bethe's Approximation for Small Values of $\gamma - 1$

Following on from the approximate treatment of spherical shock waves in the previous section, let us now turn our attention to an analysis of strong shocks as presented by Bethe [6] by using the approximation that $\gamma - 1$ is small. Bethe's analysis is based on the particular nature of the point source solution of Taylor [4] and Von Neumann [5] in which most of the material is piled up at the shock front and the density is extremely low in the inner regions. In addition, the pressure is approximately constant within a large radius of the shock front. Bethe considered the first of these facts to be most relevant for the construction of a general method for dealing with shock waves other than the exact point source solution of Taylor and Von Neumann.

Bethe noted that the concentration of the material at the shock front becomes more pronounced for values of γ close to unity and the assumption that all of the material is concentrated there is much more valid as γ approaches one. In these circumstances, the velocity of all the material will be the same as the velocity of the material directly behind the shock front. This velocity is given by Eq. (3.38b), namely, $u = [2/(\gamma + 1)]U$, which is almost identical to the velocity U of the shock front itself as γ approaches unity.

We will follow Bethe's analysis but we will adopt a slightly different notation for some of the quantities encountered: Bethe uses Y, $\dot{Y}$ and $\ddot{Y}$ for the position, velocity and acceleration of the shock front, respectively; here, we use R, $\dot{R}$ and $\ddot{R}$ instead, in order to be consistent with the notation already adopted. In addition, we will use r_0 to denote the initial position of an arbitrary mass element and $r(r_0, t)$ to denote its position at some time t later, clearly, $r(r_0, 0) = r_0$; Bethe uses r for its initial position and R for its position at a time t later, otherwise, similar notation is used for the remaining parameters encountered, such as, pressure and density.

Let us now write equations for mass and momentum conservation: the continuity or mass conservation equation in the case of spherical symmetry is given by

$$\rho(r_0, t)r^2(r_0, t)dr = \rho_0 r_0^2 dr_0,$$

where ρ_0 is the initial density and $\rho(r_0, t)$ is the density of this mass element at a later time t. This equation gives

$$\frac{\partial r(r_0, t)}{\partial r_0} = \frac{\rho_0}{\rho(r_0, t)} \frac{r_0^2}{r^2(r_0, t)} \tag{5.85}$$

The momentum equation for this specific mass element is

$$\frac{\partial^2 r(r_0, t)}{\partial t^2} = -\frac{1}{\rho(r_0, t)} \frac{\partial p(r_0, t)}{\partial r}$$

and by using the continuity equation we can write this latter equation as

$$\frac{\partial^2 r(r_0, t)}{\partial t^2} = -\frac{r^2(r_0, t)}{\rho_0 r_0^2} \frac{\partial p(r_0, t)}{\partial r_0}. \tag{5.86}$$

We also need an equation for energy conservation; this equation is determined by the relationship between pressure and density after the material element has been hit by the shock. This adiabatic relation implies that

$$\frac{p(r_0, t)}{p_s(r_0)} = \left[\frac{\rho(r_0, t)}{\rho_s(r_0)} \right]^\gamma, \tag{5.87}$$

where the subscript s denotes conditions just behind the shock. By using the strong shock relationship for the density, namely, Eq. (3.17b);

$$\frac{\rho_s(r_0)}{\rho_0} = \frac{\gamma + 1}{\gamma - 1} \tag{5.88}$$

in Eq. (5.87), we have

$$\frac{\rho_0}{\rho(r_0, t)} = \left(\frac{\gamma - 1}{\gamma + 1}\right) \left[\frac{p_s(r_0)}{p(r_0, t)}\right]^{1/\gamma} \tag{5.89}$$

and substituting this latter equation in the continuity equation, yields,

$$\frac{\partial r(r_0, t)}{\partial r_0} = \left(\frac{\gamma - 1}{\gamma + 1}\right) \frac{r_0^2}{r^2(r_0, t)} \left[\frac{p_s(r_0)}{p(r_0, t)}\right]^{1/\gamma}. \tag{5.90}$$

Assuming strong shock conditions, these equations are supplemented by the Rankine-Hugoniot equations at the shock front: one in relation to the pressure at the front in terms of the velocity of the shock, namely, Eq. (3.26b), that is

$$p_s(R) = \frac{2}{\gamma + 1} \rho_0 \dot{R}^2 \tag{5.91}$$

and the other in relation to the particle or material velocity immediately behind the shock front in terms of the shock velocity $\dot{R}$, namely, Eq. (3.38b), that is

$$\dot{r}(R, t) = \frac{2}{\gamma + 1} \dot{R} \tag{5.92}$$

where R and $\dot{R}$ denote the position and shock front velocity, respectively, with the dot above the symbols denoting differentiating with respect to time.

Using Bethe's assumption that all of the material is concentrated at the shock front, one can identify the position $r(r_0, t)$ with the position of the front R and the acceleration $\partial^2 r(r_0, t)/\partial t^2$ with the acceleration of the shock front itself, that is, $\ddot{R}$, hence Eq. (5.86) gives

$$\frac{1}{r_0^2} \frac{\partial p}{\partial r_0} = -\rho_0 \frac{\ddot{R}}{R^2}$$

and as the right-hand side of this equation is independent of r_0, it can be integrated to give

$$p(r_0, t) = -\rho_0 \frac{\ddot{R}}{R^2} \frac{r_0^3}{3} + \text{constant},$$

where the constant of integration can be determined from the condition that $p(R, t) = p_s(R)$ when $r_0 = R$ and $p_s(R)$ is the pressure at the shock front, hence,

$$p(r_0, t) = p_s(R) + \frac{\ddot{R}\rho_0}{3R^2}\left(R^3 - r_0^3\right) \tag{5.93}$$

Substituting Eq. (5.91) in this latter equation, gives

$$p(r_0, t) = \frac{2}{\gamma + 1}\rho_0\dot{R}^2 + \frac{\ddot{R}R\rho_0}{3}\left(1 - \frac{r_0^3}{R^3}\right)$$

and for γ close to unity, this latter equation can be approximated by

$$\frac{p(r_0, t)}{\rho_0} = \dot{R}^2 + \frac{\ddot{R}R}{3}\left(1 - \frac{r_0^3}{R^3}\right), \tag{5.94}$$

which gives the pressure distribution at time t in terms of the position, velocity and acceleration of the shock as indicated by Bethe.

The total energy E available for the shock comprises potential and kinetic energy. The potential energy per unit volume is $p/(\gamma - 1)$ and, therefore, the total potential energy is

$$P.E. = \frac{4\pi}{3}R^3\frac{p}{\gamma - 1}.$$

However, in the case of the point-source solution, we have already observed that the pressure is essentially uniform within the interior and approximately equal to half the pressure at the shock front, that is, $p \approx p_s(R)/2$, hence,

$$P.E. = \frac{2\pi}{3}R^3\frac{p_s(R)}{\gamma - 1}$$

and substituting Eq. (5.91) in this latter equation, gives,

$$P.E. = \frac{2\pi}{3}\rho_0\frac{R^3\dot{R}^2}{\gamma - 1} \tag{5.95}$$

and as we shall see in due course, this is the dominant contribution to the total energy. The other component contributing to the total energy comprises kinetic energy and it is given, in general, by the equation

$$K.E. = \frac{1}{2}mu^2$$

where m is the mass and u is the velocity. Since we are assuming that the total original mass is piled up in a very thin region at the shock front, then $m = 4\pi R^3\rho_0/3$ and the velocity of this mass of material is approximately equal to the velocity of the shock front, hence,

$$K.E. = \frac{2\pi}{3}\rho_0 R^3 \dot{R}^2. \tag{5.96}$$

Comparing this latter equation with the equation for the potential energy, one observes that the kinetic energy is smaller by a factor of $\gamma - 1$, and as γ is considered close to unity the total energy is largely potential energy. Nonetheless, as the total energy is conserved we obtain the result that $R^3\dot{R}^2 = \text{constant}$, hence,

$$\dot{R}^2 = AR^{-3} \tag{5.97}$$

where A is a constant which is related to the total energy generated. Integrating this equation, yields,

$$R = \left(\frac{5}{2}\right)^{2/5} A^{1/5} t^{2/5} \tag{5.98}$$

and we recover the important relationship showing how the radius of the shock front varies with time. Differentiating Eq. (5.98), gives,

$$R\ddot{R} = -\frac{3}{2}\dot{R}^2 \tag{5.99}$$

and by using this relationship in Eq. (5.94), we obtain,

$$\frac{p(r_0, t)}{\rho_0} = \dot{R}^2 - \frac{\dot{R}^2}{2}\left(1 - \frac{r_0^3}{R^3}\right) \tag{5.100}$$

and at the centre of the shock wave we have

$$p(0, t) = \frac{1}{2}p_s(R)$$

after using Eq. (5.91) and assuming that γ is close to one; hence, the pressure at the centre is approximately one half of the pressure at the shock front and the pressure distribution in terms of the pressure at the shock front is

$$p(r_0, t) = p_s(R) - \frac{1}{2}p_s(R)\left(1 - \frac{r_0^3}{R^3}\right). \tag{5.101}$$

From the pressure distribution one can obtain the density distribution by using Eq. (5.89). The quantity $p(r_0, t)$ is given by Eq. (5.101) and $p_s(r_0)$ is the pressure at r_0 after being hit by the shock. Using the Rankine-Hugoniot relationship between the shock pressure and the shock velocity in addition to the relationship between velocity and radius according to Eq. (5.97), one can write

$$p_s(r_0) = \frac{R^3}{r_0^3} p_s(R) \tag{5.102}$$

and, therefore,

$$\frac{p_s(r_0)}{p(r_0,t)} = \frac{2}{\left(\frac{r_0}{R}\right)^3 \left[1 + \left(\frac{r_0}{R}\right)^3\right]}, \tag{5.103}$$

accordingly, the equation for the density distribution becomes

$$\frac{\rho(r_0,t)}{\rho_0} = \frac{\gamma+1}{\gamma-1} \left[\frac{(r_0/R)^3 \left[1 + (r_0/R)^3\right]}{2}\right]^{1/\gamma}. \tag{5.104}$$

From this density distribution one can determine how the material's position changes with time by integrating the continuity equation. Once the material's position is known, one can proceed to ascertain the material's velocity distribution by performing a simple differentiation.

Continuing on with Bethe's analysis, let us now substitute Eq. (5.103) in Eq. (5.90), thereby obtaining

$$\frac{\partial r(r_0,t)}{\partial r_0} = \left(\frac{\gamma-1}{\gamma+1}\right) \frac{r_0^2}{r^2(r_0,t)} \left[\frac{2}{(r_0/R)^3 \left[1 + (r_0/R)^3\right]}\right]^{1/\gamma}. \tag{5.105}$$

Bethe defines $x = r_0^3/R^3$ and $y = r^3(r_0,t)/R^3$, so that Eq. (5.105) becomes

$$\frac{dy}{dx} = \left(\frac{\gamma-1}{\gamma+1}\right) \left[\frac{2}{x(1+x)}\right]^{1/\gamma}. \tag{5.106}$$

Bethe proceeds to consider the case where x is not too small while $\gamma - 1$ is small enough to make γ in the exponent is equal to 1, hence, Eq. (5.106) becomes,

$$\frac{dy}{dx} = (\gamma-1) \left(\frac{1}{x(1+x)}\right).$$

By performing a partial fraction expansion and integrating, yields,

$$y = 1 - (\gamma-1)\ln\left(\frac{1+x}{2x}\right)$$
$$= 1 - (\gamma-1)[\ln(1+x) - \ln 2 - \ln x]$$

and as x is considered not too small, the latter equation can be approximated by

$$y = 1 + (\gamma - 1) \ln x \tag{5.107}$$

Bethe then considers the case where x is small enough that the term $1 + x$ in Eq. (5.106) is approximately unity; hence, Eq. (5.106) becomes,

$$\frac{dy}{dx} = \left(\frac{\gamma - 1}{\gamma + 1}\right) \frac{2^{1/\gamma}}{x^{1/\gamma}}$$

$$\cong (\gamma - 1)x^{-\frac{1}{\gamma}},$$

since $\gamma \cong 1$. Integrating this latter equation, implies that

$$y = (\gamma - 1) \frac{x^{-\frac{1}{\gamma}+1}}{1 - \frac{1}{\gamma}} + \text{constant}$$

$$\cong x^{\frac{\gamma-1}{\gamma}} + \text{constant}. \tag{5.108}$$

By using the relation, $x = e^{\log_e x}$, one can write the latter equation as,

$$y = e^{\frac{\gamma-1}{\gamma} \log_e x} + \text{constant}$$

and by expanding this expression and just retaining terms of the order of $\gamma - 1$, we have,

$$y \cong 1 + (\gamma - 1) \ln x + \text{constant}$$

and by comparing this latter equation with Eq. (5.107) we can see that the constant of integration is zero, hence, Eq. (5.108) becomes

$$y \cong x^{\frac{\gamma-1}{\gamma}}. \tag{5.109}$$

Let us now compare these approximate solutions according to Eqs. (5.107) and (5.109) with the exact solution of Eq. (5.106) by direct numerical integration of the equation,

$$y(x) = 1 - \left(\frac{\gamma - 1}{\gamma + 1}\right) \int_x^1 \left[\frac{2}{\zeta(1 + \zeta)}\right]^{1/\gamma} d\zeta,$$

and these are shown plotted in Fig. 5.12. It can be observed that Eq. (5.109) is a better approximation to the exact solution and we will proceed to use it in the subsequent analysis. Substituting the relationships, $x = r_0^3/R^3$ and $y = r^3(r_0, t)/R^3$ in Eq. (5.109), we obtain,

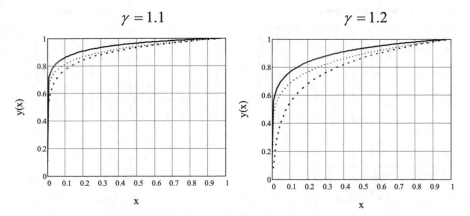

Fig. 5.12 Plots of $y(x)$ versus x for the exact solution of Eq. (5.106); (solid line) and the approximate solutions according to Eq. 5.107 (dash-dot line) and Eq. 5.109 (dotted line) for two different values of γ (see text)

$$\frac{r(r_0,t)}{R} = \left(\frac{r_0}{R}\right)^{\frac{\gamma-1}{\gamma}},\tag{5.110}$$

hence,

$$r(r_0,t) = R^{\frac{1}{\gamma}}r_0^{\frac{\gamma-1}{\gamma}}.\tag{5.111}$$

Noting that $R \equiv R(t)$, Eq. (5.111) gives the position $r(r_0, t)$ as a function of time of an arbitrary material particle or mass element whose initial position is r_0. The velocity $\dot{r}(r_0, t)$ of a specific mass element can be determined from the latter equation by using the formula;

$$\dot{r}(r_0,t) = \left(\frac{\partial r(r_0,t)}{\partial t}\right)_{r_0}.\tag{5.112}$$

Carrying out the differentiation of Eq. (5.111), we have

$$\dot{r}(r_0,t) = \frac{1}{\gamma}\left(\frac{r_0}{R}\right)^{\frac{\gamma-1}{\gamma}}\dot{R}$$

and by using Eq. (5.110) we can write the latter equation as

$$\dot{r}(r_0,t) = \frac{1}{\gamma}\frac{r(r_0,t)}{R}\dot{R}.\tag{5.113}$$

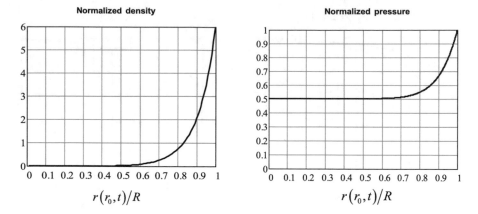

Fig. 5.13 Plots of Eqs. (5.114) and (5.115) for $\gamma = 1.4$ (see text)

Within the approximations considered, Eq. (5.113) shows that the particle or material velocity is approximately linear with Eulerian position $r(r_0, t)$ as already observed in Fig. 5.1. In fact, when Eq. (5.113) is compared with Eqs. (5.2), (5.7b) and (5.32) they are identical.

Let us now substitute Eq. (5.110) in Eqs. (5.101) and (5.104), thereby obtaining,

$$p(r_0, t) = p_s(R) - \frac{1}{2} p_s(R) \left[1 - \left(\frac{r(r_0, t)}{R} \right)^{\frac{3\gamma}{\gamma - 1}} \right] \tag{5.114}$$

and

$$\frac{\rho(r_0, t)}{\rho_0} = \frac{\gamma + 1}{\gamma - 1} \left[\frac{(r(r_0, t)/R)^{\frac{3\gamma}{\gamma - 1}} \left[1 + (r(r_0, t)/R)^{\frac{3\gamma}{\gamma - 1}} \right]}{2} \right]^{1/\gamma} \tag{5.115}$$

for the pressure and density variations as a function of Eulerian position. These are shown plotted in Fig. 5.13 and should be compared with the plots shown in Fig. 5.1.

5.17   Route to an Analytical Solution

Taylor's analysis was followed quite closely in this chapter and it was shown that three coupled ordinary differential equations (ODEs) were obtained from the partial differential equations. These ODEs were then numerically integrated to generate plots of the pressure, velocity and density corresponding to the similarity hypothesis. Following Taylor's method, approximate analytical expressions for the pressure, velocity and density were presented and were shown to be remarkably close to the numerical results. However, an analytical solution of the equations was obtained by Sedov [7] and discussed in the text by Needham [14], while Von Neumann [5], using the Lagrangian formulation, also obtained an analytical solution. Later on in 1955, J. L. Taylor [15] obtained an analytical result in a slightly different manner as described by Sachdev [16]. A route to an analytical solution, that is largely similar to the methods presented by Sachdev [16] and Lee [17], is described below. We will see in due course that the method is quite lengthy and laborious and the potential to make simple algebraic mistakes is high. In the initial stage of the calculation the energy-balance equation;

$$\left(\frac{\partial}{\partial t} + u \frac{\partial}{\partial r} \right) (p\rho^{-\gamma}) = 0$$

is replaced by the following equivalent energy equation [16];

$$\frac{\partial}{\partial t} \left(\frac{1}{2} \rho u^2 + \frac{p}{\gamma - 1} \right) + \frac{1}{r^2} \frac{\partial}{\partial r} \left[r^2 \left(\frac{1}{2} \rho u^2 + \frac{\gamma p}{\gamma - 1} \right) u \right] = 0 \tag{5.116}$$

which is Eq. (1.66b) of chapter 1. Defining [16],

$$E = \frac{1}{2} \rho u^2 + \frac{p}{\gamma - 1} \quad \text{and} \quad I = \frac{1}{2} \rho u^2 + \frac{\gamma p}{\gamma - 1}$$

then Eq. (5.116) becomes

$$\frac{\partial E}{\partial t} + \frac{1}{r^2} \frac{\partial (r^2 u I)}{\partial r} = 0. \tag{5.117}$$

But $u = AR^{-3/2}\phi(\eta)$ and $R \propto t^{2/5}$, hence, $u^2 \propto t^{-6/5}$ and similarly $p \propto t^{-6/5}$ so that $E \propto t^{-6/5}$. Hence, define,

$$E = t^{-6/5} F(\eta) = t^{-6/5} F\left(r/t^{2/5} \right).$$

Consequently,

$$\frac{\partial E}{\partial t} = -\frac{6}{5} t^{-11/5} F(\eta) - \frac{2}{5} t^{-13/5} r \frac{\partial F}{\partial \eta} \tag{5.118}$$

and

$$\frac{\partial E}{\partial r} = t^{-8/5} \frac{\partial F}{\partial \eta}.$$

Substituting for $\partial F/\partial \eta$ in Eq. (5.118) gives

$$\frac{\partial E}{\partial t} = -\frac{6}{5}\frac{E}{t} - \frac{2}{5}\frac{r}{t}\frac{\partial E}{\partial r}$$

and using this value for $\partial E/\partial t$ in Eq. (5.117) gives

$$\frac{\partial (r^2 uI)}{\partial r} = \frac{6}{5}\frac{r^2}{t}E + \frac{2}{5}\frac{r^3}{t}\frac{\partial E}{\partial r}$$

$$= \frac{2}{5t}\frac{\partial (r^3 E)}{\partial r}.$$

Integrating this latter equation yields,

$$uI = \frac{2}{5}\frac{r}{t}E + \text{constant of integration}. \tag{5.119}$$

We will see from the boundary condition below that the constant of integration is equal to zero. Using Eq. (5.2) and Eqs. (5.7a), (5.7b) and (5.7c) we can write Eq. (5.119) as

$$\phi I = \eta E.$$

Substituting the expressions for I and E, and by using Eqs. (5.7a), (5.7b) and (5.7c), we obtain

$$\phi = \eta \left[\frac{\rho_0 \psi U^2 \phi^2 + \frac{2\rho_0 U^2 f}{\gamma(\gamma-1)}}{\rho_0 \psi U^2 \phi^2 + \frac{2\rho_0 U^2 f}{(\gamma-1)}}\right].$$

Multiplying across and solving for f gives

$$f = \frac{\gamma(\gamma-1)}{2}\psi\left[\frac{\phi-\eta}{\eta-\gamma\phi}\right]\phi^2. \tag{5.120}$$

This latter equation gives a simple relationship between the pressure, density and velocity. Before proceeding further let us check to see if the latter equation satisfies the boundary condition at the shock front according to Eq. (5.8); at $\eta = 1$ we have

$$\psi(1) = \frac{\gamma+1}{\gamma-1} \quad \text{and} \quad \phi(1) = \frac{2}{\gamma+1}$$

and by substituting these in Eq. (5.120) one can verify that $f(1) = 2\gamma/(\gamma+1)$ as given by Eq. (5.8). Let us define φ according to the equation,

$$\varphi = \phi/\eta, \tag{5.121}$$

consequently, we can write Eq. (5.120) as

$$f = \frac{\gamma(\gamma-1)}{2}\psi\eta^2\varphi^2\left[\frac{1-\varphi}{\gamma\varphi-1}\right] \tag{5.122}$$

$$= \eta^2\Psi\psi \tag{5.123}$$

where

$$\Psi = \frac{\gamma(\gamma-1)}{2}\varphi^2\frac{1-\varphi}{\gamma\varphi-1}. \tag{5.124}$$

and, as a result, Ψ is expressed in terms of the velocity variable φ. We will now dispense with the momentum equation, namely Eq. (5.13), and, instead, use Eq. (5.122) in addition the continuity and energy equations. Firstly, the continuity equation, namely Eq. (5.10), is

$$\frac{\psi'}{\psi} = \frac{\phi' + 2\phi/\eta}{\eta - \phi},$$

which can be written as,

$$\frac{(\phi-\eta)d\ln\psi}{d\eta} = -\frac{1}{\eta^2}\frac{d(\eta^2\phi)}{d\eta},$$

where, for example, $\ln\psi$ implies $\log_e\psi$, and in view of the above definitions we can write the latter equation as,

$$\frac{\eta(\varphi-1)d\ln\psi}{d\eta} = -\frac{1}{\eta^2}\frac{d(\eta^3\varphi)}{d\eta},$$

so that

$$\frac{d\ln\psi}{d\ln\eta} = -\frac{1}{\varphi-1}\frac{d\varphi}{d\ln\eta} - \frac{3\varphi}{\varphi-1}, \tag{5.125}$$

which is an alternative form of the continuity equation. The energy equation, namely, Eq. (5.12), is

$$3f + \eta f' + \gamma\frac{\psi'}{\psi}f(\phi-\eta) - \phi f' = 0,$$

with $\phi = \eta\varphi$, this latter equation becomes

$$3 + \gamma\eta(\varphi - 1)\frac{\psi'}{\psi} - \eta\frac{f'}{f}(\varphi - 1) = 0. \tag{5.126}$$

From Eq. (5.123) we have

$$f' = 2\eta\Psi\psi + \eta^2\left(\psi\frac{d\Psi}{d\eta} + \Psi\frac{d\psi}{d\eta}\right),$$

and dividing across by f, gives,

$$\frac{f'}{f} = \frac{2}{\eta} + \frac{1}{\Psi}\frac{d\Psi}{d\eta} + \frac{1}{\psi}\frac{d\psi}{d\eta}$$

and substituting this latter equation in Eq. (5.126), we have

$$3 + \gamma\eta(\varphi - 1)\frac{\psi'}{\psi} - \eta(\varphi - 1)\left(\frac{2}{\eta} + \frac{1}{\Psi}\frac{d\Psi}{d\eta} + \frac{1}{\psi}\frac{d\psi}{d\eta}\right) = 0.$$

Simplifying, yields

$$3 + \gamma(\varphi - 1)\frac{d\ln\psi}{d\ln\eta} - 2(\varphi - 1) - (\varphi - 1)\frac{d\ln\Psi}{d\ln\eta} - (\varphi - 1)\frac{d\ln\psi}{d\ln\eta} = 0,$$

therefore,

$$(\varphi - 1)\frac{d\ln\Psi}{d\ln\eta} = 3 + (\varphi - 1)(\gamma - 1)\frac{d\ln\psi}{d\ln\eta} - 2(\varphi - 1),$$

hence,

$$\frac{d\ln\Psi}{d\ln\eta} = (\gamma - 1)\frac{d\ln\psi}{d\ln\eta} - 2 + \frac{3}{\varphi - 1}$$

and after further simplifications we finally obtain

$$\frac{d\ln\Psi}{d\ln\eta} - (\gamma - 1)\frac{d\ln\psi}{d\ln\eta} + 2\left(\frac{\varphi - \frac{5}{2}}{\varphi - 1}\right) = 0. \tag{5.127}$$

This is the final form of the energy equation. Accordingly, the following three equations will be used to obtain an analytical solution;

$$\Psi = \frac{\gamma(\gamma-1)}{2}\varphi^2 \frac{1-\varphi}{\gamma\varphi-1}$$

$$\frac{d\ln\psi}{d\ln\eta} = -\frac{1}{\varphi-1}\frac{d\varphi}{d\ln\eta} - \frac{3\varphi}{\varphi-1}$$

$$\frac{d\ln\Psi}{d\ln\eta} - (\gamma-1)\frac{d\ln\psi}{d\ln\eta} + 2\left(\frac{\varphi-\frac{5}{2}}{\varphi-1}\right) = 0,$$

with the first of these equations replacing the momentum equation.

5.18 Analytical Solution Method

Let us now proceed to obtain analytical solutions for the velocity, density and pressure. As we will see, the method is straightforward but, nonetheless, quite lengthy and one has to be careful in performing the many algebraic manipulations.

5.18.1 The Analytical Expression for the Velocity

Substituting the continuity equation in the energy equation gives

$$\frac{d\ln\Psi}{d\ln\eta} + \frac{(\gamma-1)}{\varphi-1}\frac{d\varphi}{d\ln\eta} + 3\varphi\frac{(\gamma-1)}{\varphi-1} + 2\frac{(\varphi-\frac{5}{2})}{\varphi-1} = 0$$

hence,

$$\left(\frac{\varphi-1}{\gamma-1}\right)\frac{d\ln\Psi}{d\ln\eta} + \frac{d\varphi}{d\ln\eta} + 3\varphi + 2\left(\frac{\varphi-\frac{5}{2}}{\gamma-1}\right) = 0$$

and, accordingly,

$$\frac{d\varphi}{d\ln\eta} - \left(\frac{1-\varphi}{\gamma-1}\right)\frac{d\ln\Psi}{d\ln\eta} - \left(\frac{5-(3\gamma-1)\varphi}{\gamma-1}\right) = 0.$$

Multiplying the latter equation by $d\ln\eta/d\varphi$ gives

$$1 - \left(\frac{1-\varphi}{\gamma-1}\right)\frac{d\ln\Psi}{d\varphi} - \left(\frac{5-(3\gamma-1)\varphi}{\gamma-1}\right)\frac{d\ln\eta}{d\varphi} = 0$$

and solving for $d\ln\eta/d\varphi$ gives

$$\frac{d\ln\eta}{d\varphi} = \frac{(\gamma - 1)}{5 - (3\gamma - 1)\varphi} - \frac{(1 - \varphi)}{5 - (3\gamma - 1)\varphi}\frac{d\ln\Psi}{d\varphi}. \tag{5.128}$$

However,

$$\Psi = \frac{\gamma(\gamma - 1)}{2}\frac{\varphi^2(1 - \varphi)}{\gamma\varphi - 1}$$

and taking logs of both sides gives

$$\ln\Psi = \ln\gamma\frac{(\gamma - 1)}{2} + 2\ln\varphi + \ln(1 - \varphi) - \ln(\gamma\varphi - 1)$$

and differentiating the latter equation yields,

$$\frac{d\ln\Psi}{d\varphi} = \frac{2}{\varphi} - \frac{1}{1 - \varphi} - \frac{\gamma}{\gamma\varphi - 1}.$$

Hence, Eq. (5.128) becomes,

$$\frac{d\ln\eta}{d\varphi} = \frac{(\gamma - 1)}{5 - (3\gamma - 1)\varphi} - \frac{(1 - \varphi)}{5 - (3\gamma - 1)\varphi}\left[\frac{2}{\varphi} - \frac{1}{1 - \varphi} - \frac{\gamma}{\gamma\varphi - 1}\right].$$

Gathering similar terms together implies that we can write the latter equation as,

$$\frac{d\ln\eta}{d\varphi} = \frac{\gamma}{5 - (3\gamma - 1)\varphi} - \frac{2(1 - \varphi)}{[5 - (3\gamma - 1)\varphi]\varphi} + \frac{\gamma(1 - \varphi)}{(\gamma\varphi - 1)[5 - (3\gamma - 1)\varphi]}. \tag{5.129}$$

At this stage we need to carry out some partial fraction expansions in order to integrate this latter equation. Firstly, let

$$\frac{2(1 - \varphi)}{[5 - (3\gamma - 1)\varphi]\varphi} \equiv \frac{A_1}{\varphi} + \frac{B_1}{[5 - (3\gamma - 1)\varphi]},$$

hence,

$$5A_1 = 2 \quad \text{and} \quad (3\gamma - 1)A_1 - B_1 = 2.$$

Solving for A_1 and B_1 gives,

$$A_1 = \frac{2}{5} \quad \text{and} \quad B_1 = \frac{6}{5}(\gamma - 2).$$

Similarly, let

$$\frac{\gamma(1-\varphi)}{(\gamma\varphi-1)[5-(3\gamma-1)\varphi]} \equiv \frac{A_2}{\gamma\varphi-1} + \frac{B_2}{[5-(3\gamma-1)\varphi]}$$

and by solving this partial fraction expansion we find that

$$A_2 = \frac{\gamma(\gamma-1)}{2\gamma+1} \quad \text{and} \quad B_2 = \frac{3\gamma(\gamma-2)}{2\gamma+1}.$$

Collecting all terms having denominator $[5-(3\gamma-1)\varphi]$ we have

$$\frac{1}{[5-(3\gamma-1)\varphi]} \left[\gamma - \frac{6(\gamma-2)}{5} + \frac{3\gamma(\gamma-2)}{2\gamma+1}\right] = \frac{(13\gamma^2-7\gamma+12)/5(2\gamma+1)}{[5-(3\gamma-1)\varphi]}.$$

Substituting these results back in Eq. (5.129) gives,

$$5\frac{d\ln\eta}{d\varphi} = \frac{(13\gamma^2-7\gamma+12)/(2\gamma+1)}{[5-(3\gamma-1)\varphi]} - \frac{2}{\varphi} + \frac{5\gamma(\gamma-1)/(2\gamma+1)}{\gamma\varphi-1} \tag{5.130}$$

We are now in a position to integrate the latter equation; noting that

$$\int \frac{d\varphi}{[5-(3\gamma-1)\varphi]} = -\frac{1}{(3\gamma-1)} \ln[5-(3\gamma-1)\varphi] + \text{constant}$$

and

$$\int \frac{d\varphi}{\gamma\varphi-1} = \frac{1}{\gamma} \ln(\gamma\varphi-1) + \text{constant}.$$

Carrying out the integration we obtain

$$5\ln\eta = -\frac{(13\gamma^2-7\gamma+12)}{(3\gamma-1)(2\gamma+1)} \ln[5-(3\gamma-1)\varphi] - 2\ln\varphi$$

$$+ \frac{5(\gamma-1)}{2\gamma+1} \ln(\gamma\varphi-1) + C_1. \tag{5.131}$$

The constant of integration C_1 is determined by noting that at $\eta=1$, $\varphi=2/(\gamma+1)$, hence,

$$0 = -\frac{(13\gamma^2-7\gamma+12)}{(3\gamma-1)(2\gamma+1)} \ln\left[5-(3\gamma-1)\frac{2}{\lambda+1}\right] - 2\ln\left(\frac{2}{\gamma+1}\right)$$

$$+ \frac{5(\gamma-1)}{2\gamma+1} \ln\left(\frac{2\gamma}{\gamma+1}-1\right) + C_1$$

Hence,

$$C_1 = \frac{(13\gamma^2 - 7\gamma + 12)}{(3\gamma - 1)(2\gamma + 1)} \ln\left(\frac{7-\gamma}{\gamma+1}\right) + 2\ln\left(\frac{2}{\gamma+1}\right) - \frac{5(\gamma-1)}{2\gamma+1} \ln\left(\frac{\gamma-1}{\gamma+1}\right).$$

Substituting this back in Eq. (5.131) gives

$$5\ln\eta = -\frac{(13\gamma^2 - 7\gamma + 12)}{(3\gamma - 1)(2\gamma + 1)} \ln\left[\left(\frac{\gamma+1}{7-\gamma}\right)\{5 - (3\gamma - 1)\varphi\}\right] - 2\ln\left[\frac{\gamma+1}{2}\varphi\right]$$
$$+ \frac{5(\gamma-1)}{2\gamma+1} \ln\left[\left(\frac{\gamma+1}{\gamma-1}\right)(\gamma\varphi - 1)\right]$$

hence,

$$\eta = \left[\left(\frac{\gamma+1}{7-\gamma}\right)\left\{5 - (3\gamma - 1)\frac{\phi}{\eta}\right\}\right]^{\frac{\alpha_1}{5}} \left[\frac{\gamma+1}{2}\frac{\phi}{\eta}\right]^{\frac{\alpha_2}{5}} \left[\left(\frac{\gamma+1}{\gamma-1}\right)\left(\gamma\frac{\phi}{\eta} - 1\right)\right]^{\frac{\alpha_3}{5}},$$

$$(5.132)$$

where we have substituted the relation, $\phi/\eta = \varphi$ and where

$$\alpha_1 = -\frac{(13\gamma^2 - 7\gamma + 12)}{(3\gamma - 1)(2\gamma + 1)}, \quad \alpha_2 = -2, \quad \alpha_3 = \frac{5(\gamma-1)}{(2\gamma+1)}. \qquad (5.133)$$

Eq. (5.132) gives an implicit relationship between the velocity ϕ and η.

5.18.2   The Analytical Expression for the Density

Let us now determine an analytical expression for the density. The continuity equation is

$$(\varphi - 1)\frac{d\ln\psi}{d\ln\eta} + \frac{d\varphi}{d\ln\eta} + 3\varphi = 0,$$

hence,

$$\frac{d\ln\psi}{d\ln\eta} = \frac{1}{(1-\varphi)}\frac{d\varphi}{d\ln\eta} + \frac{3\varphi}{1-\varphi}.$$

Multiplying across by $d\ln\eta/d\varphi$ we obtain

$$\frac{d \ln \psi}{d\varphi} = \frac{1}{(1 - \varphi)} + \frac{3\varphi}{1 - \varphi} \frac{d \ln \eta}{d\varphi}$$

and substituting for $d \ln \eta / d\varphi$ from Eq. (5.129) we have

$$\frac{d \ln \psi}{d\varphi} = \frac{1}{1 - \varphi} + \frac{3\varphi}{1 - \varphi}$$
$$\times \left[\frac{(13\gamma^2 - 7\gamma + 12)/5(2\gamma + 1)}{[5 - (3\gamma - 1)\varphi]} - \frac{2/5}{\varphi} + \frac{\gamma(\gamma - 1)/(2\gamma + 1)}{\gamma\varphi - 1} \right]. \quad (5.134)$$

By carrying out the multiplication in relation to the second term on the right-hand side of the latter equation we can identify again the need to consider partial fraction expansions. In this context we, firstly, deal the quantity,

$$\frac{\frac{3\varphi}{5(2\gamma+1)}(13\gamma^2 - 7\gamma + 12)}{(1 - \varphi)[5 - (3\gamma - 1)\varphi]} \equiv \frac{A_3}{1 - \varphi} + \frac{B_3}{[5 - (3\gamma - 1)\varphi]}.$$

Solving this latter equation in the usual manner we obtain

$$A_3 = \frac{(13\gamma^2 - 7\gamma + 12)}{5(2 - \gamma)(2\gamma + 1)} \quad \text{and} \quad B_3 = -\frac{(13\gamma^2 - 7\gamma + 12)}{(2 - \gamma)(2\gamma + 1)}.$$

Similarly, in relation to the other quantity appearing in Eq. (5.134) we have

$$\frac{\frac{3\gamma(\gamma-1)\varphi}{2\gamma+1}}{(1 - \varphi)(\gamma\varphi - 1)} \equiv \frac{A_4}{1 - \varphi} + \frac{B_4}{\gamma\varphi - 1}$$

and one finds that

$$A_4 = \frac{3\gamma}{2\gamma + 1} \quad \text{and} \quad B_4 = \frac{3\gamma}{2\gamma + 1}.$$

Substituting these results in Eq. (5.134) gives

$$\frac{d \ln \psi}{d\varphi} = \frac{1}{1 - \varphi} + \frac{(13\gamma^2 - 7\gamma + 12)/5(2 - \gamma)(2\gamma + 1)}{1 - \varphi} - \frac{(13\gamma^2 - 7\gamma + 12)/(2 - \gamma)(2\gamma + 1)}{[5 - (3\gamma - 1)\varphi]}$$
$$- \frac{6/5}{1 - \varphi} + \frac{3\gamma/(2\gamma + 1)}{1 - \varphi} + \frac{3\gamma/(2\gamma + 1)}{\gamma\varphi - 1}$$

Collecting all numerator terms having denominator $(1 - \varphi)$ in this latter equation gives the following expression for the numerator;

$$1 + \frac{13\gamma^2 - 7\gamma + 12}{5(2 - \gamma)(2\gamma + 1)} - \frac{6}{5} + \frac{3\gamma}{2\gamma + 1}$$

and by simplifying this expression one can show that it reduces to $2/(2 - \gamma)$. Substituting this back in the equation for $d \ln \psi / d\varphi$ gives

$$\frac{d \ln \psi}{d\varphi} = \frac{2/(2 - \gamma)}{1 - \varphi} - \frac{(13\gamma^2 - 7\gamma + 12)/(2 - \gamma)(2\gamma + 1)}{[5 - (3\gamma - 1)\varphi]} + \frac{3\gamma/(2\gamma + 1)}{\gamma\varphi - 1}.$$

Integrating, we obtain,

$$\ln \psi = -\frac{2}{2 - \gamma} \ln (1 - \varphi) + \frac{(13\gamma^2 - 7\gamma + 12)}{(2 - \gamma)(2\gamma + 1)(3\gamma - 1)} \ln [5 - (3\gamma - 1)\varphi]$$

$$+ \frac{3}{2\gamma + 1} \ln (\gamma\varphi - 1) + C_2$$

where the constant of integration C_2 is determined from the boundary condition;

$$\psi(1) = \frac{\gamma + 1}{\gamma - 1} \quad \text{and} \quad \varphi(1) = \frac{2}{\gamma + 1} \quad \text{at } \eta = 1,$$

this gives C_2 as

$$C_2 = \ln \left(\frac{\gamma + 1}{\gamma - 1}\right) + \frac{2}{2 - \gamma} \ln \left(\frac{\gamma - 1}{\gamma + 1}\right) - \frac{(13\gamma^2 - 7\gamma + 12)}{(2 - \gamma)(2\gamma + 1)(3\gamma - 1)} \ln \left(\frac{7 - \gamma}{\gamma + 1}\right)$$

$$- \frac{3}{2\gamma + 1} \ln \left(\frac{\gamma - 1}{\gamma + 1}\right)$$

Substituting this value of C_2 back in the equation for $\ln \psi$ and simplifying yields,

$$\ln \psi = \ln \left(\frac{\gamma + 1}{\gamma - 1}\right) + \frac{(13\gamma^2 - 7\gamma + 12)}{(2 - \gamma)(2\gamma + 1)(3\gamma - 1)} \ln \left(\frac{\gamma + 1}{7 - \gamma}\right) [5 - (3\gamma - 1)\varphi]$$

$$+ \frac{3}{2\gamma + 1} \ln \left(\frac{\gamma + 1}{\gamma - 1}\right)(\gamma\varphi - 1) - \frac{2}{2 - \gamma} \ln \left(\frac{\gamma + 1}{\gamma - 1}\right)(1 - \varphi)$$

hence,

$$\psi = \left(\frac{\gamma+1}{\gamma-1}\right)\left[\left(\frac{\gamma+1}{7-\gamma}\right)\{5-(3\gamma-1)\varphi\}\right]^{\beta_1}\left[\left(\frac{\gamma+1}{\gamma-1}\right)(\gamma\varphi-1)\right]^{\beta_2}\left[\left(\frac{\gamma+1}{\gamma-1}\right)(1-\varphi)\right]^{\beta_3},$$

$$(5.135)$$

where

$$\beta_1 = \frac{13\gamma^2 - 7\gamma + 12}{(2-\gamma)(2\gamma+1)(3\gamma-1)}, \quad \beta_2 = \frac{3}{2\gamma+1} \quad \text{and} \quad \beta_3 = -\frac{2}{2-\gamma}. \qquad (5.136)$$

5.18.3 The Analytical Expression for the Pressure

By substituting Eq. (5.135) in Eq. (5.123) we obtain the following expression for the pressure f,

$$f = \frac{\gamma(\gamma+1)}{2}\eta^2\varphi^2\left[\left(\frac{\gamma+1}{7-\gamma}\right)\{5-(3\gamma-1)\varphi\}\right]^{\beta_1}\left[\left(\frac{\gamma+1}{\gamma-1}\right)(\gamma\varphi-1)\right]^{\beta_2-1}\left[\left(\frac{\gamma+1}{\gamma-1}\right)(1-\varphi)\right]^{\beta_3+1}$$

and substituting the expression for η from Eq. (5.132), gives

$$f = \frac{2\gamma}{(\gamma+1)}\left[\frac{\gamma+1}{2}\varphi\right]^2\left[\left(\frac{\gamma+1}{7-\gamma}\right)\{5-(3\gamma-1)\varphi\}\right]^{\beta_1}\left[\left(\frac{\gamma+1}{\gamma-1}\right)(\gamma\varphi-1)\right]^{\beta_2-1}\left[\left(\frac{\gamma+1}{\gamma-1}\right)(1-\varphi)\right]^{\beta_3+1} \times$$

$$\left[\left(\frac{\gamma+1}{7-\gamma}\right)\{5-(3\gamma-1)\varphi\}\right]^{\frac{2\alpha_1}{5}}\left[\frac{\gamma+1}{2}\varphi\right]^{\frac{2\alpha_2}{5}}\left[\left(\frac{\gamma+1}{\gamma-1}\right)(\gamma\varphi-1)\right]^{\frac{2\alpha_3}{5}}$$

Hence,

$$f = \frac{2\gamma}{(\gamma+1)}\left[\left(\frac{\gamma+1}{7-\gamma}\right)\{5-(3\gamma-1)\varphi\}\right]^{\frac{2\alpha_1}{5}+\beta_1}\left[\frac{\gamma+1}{2}\varphi\right]^{\frac{2\alpha_2}{5}+2}\left[\left(\frac{\gamma+1}{\gamma-1}\right)(\gamma\varphi-1)\right]^{\frac{2\alpha_3}{5}+\beta_2-1} \times$$

$$\left[\left(\frac{\gamma+1}{\gamma-1}\right)(1-\varphi)\right]^{\beta_3+1}$$

the exponent terms in this latter equation are obtained from Eqs. (5.133) and (5.136), hence,

$$\frac{2}{5}\alpha_1 + \beta_1 = -\frac{2}{5}\frac{(13\gamma^2 - 7\gamma + 12)}{(3\gamma-1)(2\gamma+1)} + \frac{(13\gamma^2 - 7\gamma + 12)}{(3\gamma-1)(2\gamma+1)(2-\gamma)}$$

$$= \frac{(13\gamma^2 - 7\gamma + 12)}{5(3\gamma - 1)(2 - \gamma)};$$

$$\frac{2\alpha_2}{5} + 2 = \frac{6}{5};$$

$$\frac{2\alpha_3}{5} + \beta_2 - 1 = \frac{2(\gamma - 1)}{2\gamma + 1} + \frac{3}{2\gamma + 1} - 1 = 0;$$

and

$$\beta_3 + 1 = -\frac{\gamma}{2 - \gamma}.$$

Therefore, the terms $\gamma\varphi - 1$ disappears and we finally have the following equation for the pressure f;

$$f = \frac{2\gamma}{(\gamma + 1)} \left[\left(\frac{\gamma + 1}{7 - \gamma} \right) \{5 - (3\gamma - 1)\varphi\} \right]^{\frac{13\gamma^2 - 7\gamma + 12}{5(3\gamma - 1)(2 - \gamma)}} \left[\left(\frac{\gamma + 1}{\gamma - 1} \right) (1 - \varphi) \right]^{\frac{-\gamma}{2 - \gamma}} \left[\frac{\gamma + 1}{2} \varphi \right]^{\frac{6}{5}}$$

(5.137)

It can be seen that analytical solutions have been obtained for all three quantities but the final results are not very revealing as the expressions obtained are implicit. Accordingly, it is clear that Taylor's method in solving the coupled equations numerically was perhaps the best approach to take.

Nonetheless, it is interesting to note that all three equations, namely, Eqs. (5.132), (5.135) and (5.137) contain some of the following terms in square brackets that are raised to various powers;

$$\left[\left(\frac{\gamma + 1}{7 - \gamma} \right) \left\{ 5 - (3\gamma - 1) \frac{\phi}{\eta} \right\} \right], \quad \left[\left(\frac{\gamma + 1}{\gamma - 1} \right) \left(1 - \frac{\phi}{\eta} \right) \right], \quad \left[\frac{\gamma + 1}{2} \frac{\phi}{\eta} \right] \quad \text{and} \left[\left(\frac{\gamma + 1}{\gamma - 1} \right) \left(\gamma \frac{\phi}{\eta} - 1 \right) \right],$$

and all these terms are equal to 1 at the shock front ($\eta = 1$) where $\phi = 2/(\gamma + 1)$ as can be verified by direct substitution. Consequently, Eqs. (5.135) and (5.137) for the density and pressure reduce to

$$\left(\frac{\gamma + 1}{\gamma - 1} \right) \quad \text{and} \quad \frac{2\gamma}{\gamma + 1},$$

respectively, which are just the first multiplying factors that appear on the right-hand side of each equation and these factors are consistent with those expected for the density and pressure at the shock front.

In order to generate plots for the velocity, pressure and density, one can use any standard mathematical *root-solver* to find the values of ϕ for specific values of η according to Eq. (5.132); however, one encounters a problem in carrying out this procedure due to the term, $\left(\gamma\frac{\phi}{\eta} - 1\right)$, since we already know that $\phi \cong \eta/\gamma$ (see Eq. 5.32) for a substantial part of the velocity profile and imaginary roots emerge when using the *root-solver*. One can overcome this difficulty by writing Eq. (5.132) in the following form,

$$\eta^{\frac{5}{\alpha_3}} = \left[\left(\frac{\gamma+1}{7-\gamma}\right)\left\{5 - (3\gamma - 1)\frac{\phi}{\eta}\right\}\right]^{\frac{\alpha_1}{\alpha_3}}\left[\frac{\gamma+1}{2}\frac{\phi}{\eta}\right]^{\frac{\alpha_2}{\alpha_3}}\left[\left(\frac{\gamma+1}{\gamma-1}\right)\left(\gamma\frac{\phi}{\eta} - 1\right)\right]$$

so that

$$\left[\left(\frac{\gamma+1}{\gamma-1}\right)\left(\gamma\frac{\phi}{\eta} - 1\right)\right] = \eta^{\frac{5}{\alpha_3}}\left[\left(\frac{\gamma+1}{7-\gamma}\right)\left\{5 - (3\gamma - 1)\frac{\phi}{\eta}\right\}\right]^{\frac{-\alpha_1}{\alpha_3}}\left[\frac{\gamma+1}{2}\frac{\phi}{\eta}\right]^{\frac{-\alpha_2}{\alpha_3}}$$

and, hence,

$$\phi = \frac{\eta}{\gamma}\left[1 + \frac{\eta^{\frac{5}{\alpha_3}}\left[\left(\frac{\gamma+1}{7-\gamma}\right)\left\{5 - (3\gamma - 1)\frac{\phi}{\eta}\right\}\right]^{\frac{-\alpha_1}{\alpha_3}}\left[\frac{\gamma+1}{2}\frac{\phi}{\eta}\right]^{\frac{-\alpha_2}{\alpha_3}}}{\left(\frac{\gamma+1}{\gamma-1}\right)}\right].$$

We now form the expression,

$$F(\phi) = \phi - \frac{\eta}{\gamma}\left[1 + \frac{\eta^{\frac{5}{\alpha_3}}\left[\left(\frac{\gamma+1}{7-\gamma}\right)\left\{5 - (3\gamma - 1)\frac{\phi}{\eta}\right\}\right]^{\frac{-\alpha_1}{\alpha_3}}\left[\frac{\gamma+1}{2}\frac{\phi}{\eta}\right]^{\frac{-\alpha_2}{\alpha_3}}}{\left(\frac{\gamma+1}{\gamma-1}\right)}\right]$$

and by using Mathcad's *root-solver* this latter equation will return the value for ϕ (after an initial guess value) that results in $F(\phi) = 0$ for a specified value of η. The velocity plot shown in Fig. 5.14 represents the typical output obtained following this procedure. One can then proceed to use those values of η and ϕ to generate plots of the density and pressure by using Eqs. (5.135) and (5.137) and these results are also shown in Fig. 5.14 and these plots should be compared with the plots shown in Fig. 5.1 which are obtained numerically.

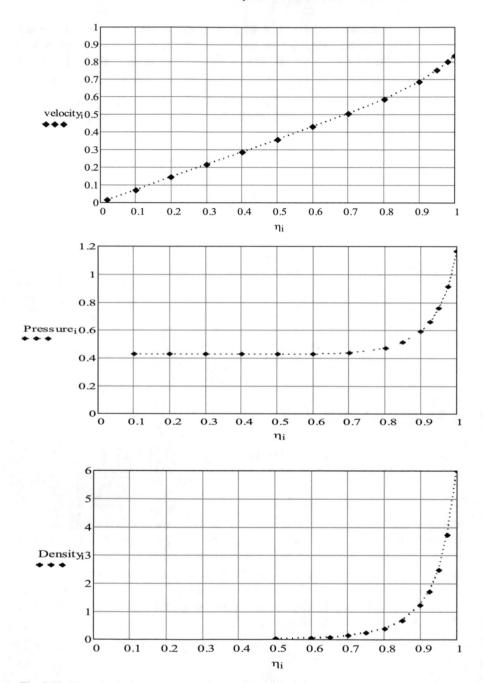

Fig. 5.14 Plots of velocity, pressure and density for specific values of η according to the analytical solution with $\gamma = 1.4$ (see text)

References

1. R. Serber, *The Los Alamos Primer: The First Lectures on How to Build an Atomic Bomb* (University of California Press, Berkeley, 1992)
2. B. Cameron Reed, *The Physics of the Manhattan Project*, 3rd edn. (Springer, Heidelberg, 2015)
3. J.D. Logan, *Applied Mathematics; A Contemporary Approach*, (John Wiley & Sons, Inc., New York, 1987), Chapter 7
4. G.I. Taylor, The formation of a blast wave by a very intense explosion, I. Theor. Discuss., Proc. R. Soc. A **201**, 159 (1950)
5. J. von Neumann, *The Point Source Solution*, Collected Works, vol VI (Pergamon Press, New York, 1976), p. 219
6. H.A. Bethe et al., *LA-2000 Report* (Los Alamos Scientific Laboratory of the University of California, Los Alamos, 1958)
7. L.I. Sedov, *Similarity and Dimensional Methods in Mechanics* (Academic, New York/London, 1959)
8. L.D. Landau, E.M. Lifshitz, *Fluid Mechanics*, (Pergamon Press, London, 1966), Chapter 9
9. R.E. Scraton, *Basic Numerical Methods* (Edward Arnold Pub. Ltd., London, 1984)
10. H. Bethe, J. Hirschfelder, V. Waters, *LA-213 Report* (Los Alamos Scientific Laboratory of the University of California, Los Alamos, 1946)
11. G.I. Taylor, The formation of a blast wave by a very intense explosion, II. The atomic explosion of 1945. Proc. R. Soc. A **201**, 175 (1950)
12. G.G. Chernyi, The problem of a point explosion. Dokl. Akad. Nauk SSSR **112**, 213 (1957)
13. Ya. B. Zel'dovich, Yu. P. Raizer, *Physics of Shock Waves and High-Temperature Hydrodynamic Phenomena*, (Dover Publications Inc., Mineola, 2002), Section 26
14. C.E. Needham, *Blast Waves*, (Springer-Verlag Berlin, Heidelberg, 2010), Section 4.2
15. J.L. Taylor, An exact solution of the spherical blast wave problem. Philos. Magn. **46**, 317–320 (1955)
16. P.L. Sachdev, *Shock Waves and Explosions*, (Chapman & Hall, London, 2004), Chapter 3
17. J.H.S. Lee, *The Gas Dynamics of Explosions*, (Cambridge University Press, New York, 2016), Chapter 4

Chapter 6
Numerical Treatment of Spherical Shock Waves

6.1 Introduction

The development of nuclear weapons heralded the need for numerical methods for predicting the hydrodynamic effects of these devices outside the very strong shock regime. The similarity solution to the intense point-source explosion in air only applies to the early phases of the explosion where the pressures generated are very much greater than the ambient air pressure. The blast wave becomes progressively weaker at later stages of the expansion and the pressure behind the shock front will eventually becomes comparable with the atmospheric pressure. The *self-similar solution* no longer applies when the pressure drops below about 20 atmospheres [1]. Consequently, it becomes necessary to take into account the counter-pressure which has so far been neglected and, when this is included, the partial differential equations describing the flow must be integrated numerically. Von Neumann, who provided an analytical solution to the point-source strong shock problem, [2] pioneered the application of numerical techniques for blast wave problems and was instrumental in the development of high-speed computing machines for performing numerical calculations.

As the blast wave travels further away from the point of detonation the over pressure, $p - p_0$, steadily decreases. Once the air has crossed the shock front and been compressed, it expands again to a pressure even lower than the pre-shock ambient pressure, p_0; this so-called "suction phase" is an important feature encountered in explosions. In Fig. 6.1 we illustrate the overpressure at four successive times. In relation to the plot corresponding to t_4 it can be seen that the overpressure has a negative value at some distance behind the shock front. During this phase, a partial vacuum is created and the surrounding air is sucked in, which results in a reversal in the air flow towards the centre as opposed to being pushed away from the centre during the positive phase ($p > p_0$). The duration of the negative phase is, in general, larger than the positive phase and the air eventually returns to atmospheric pressure. This particular feature of explosive behaviour cannot be accounted for with

S. Prunty, *Introduction to Simple Shock Waves in Air*, Shock Wave and High Pressure Phenomena, https://doi.org/10.1007/978-3-030-63606-7_6

Fig. 6.1 Typical variation
of overpressure with
distance from the centre of
the expanding spherical
shock wave at successive
times is illustrated

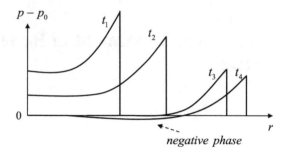

the *similarity solution* and we must resort to numerical techniques to solve the
equations of fluid flow.

In the analysis to follow we will use the Lagrangian form of the equations in
spherical geometry similar to that presented in Chap. 4 in the case of plane shocks; in
addition, artificial viscosity will be introduced to avoid discontinuities.

6.2 Lagrangian Equations in Spherical Geometry

The equations are written in Lagrangian form, that is, in a coordinate system that
moves with the particles so that each particle or infinitesimal cross-section carries a
label as its position changes with time. We will follow the Lagrangian notation of
Zel'dovich and Raizer [1] and in the article by Brode [3].

We have already seen (in Chap. 5) that the total blast energy can be written as

$$E_0 = \int_0^R \left(\frac{1}{2}\rho u^2 + \frac{p}{\gamma - 1} \right) 4\pi r^2 dr - \frac{4\pi R^3}{3} \frac{p_0}{\gamma - 1} \tag{6.1}$$

where R is the radius of the shock front. The subtracted term is the internal energy of
the air prior to being engulfed by the shock of radius R. Dividing across by p_0, the
ambient air pressure ($p_0 = 1.013 \times 10^5 Nm^{-2}$), we have

$$\frac{E_0}{p_0} = \frac{4\pi}{p_0} \int_0^R r^2 \left(\frac{1}{2}\rho u^2 + \frac{p}{\gamma - 1} \right) dr - \frac{4\pi R^3}{3(\gamma - 1)}. \tag{6.2}$$

Noting that the last term in the latter equation has dimensions of length cubed, so
let us define a length ε in terms of the energy E_0 and the ambient pressure p_0, such
that,

$$\varepsilon^3 = \frac{E_0}{p_0} = \frac{4\pi}{p_0} \int_0^R r^2 \left(\frac{1}{2} \rho u^2 + \frac{p}{\gamma - 1} \right) dr - \frac{4\pi R^3}{3(\gamma - 1)}, \tag{6.3}$$

accordingly, radial distances will be expressed in dimensionless units (that is, the radial distance will be divided by ε as indicated below). Let us now proceed to establish the equations for momentum, mass and energy conservation.

6.2.1 Momentum Equation

The Lagrangian formulation (by following specific particles) for the equation of motion is considered here; the particle or "mass packet" has spherical symmetry in this specific case. At some instant, $t = t_0$, the air between radius r_0 and $r_0 + dr_0$ carries the label r_0 with mass of $4\pi r_0^2 \rho(r_0, t_0) dr_0$. Let us apply Newton's equation of motion to this "particle" as it moves with velocity u in the direction r; hence,[1]

$$4\pi r_0^2 \rho(r_0, t_0) dr_0 \frac{\partial u}{\partial t} = 4\pi r^2 p(r_0, t) - 4\pi r^2 p(r_0 + dr_0, t)$$
$$= -4\pi r^2 dr_0 \frac{\partial p}{\partial r_0}, \tag{6.4}$$

where $r \equiv r(r_0, t)$ and clearly $r_0 \equiv r(r_0, t_0)$, where t_0 represents some initial time which is usually taken as $t_0 = 0$ when the particle had position r_0. In this case $r_0 \equiv r(r_0, 0)$ and the particle's initial density is $\rho(r_0, 0)$, therefore, Eq. (6.4) becomes;

$$\frac{\partial u}{\partial t} = -\frac{r^2}{r_0^2 \rho(r_0, 0)} \frac{\partial p}{\partial r_0}. \tag{6.5}$$

Defining the following dimensionless variables; $\lambda = r/\varepsilon$ and $\lambda_0 = r_0/\varepsilon$ in terms of ε above, then the latter equation becomes;

$$\frac{\partial u}{\partial t} = -\frac{\varepsilon^2 \lambda^2}{r_0^2 \rho(\lambda_0, 0)} \frac{\partial p}{\partial r_0}, \tag{6.6}$$

and by defining, $\chi \equiv (r_0/\varepsilon)^3/3$, Eq. (6.6) can be written as

---

[1]Partial derivatives are used here to indicate the changes in position and time of *specific particles*; nonetheless, it should be understood that these partial derivatives imply that we are in fact following the path taken by a *specific particle* according to the Lagrangian description.

$$\frac{\partial u}{\partial t} = -\frac{\varepsilon^2 \lambda^2}{r_0^2 \rho(\lambda_0, 0)} \frac{\partial p}{\partial \chi} \frac{\partial \chi}{\partial r_0} = -\frac{\lambda^2}{\rho(\lambda_0, 0)\varepsilon} \frac{\partial p}{\partial \chi}. \tag{6.7}$$

Let us now measure the particle velocity u in terms of the ambient sound speed c_0, the pressure p in terms of atmospheric pressure p_0, and the density ρ normalized to ambient air density ρ_0, then the following dimensionless quantities are obtained;
$\tilde{u} \equiv u/c_0, \tilde{p} \equiv p/p_0, \tilde{\rho}(\lambda_0, 0) = \rho(\lambda_0, 0)/\rho_0$, hence, Eq. (6.7) becomes

$$c_0 \frac{\partial \tilde{u}}{\partial t} = -\frac{\lambda^2 p_0}{\rho_0 \tilde{\rho}(\lambda_0, 0)\varepsilon} \frac{\partial \tilde{p}}{\partial \chi}$$

so that

$$\frac{\partial \tilde{u}}{\partial t} = -\frac{\lambda^2 p_0}{\rho_0 \tilde{\rho}(\lambda_0, 0)c_0\varepsilon} \frac{\partial \tilde{p}}{\partial \chi}.$$

However, the speed of sound c_0 in ambient air is given by, $c_0^2 = \gamma p_0/\rho_0$, so that the latter equation becomes

$$\frac{\partial \tilde{u}}{\partial t} = -\frac{\lambda^2 c_0}{\gamma \varepsilon \tilde{\rho}(\lambda_0, 0)} \frac{\partial \tilde{p}}{\partial \chi}. \tag{6.8}$$

Clearly, the ratio, ε/c_0, has dimensions of time, so defining a dimensionless time according to

$$\tau = c_0 t/\varepsilon, \tag{6.9}$$

then Eq. (6.8) can be written in the form;

$$\frac{\partial \tilde{u}}{\partial \tau} = -\frac{\lambda^2}{\gamma \tilde{\rho}(\lambda_0, 0)} \frac{\partial \tilde{p}}{\partial \chi}, \tag{6.10}$$

or writing the latter equation in terms of the normalized specific volume $\tilde{v}(\lambda_0, 0)$ rather than the normalized density $\tilde{\rho}(\lambda_0, 0)$, where $\tilde{v}(\lambda_0, 0) = 1/\tilde{\rho}(\lambda_0, 0)$, then Eq. (6.10) becomes,

$$\frac{\partial \tilde{u}}{\partial \tau} = -\frac{\lambda^2 \tilde{v}(\lambda_0, 0)}{\gamma} \frac{\partial \tilde{p}}{\partial \chi} \tag{6.11}$$

When artificial viscosity is included, however, the previous equation becomes

$$\frac{\partial \tilde{u}}{\partial \tau} = -\frac{\lambda^2 \tilde{v}(\lambda_0, 0)}{\gamma} \frac{\partial}{\partial \chi}(\tilde{p} + \tilde{q}),$$ (6.12)

where the artificial viscosity is also in units of the ambient pressure p_0. Clearly, the particle velocity is given by

$$u = \frac{\partial r}{\partial t}$$ (6.13)

and with the variables defined as above we have the following equation for the normalized particle velocity,

$$\tilde{u} = \frac{\partial \lambda}{\partial \tau}.$$ (6.14)

6.2.2   Continuity Equation

Mass conservation gives the following equation,

$$4\pi r^2 \rho(r_0, t)dr = 4\pi r_0^2 \rho(r_0, 0)dr_0$$ (6.15)

where

$$dr = r(r_0 + dr_0, t) - r(r_0, t),$$

and it follows that

$$\frac{\partial r}{\partial r_0} = \frac{\rho(\lambda_0, 0)}{\rho} \frac{\lambda_0^2}{\lambda^2}.$$ (6.16)

Writing this in the form;

$$\frac{\partial r}{\partial \chi} \frac{\partial \chi}{\partial r_0} = \frac{\rho(\lambda_0, 0)}{\rho} \frac{\lambda_0^2}{\lambda^2};$$

noting that $d\chi = (r_0^2/\varepsilon^3)dr_0$, then the latter equation becomes

$$\frac{\partial r}{\partial \chi} \frac{\lambda_0^2}{\varepsilon} = \frac{\rho(\lambda_0, 0)}{\rho} \frac{\lambda_0^2}{\lambda^2}$$

and using $\lambda = r/\varepsilon$ we have

$$\frac{\partial \lambda}{\partial \chi} = \frac{\rho(\lambda_0, 0)}{\rho} \frac{1}{\lambda^2}, \tag{6.17}$$

and with density measured in units of ambient density ρ_0 according to $\tilde{\rho} = \rho/\rho_0$, then the latter equation becomes

$$\frac{\partial \lambda}{\partial \chi} = \frac{\tilde{\rho}(\lambda_0, 0)}{\tilde{\rho}\lambda^2}. \tag{6.18}$$

Differentiating this equation with respect to τ gives,

$$\begin{aligned}
\frac{\partial^2 \lambda}{\partial \chi \partial \tau} &= \tilde{\rho}(\lambda_0, 0) \frac{\partial}{\partial \tau}\left(\frac{1}{\tilde{\rho}\lambda^2}\right) = -\tilde{\rho}(\lambda_0, 0) \frac{\frac{\partial}{\partial \tau}(\tilde{\rho}\lambda^2)}{\tilde{\rho}^2 \lambda^4} \\
&= -\frac{\tilde{\rho}(\lambda_0, 0)}{\tilde{\rho}^2 \lambda^4}\left(\lambda^2 \frac{\partial \tilde{\rho}}{\partial \tau} + 2\lambda\tilde{\rho}\frac{\partial \lambda}{\partial \tau}\right).
\end{aligned}$$

Noting that $\tilde{u} = \partial \lambda/\partial \tau$, then the latter equation becomes

$$\begin{aligned}
\frac{\partial \tilde{u}}{\partial \chi} &= -\frac{\tilde{\rho}(\lambda_0, 0)}{\tilde{\rho}^2 \lambda^4}\left(\lambda^2 \frac{\partial \tilde{\rho}}{\partial \tau} + 2\lambda\tilde{\rho}\tilde{u}\right) \\
&= -\frac{\tilde{\rho}(\lambda_0, 0)}{\tilde{\rho}^2 \lambda^2} \frac{\partial \tilde{\rho}}{\partial \tau} - \tilde{\rho}(\lambda_0, 0)\frac{2\tilde{u}}{\tilde{\rho}\lambda^3}.
\end{aligned}$$

Hence,

$$\frac{\partial \tilde{\rho}}{\partial \tau} = -\tilde{\rho}\left(\frac{2\tilde{u}}{\lambda} + \frac{\tilde{\rho}\lambda^2}{\tilde{\rho}(\lambda_0, 0)} \frac{\partial \tilde{u}}{\partial \chi}\right)$$

or

$$\frac{\partial \tilde{\rho}}{\partial \tau} = -\tilde{\rho}\left(\frac{2\tilde{u}}{\lambda} + \frac{\partial \tilde{u}/\partial \chi}{\partial \lambda/\partial \chi}\right), \tag{6.19}$$

after using Eq. (6.18). This mass conservation equation corresponds to Eq. (3) in the article by Brode [3].

An alternative form of this latter equation can be obtained by using the specific volume rather than the density. In order to see this let us write Eq. (6.18) in terms of the specific volume v as

$$\frac{\partial \lambda}{\partial \chi} = \frac{v}{v(\lambda_0, 0)} \frac{1}{\lambda^2}.$$

If the latter equation is written in terms of the normalized specific volume of the ambient air, namely, $v_0 = 1/\rho_0$, such that, $\tilde{v} = v/v_0$ and $\tilde{v}(\lambda_0, 0) = v(\lambda_0, 0)/v_0$, we have

$$\tilde{v} = \tilde{v}(\lambda_0, 0)\lambda^2 \frac{\partial \lambda}{\partial \chi},$$

and differentiating this with respect to τ yields,

$$\frac{\partial \tilde{v}}{\partial \tau} = \tilde{v}(\lambda_0, 0)\lambda \left(2\tilde{u} \frac{\partial \lambda}{\partial \chi} + \lambda \frac{\partial \tilde{u}}{\partial \chi} \right), \tag{6.20}$$

which is an alternative form of the continuity equation.

6.2.3 Energy Equation

The energy equation (Eq. 4.12) in Lagrangian form with artificial viscosity included is

$$\frac{\partial p}{\partial t} = \frac{1}{\rho}[\gamma p + (\gamma - 1)q]\frac{\partial \rho}{\partial t},$$

and it is straightforward to show that it reduces to

$$\frac{\partial \tilde{p}}{\partial \tau} = \frac{1}{\tilde{\rho}}[\gamma \tilde{p} + (\gamma - 1)\tilde{q}]\frac{\partial \tilde{\rho}}{\partial \tau},$$

where all quantities are now written in dimensionless units. When the latter equation is written in terms of the specific volume rather than the density it is easy to show that it becomes;

$$[\gamma \tilde{p} + (\gamma - 1)\tilde{q}]\frac{\partial \tilde{v}}{\partial \tau} + \tilde{v}\frac{\partial \tilde{p}}{\partial \tau} = 0. \tag{6.21}$$

6.3 Conservation Equations in Spherical Geometry: A Summary

Let us now bring together our conservation equation in the case of spherical geometry; they are;

$$\frac{\partial \tilde{u}}{\partial \tau} = -\frac{\lambda^2 \tilde{v}(\lambda_0, 0)}{\gamma} \frac{\partial}{\partial \chi}(\tilde{p} + \tilde{q}) \quad \text{(Momentum)}$$

$$\tilde{u} = \frac{\partial \lambda}{\partial \tau} \quad \text{(Normalized particle velocity)}$$

$$\frac{\partial \tilde{v}}{\partial \tau} = \tilde{v}(\lambda_0, 0)\lambda \left(2\tilde{u}\frac{\partial \lambda}{\partial \chi} + \lambda \frac{\partial \tilde{u}}{\partial \chi}\right) \quad \text{(Continuity)}$$

$$[\gamma \tilde{p} + (\gamma - 1)\tilde{q}]\frac{\partial \tilde{v}}{\partial \tau} + \tilde{v}\frac{\partial \tilde{p}}{\partial \tau} = 0 \quad \text{(Energy)}$$

The particular form of the artificial viscosity $\tilde{q}$ chosen by Brode [3] for an outward moving spherical shock wave is

$$\tilde{q} = \frac{9\gamma(\gamma + 1)}{4} \left(\frac{M}{3\pi}\right)^2 \rho(\Delta\chi)^2 \frac{\partial \tilde{u}}{\partial \chi}\left(\frac{\partial \tilde{u}}{\partial \chi} - \left|\frac{\partial \tilde{u}}{\partial \chi}\right|\right),$$

where M is the number of grid zones in the shock front. Here, we take $\tilde{q}$ to have the following form for the numerical procedure;

$$\tilde{q} = (\kappa\Delta\chi)^2(1/\tilde{v}) \frac{\partial \tilde{u}}{\partial \chi}\left(\frac{\partial \tilde{u}}{\partial \chi} - \left|\frac{\partial \tilde{u}}{\partial \chi}\right|\right), \tag{6.22}$$

where $\kappa \approx 1.2$ to 1.5 corresponding to approximately 4 to 5 grid zones (with $\gamma = 1.4$) and with the density replaced by the specific volume.

6.4 Difference Equations

The differential equations above are approximated by the following difference equations for the numerical procedure;

$$\tilde{u}_{n+1,j} = \tilde{u}_{n,j} - \frac{\Delta\tau\left(\lambda_{n,j}\right)^2 \tilde{v}_{0,j}}{\gamma\left(\Delta\chi_{0,j}\right)}\left(\tilde{p}_{n,j+1} - \tilde{p}_{n,j} + \tilde{q}_{n,j+1} - \tilde{q}_{n,j}\right). \tag{6.23}$$

$$\lambda_{n+1,j} = \lambda_{n,j} + \Delta\tau\left(\frac{\tilde{u}_{n+1,j} + \tilde{u}_{n,j}}{2}\right) \tag{6.24}$$

$$\tilde{v}_{n+1,j} = \tilde{v}_{n,j}$$
$$+ \tilde{v}_{0,j}\lambda_{n+1,j}\Delta\tau\left[2\tilde{u}_{n+1,j}\left(\frac{\lambda_{n+1,j} - \lambda_{n+1,j-1}}{\Delta\chi_{0,j}}\right) + \lambda_{n+1,j}\left(\frac{\tilde{u}_{n+1,j} - \tilde{u}_{n+1,j-1}}{\Delta\chi_{0,j}}\right)\right]. \tag{6.25}$$

$$\tilde{q}_{n+1,j} = \left(\kappa\Delta\chi_{0,j}\right)^2\left(\frac{2}{\tilde{v}_{n+1,j} + \tilde{v}_{n,j}}\right)\left(\frac{\tilde{u}_{n+1,j} - \tilde{u}_{n+1,j-1}}{\Delta\chi_{0,j}}\right)$$
$$\times\left[\left(\frac{\tilde{u}_{n+1,j} - \tilde{u}_{n+1,j-1}}{\Delta\chi_{0,j}}\right) - \left|\frac{\tilde{u}_{n+1,j} - \tilde{u}_{n+1,j-1}}{\Delta\chi_{0,j}}\right|\right] \tag{6.26}$$

$$\tilde{p}_{n+1,j} = \frac{\left(\frac{\gamma+1}{\gamma-1}\tilde{v}_{n,j} - \tilde{v}_{n+1,j}\right)\tilde{p}_{n,j} + 2\tilde{q}_{n+1,j}\left(\tilde{v}_{n,j} - \tilde{v}_{n+1,j}\right)}{\frac{\gamma+1}{\gamma-1}\tilde{v}_{n+1,j} - \tilde{v}_{n,j}}. \tag{6.27}$$

where $\lambda_{0,j} = j\Delta\lambda$ and $\Delta\chi_{0,j} = \lambda_{0,j}^2\Delta\lambda$. An alternative form of the equation for the pressure when written in terms of the density rather than the specific volume is

$$\tilde{p}_{n+1,j} = \frac{\left[\left(\frac{\gamma+1}{\gamma-1}\right)\tilde{\rho}_{n+1,j} - \tilde{\rho}_{n,j}\right]\tilde{p}_{n,j} + 2\tilde{q}_{n+1,j}\left(\tilde{\rho}_{n+1,j} - \tilde{\rho}_{n,j}\right)}{\left(\frac{\gamma+1}{\gamma-1}\right)\tilde{\rho}_{n,j} - \tilde{\rho}_{n+1,j}}. \tag{6.28}$$

Brode [3] has also approximated the differential equations for the air motion by a set of finite difference equations and these are presented below. These equations are correct to second order for small quantities $\Delta\chi$ and $\Delta\tau$. The momentum equation, that is, Eq. (6.12), is written in the following difference form (with the normalized density replacing the normalized specific volume);

$$\frac{\tilde{u}_j^{n+\frac{1}{2}} - \tilde{u}_j^{n-\frac{1}{2}}}{\Delta\tau} = -\frac{\left(\lambda_j^n\right)^2}{\gamma\Delta\chi_j^0\tilde{\rho}_j^0}\left[\tilde{p}_{j+\frac{1}{2}}^n - \tilde{p}_{j-\frac{1}{2}}^n + \tilde{q}_{j+\frac{1}{2}}^{n-\frac{1}{2}} - \tilde{q}_{j-\frac{1}{2}}^{n-\frac{1}{2}}\right],$$

and where, for convenience, the time-stepping index n is written here as a superscript, hence,

$$\widetilde{u}_j^{n+\frac{1}{2}} = \widetilde{u}_j^{n-\frac{1}{2}} - \frac{\Delta\tau\left(\lambda_j^n\right)^2}{\gamma\Delta\chi_j^0\widetilde{\rho}_j^0}\left[\widetilde{p}_{j+\frac{1}{2}}^n - \widetilde{p}_{j-\frac{1}{2}}^n + \widetilde{q}_{j+\frac{1}{2}}^{n-\frac{1}{2}} - \widetilde{q}_{j-\frac{1}{2}}^{n-\frac{1}{2}}\right], \tag{6.29}$$

which is Eq. (10) in the article by Brode. The equation for the normalized particle velocity, namely, Eq. (6.14), in discrete form gives,

$$\widetilde{u}_j^{n+\frac{1}{2}} = \frac{\lambda_j^{n+1} - \lambda_j^n}{\Delta\tau},$$

hence,

$$\lambda_j^{n+1} = \lambda_j^n + \widetilde{u}_j^{n+\frac{1}{2}}\Delta\tau, \tag{6.30}$$

and this is Brode's Eq. (11). The mass conservation equation, namely, Eq. (6.19), in discrete form becomes,

$$\frac{\widetilde{\rho}_{j-\frac{1}{2}}^{n+1} - \widetilde{\rho}_{j-\frac{1}{2}}^n}{\Delta\tau} = -\widetilde{\rho}_{j-\frac{1}{2}}^{n+\frac{1}{2}}\left[\frac{2\widetilde{u}_{j-\frac{1}{2}}^{n+\frac{1}{2}}}{\lambda_{j-\frac{1}{2}}^{n+\frac{1}{2}}} + \frac{\widetilde{u}_j^{n+\frac{1}{2}} - \widetilde{u}_{j-1}^{n+\frac{1}{2}}}{\lambda_j^{n+\frac{1}{2}} - \lambda_{j-1}^{n+\frac{1}{2}}}\right],$$

and, by using the expansions for the intermediate points, one obtains,

$$\frac{\widetilde{\rho}_{j-\frac{1}{2}}^{n+1} - \widetilde{\rho}_{j-\frac{1}{2}}^n}{\Delta\tau} = -\frac{1}{2}\left(\widetilde{\rho}_{j-\frac{1}{2}}^{n+1} + \widetilde{\rho}_{j-\frac{1}{2}}^n\right)\left[\frac{2\left(\widetilde{u}_j^{n+\frac{1}{2}} + \widetilde{u}_{j-1}^{n+\frac{1}{2}}\right)}{\lambda_j^{n+\frac{1}{2}} + \lambda_{j-1}^{n+\frac{1}{2}}} + \frac{\widetilde{u}_j^{n+\frac{1}{2}} - \widetilde{u}_{j-1}^{n+\frac{1}{2}}}{\lambda_j^{n+\frac{1}{2}} - \lambda_{j-1}^{n+\frac{1}{2}}}\right]$$

$$= -\left(\widetilde{\rho}_{j-\frac{1}{2}}^{n+1} + \widetilde{\rho}_{j-\frac{1}{2}}^n\right)\left[\frac{2\left(\widetilde{u}_j^{n+\frac{1}{2}} + \widetilde{u}_{j-1}^{n+\frac{1}{2}}\right)}{\lambda_j^{n+1} + \lambda_j^n + \lambda_{j-1}^{n+1} + \lambda_{j-1}^n} + \frac{\widetilde{u}_j^{n+\frac{1}{2}} - \widetilde{u}_{j-1}^{n+\frac{1}{2}}}{\lambda_j^{n+1} + \lambda_j^n - \lambda_{j-1}^{n+1} - \lambda_{j-1}^n}\right].$$

In relation to the quantity within the square brackets in this latter equation, Brode defines,

$$W = \Delta\tau\left[\frac{2\left(\widetilde{u}_j^{n+\frac{1}{2}} + \widetilde{u}_{j-1}^{n+\frac{1}{2}}\right)}{\lambda_j^{n+1} + \lambda_j^n + \lambda_{j-1}^{n+1} + \lambda_{j-1}^n} + \frac{\widetilde{u}_j^{n+\frac{1}{2}} - \widetilde{u}_{j-1}^{n+\frac{1}{2}}}{\lambda_j^{n+1} + \lambda_j^n - \lambda_{j-1}^{n+1} - \lambda_{j-1}^n}\right],$$

which is his Eq. (13), hence, with this substitution the mass conservation equation becomes,

$$\widetilde{\rho}^{\,n+1}_{j-\frac{1}{2}} - \widetilde{\rho}^{\,n}_{j-\frac{1}{2}} = -\left(\widetilde{\rho}^{\,n+1}_{j-\frac{1}{2}} + \widetilde{\rho}^{\,n}_{j-\frac{1}{2}}\right)W,$$

and by solving for $\widetilde{\rho}^{\,n+1}_{j-\frac{1}{2}}$ we obtain,

$$\widetilde{\rho}^{\,n+1}_{j-\frac{1}{2}} = \widetilde{\rho}^{\,n}_{j-\frac{1}{2}}\left(\frac{1-W}{1+W}\right), \tag{6.31}$$

which is Eq. (12) in the article by Brode. As regards the artificial viscosity term, Brode gives the following equation for $\widetilde{q}$;

$$
\begin{aligned}
\widetilde{q}^{\,n+\frac{1}{2}}_{j-\frac{1}{2}} &= 9\frac{\gamma(\gamma+1)}{2}\left(\frac{M}{3\pi}\right)^2 \widetilde{\rho}^{\,n+1}_{j-\frac{1}{2}}\left[\widetilde{u}^{\,n+\frac{1}{2}}_{j-1} - \widetilde{u}^{\,n+\frac{1}{2}}_{j}\right]^2 \quad \text{for} \ \widetilde{u}^{\,n+\frac{1}{2}}_{j-1} > \widetilde{u}^{\,n+\frac{1}{2}}_{j} \\
&= 0 \quad \text{for} \ \widetilde{u}^{\,n+\frac{1}{2}}_{j-1} \le \widetilde{u}^{\,n+\frac{1}{2}}_{j},
\end{aligned}
\tag{6.32}
$$

which is Eq. (6.14) in his article. Similarly, the energy equation (see Sect. 6.2.3 when expressed in terms of the normalized density) has the following finite difference form,

$$\frac{\widetilde{p}^{\,n+1}_{j-\frac{1}{2}} - \widetilde{p}^{\,n}_{j-\frac{1}{2}}}{\Delta\tau} = \frac{1}{\widetilde{\rho}^{\,n+\frac{1}{2}}_{j-\frac{1}{2}}}\left[\gamma\widetilde{p}^{\,n+\frac{1}{2}}_{j-\frac{1}{2}} + (\gamma-1)\widetilde{q}^{\,n+\frac{1}{2}}_{j-\frac{1}{2}}\right]\left(\frac{\widetilde{\rho}^{\,n+1}_{j-\frac{1}{2}} - \widetilde{\rho}^{\,n}_{j-\frac{1}{2}}}{\Delta\tau}\right),$$

and by solving this latter equation for $\widetilde{p}^{\,n+1}_{j-\frac{1}{2}}$, one finds that,

$$\widetilde{p}^{\,n+1}_{j-\frac{1}{2}} = \frac{\left[\left(\frac{\gamma+1}{\gamma-1}\right)\widetilde{\rho}^{\,n+1}_{j-\frac{1}{2}} - \widetilde{\rho}^{\,n}_{j-\frac{1}{2}}\right]\widetilde{p}^{\,n}_{j-\frac{1}{2}} + 2\widetilde{q}^{\,n+\frac{1}{2}}_{j-\frac{1}{2}}\left(\widetilde{\rho}^{\,n+1}_{j-\frac{1}{2}} - \widetilde{\rho}^{\,n}_{j-\frac{1}{2}}\right)}{\left(\frac{\gamma+1}{\gamma-1}\right)\widetilde{\rho}^{\,n}_{j-\frac{1}{2}} - \widetilde{\rho}^{\,n+1}_{j-\frac{1}{2}}}, \tag{6.33}$$

which is Eq. (15) in the article by Brode. When implemented, these finite difference equations were shown to produce similar numerical results to those produced by Eqs. (6.23, 6.24, 6.25, 6.26 and 6.27).

6.5   Numerical Solution of Spherical Shock Waves: The Point Source Solution

Brode [3] was one of the first to numerically solve the differential equations of gas motion for spherical shock waves by taking counter-pressure into account and he used the Von Neumann [2] point-source solution as initial conditions. The equations were written in Lagrangian form and the von Neumann-Richtmyer artificial viscosity technique was employed to avoid shock discontinuities.

Brode [3] obtained detailed results of pressure, particle velocity and density as functions of position and time outside the very strong shock region, so only a brief account of some numerical result that correspond to the later stages of the spherical expansion into the surrounding atmosphere are presented here.

6.6   Initial Conditions Using the Strong-Shock, Point-Source Solution

The point-source solution for the pressure, particle velocity and density as outlined in Chap. 5 is taken as the initial conditions for the numerical integration of the difference equations in Lagrangian form. The point-source solution starting at 1000 atmospheres pressure at the shock front is used as the initial condition. This pressure is large enough to ensure that the similarity solution can be used as the initial condition for the numerical procedure.

6.6.1   The Pressure

From Eq. (5.25) we have the following relation for the strong shock wave pressure;

$$p = \left(\frac{1}{\gamma B(\gamma)}\right)\frac{E_0 f}{R^3}.$$

Dividing across by p_0, the ambient air pressure, we have

$$\frac{p}{p_0} = \left(\frac{1}{\gamma B(\gamma)}\right)\frac{E_0}{p_0}\frac{f}{R^3},$$

and in terms of energy-reduced dimensionless units the latter equation becomes

$$\frac{p}{p_0} = \left(\frac{1}{\gamma B(\gamma)}\right)\left(\frac{\varepsilon}{R}\right)^3 f,$$

where $\varepsilon^3 = E_0/p_0$, hence, the normalized pressure (in Atmospheres) is

$$\tilde{p} = \left(\frac{1}{\gamma B(\gamma)}\right)\left(\frac{R}{\varepsilon}\right)^{-3} f(\eta),$$

and with $\eta = r/R = (r/\varepsilon)/(R/\varepsilon)$, so that $\lambda_s = R/\varepsilon$ and $\lambda = r/\varepsilon$, hence, $\eta = \lambda/\lambda_s$, then

$$\tilde{p} = \left(\frac{1}{\gamma B(\gamma)}\right)\lambda_s^{-3} f\left(\frac{\lambda}{\lambda_s}\right). \tag{6.34}$$

6.6.2   The Velocity

Also, from Eq. (5.7b) we have the following equation for the particle velocity,

$$u = AR^{-3/2}\phi(\eta)$$

and using Eq. (5.22), namely, $E_0 = \rho_0 A^2 B(\gamma)$, and substituting for A yields

$$u = \frac{E_0^{1/2}}{[\rho_0 B(\gamma)]^{1/2}} R^{-3/2}\phi(\eta).$$

Dividing above and below by p_0 (the ambient air pressure) on the right-hand side of the latter equation gives,

$$u = \frac{(E_0/p_0)^{1/2}}{\left[\frac{\rho_0 B(\gamma)}{p_0}\right]^{1/2}} R^{-3/2}\phi(\eta).$$

As $\varepsilon^3 = E_0/p_0$, hence,

$$u = \frac{1}{\left[\gamma B(\gamma)/\left(\frac{\gamma p_0}{\rho_0}\right)\right]^{1/2}} \lambda_s^{-3/2}\phi(\eta).$$

We identify the quantity $(\gamma p_0/\rho_0)^{1/2}$ as the ambient speed of sound, c_0, hence,

$$\tilde{u} = \frac{1}{[\gamma B(\gamma)]^{1/2}} \lambda_s^{-3/2}\phi\left(\frac{\lambda}{\lambda_s}\right), \tag{6.35}$$

which is the initial normalized particle velocity $(\tilde{u} = u/c_0)$ for the numerical integration.

6.6.3   The Density

The corresponding initial normalized density is

$$\widetilde{\rho} = \psi\left(\frac{\lambda}{\lambda_s}\right). \tag{6.36}$$

6.7   Specification of Initial Conditions

Now, once $\widetilde{p}$ is specified, such as, 1000 atmospheres at the shock front (and γ and B (γ) are known), then λ_s is known since $f(\eta)$ is a fixed function according to the point-source solution. Hence, with $\widetilde{p} = 1000$ then Eq. (6.34) gives

$$1000 = \left(\frac{1}{\gamma B(\gamma)}\right)\lambda_s^{-3}f(1),$$

as $\gamma = 1.4$, $B(\gamma) = 5.31$ and $f(1) = 1.166$, hence,

$$\lambda_s = 0.054, \tag{6.37}$$

so λ_s is known once the pressure at the shock front has been specified.

6.8   Results of the Numerical Integration

The discrete from of the equations for the numerical integration have already been presented in Sect. 6.4. Plots of the pressure, particle velocity and density as a function of normalized Lagrangian radius are shown in Figs. 6.2, 6.3 and 6.4 for three different times during the expansion into the surrounding atmosphere. The plots correspond to times when the peak pressure has dropped to just a few atmospheres. The pressure at slightly later times is also shown in Fig. 6.5 and one can observe that the profile develops a negative phase with the pressure dropping below its ambient value beyond about 3 atmospheres peak pressure.

Let us now investigate in some detail what these plots predict and let us choose the plots corresponding to $\tau = 0.1$. In the first instance, let us make an estimate of the velocity of the shock front. In order to do this we will consider plots on either side of $\tau = 0.1$ and take, for example, plots at $\tau = 0.09$ and at $\tau = 0.11$ and determine the radial distance the shock front advances in this dimensionless time interval of $\delta\tau = 0.02$. This should provide a good estimate of the velocity of the shock front at $\tau = 0.1$. As before, we can estimate the approximate positions of the shock front in

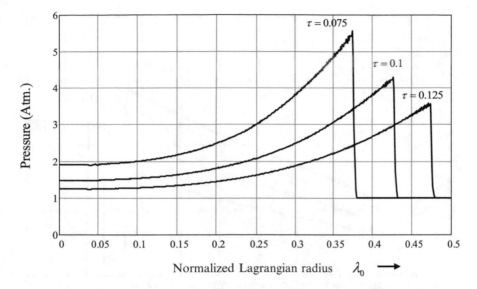

Fig. 6.2 Pressure as a function of normalized Lagrangian radius is shown for the point source explosion at the times indicated. For the numerical procedure the following parameters apply; $\lambda_s = 0.054$, $\Delta\lambda = 1.08 \times 10^{-3}$, $\Delta\tau = 5 \times 10^{-6}$ and $\kappa = 1.5$

each case, namely, at $\tau = 0.09$ and at $\tau = 0.11$, by finding the positions where the maximum value of q occurs. For this we generate plots of q at $\tau = 0.09$ and at $\tau = 0.11$; noting that $\Delta\tau = 5 \times 10^{-6}$ which implies that plots of $q_{18000, j}$ and $q_{22000, j}$ are required as functions of j and these are shown in Fig. 6.6. For these plots we find that the maxima occur at $j = 379$ and 415, respectively (see Fig. 6.6). Since $\lambda_{0, j} = j\Delta\lambda$ we find that the maxima occur at $\lambda_0 = 0.409$ and 0.448, respectively, which gives a separation of $\delta\lambda = 0.039$, hence, $\delta\lambda/\delta\tau = 1.95$. However,

$$\delta\lambda = \frac{\delta R}{\varepsilon} \quad \text{and} \quad \delta\tau = \frac{c_0 \delta t}{\varepsilon}, \text{(see Sect.6.2)}$$

where δR represents the amount by which the shock front advances in a time interval of δt, while ε is the length expressed in terms of the energy yield and the ambient air pressure and c_0 is the ordinary sonic speed ahead of the shock ($c_0 \cong 340 ms^{-1}$). Consequently,

$$\frac{\delta\lambda}{\delta\tau} = \frac{1}{c_0} \frac{\delta R}{\delta t} = \frac{U_s}{c_0},$$

so that $\delta\lambda/\delta\tau$ is just the Mach number M, hence, $M = 1.95$.

By taking $c_0 = 340 ms^{-1}$ we find that $U_s = 663 ms^{-1}$ for the velocity of the shock wave at $\tau = 0.1$. For a point source explosion with an energy yield of 20 kTons of TNT (equal to $8.4 \times 10^{13} Joules$) and with the ambient air pressure

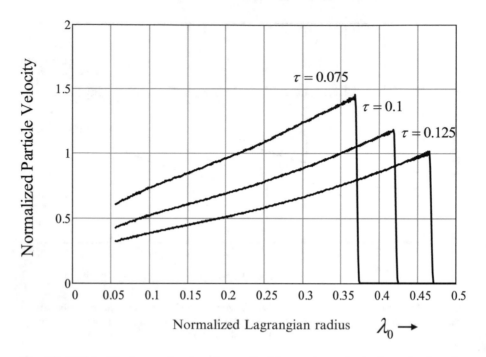

Fig. 6.3 Particle velocity as a function of normalized Lagrangian radius is shown for the point source explosion at the times indicated. For the numerical procedure the following parameters apply; $\lambda_s = 0.054$, $\Delta\lambda = 1.08 \times 10^{-3}$, $\Delta\tau = 5 \times 10^{-6}$ and $\kappa = 1.5$

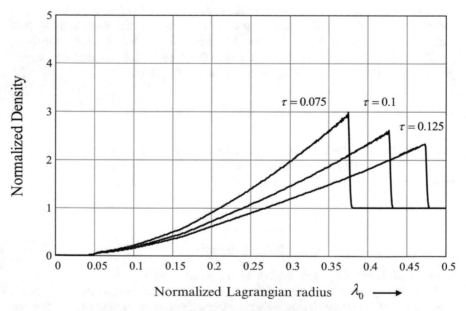

Fig. 6.4 Density as a function of normalized Lagrangian radius is shown for the point source explosion at the times indicated. For the numerical procedure the following parameters apply; $\lambda_s = 0.054$, $\Delta\lambda = 1.08 \times 10^{-3}$, $\Delta\tau = 5 \times 10^{-6}$ and $\kappa = 1.5$

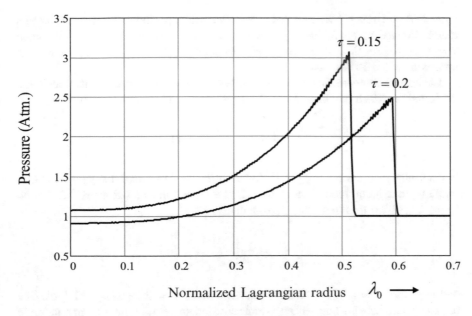

Fig. 6.5 Pressure as a function of normalized Lagrangian radius is shown for the point source explosion at the times indicated. For the numerical procedure the following parameters apply; $\lambda_s = 0.054$, $\Delta\lambda = 2.16 \times 10^{-3}$, $\Delta\tau = 1 \times 10^{-5}$ and $\kappa = 1.5$

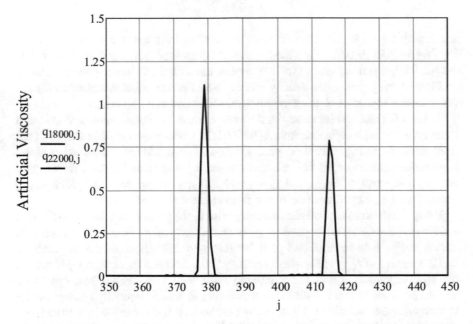

Fig. 6.6 Artificial viscosity as a function of j for the point source explosion at times $\tau = 0.09$ and at $\tau = 0.11$. For the numerical procedure the following parameters apply; $\lambda_s = 0.054$, $\Delta\lambda = 1.08 \times 10^{-3}$, $\Delta\tau = 5 \times 10^{-6}$ and $\kappa = 1.5$ (see text)

$p_0 = 1.01 \times 10^5 Nm^{-2}$ we find that $\varepsilon \cong 940m$ which positions the shock front at a radial distance of approximately $400m$ (the shock front is at $\lambda_0 = 0.426$ when $\tau = 0.1$) and its time to arrive at this distance is given by $t = \varepsilon\tau/c_0$ which is approximately 0.28 seconds.

Let us now compare the numerical results with theoretical predictions. Recall that we found the Mach number, $M = 1.95$, so let us use Eq. (3.25), namely,

$$\frac{p_2}{p_1} = 1 + \frac{2\gamma}{\gamma+1}(M^2 - 1)$$

to determine p_2 (noting that $p_1 = 1$). Substituting for M we find that $p_2 = 4.27 Atm$. while the maximum tabular value for the plot shown in Fig. 6.2 gives 4.283 Atm. Using Eq. (3.28) for the density ratio, namely,

$$\frac{\rho_2}{\rho_1} = \frac{(\gamma+1)M^2}{\left[(\gamma-1)M^2 + 2\right]},$$

and inserting $M = 1.95$ in this latter equation gives $\rho_2 = 2.59$ as $\rho_1 = 1$ (note that density values here are normalized to ambient air density ρ_0). Similarly, in relation to the particle velocity (normalized to the sonic velocity) we use Eq. (3.39b), namely,

$$u_p = \frac{2c_0}{(\gamma+1)}\left(M - \frac{1}{M}\right)$$

and find that $u_p/c_0 \cong 1.2$. The estimated value of ρ_2 from the density plot shown in Fig. 6.4 is 2.58 ± 0.02 while the estimated normalized velocity from Fig. 6.3 is 1.17 ± 0.02, and these are in excellent agreement with the theoretical values above.

Plots of the pressure and density arising from the numerical calculations for the point source explosion at much greater times are shown in Fig. 6.6a.

In Figs. 6.2 and 6.4, for example, the pressure and density are shown plotted as a function of the Lagrangian variable λ_0 which is in terms of the initial positions of the fluid elements. In Fig. 6.6b these parameters are shown plotted as a function of the Lagrangian position of the fluid elements at the times indicated and, therefore, they are given in terms of $\lambda_{15000,\,j}$, $\lambda_{20000,\,j}$ and $\lambda_{25000,\,j}$, where, for example, $\lambda_{15000,\,j}$ denotes the Lagrangian position of the fluid element j at $\tau = 15000\Delta\tau$.

Using the same numerical data that gave rise to the plots shown in Fig. 6.6b, one can also use the same data to produce plots of some typical particle paths. This plot is shown in Fig. 6.6c for different particles that were initially located at normalized radial distances of $\lambda_{0,\,j} = j\Delta\lambda$, where $j = 100, 150, ...350$ and the subscript j identifies that specific particle or fluid element being considered. One can see from this figure that the particles remain at their initial locations and are subsequently driven into the motion once they have been impacted by the shock. It is clear to identify from these plots when the impacts occur, and the solid line (S_k vs. t_k) has been drawn between adjacent impact points to show the trajectory of the shock wave as it arrives at the different particle locations.

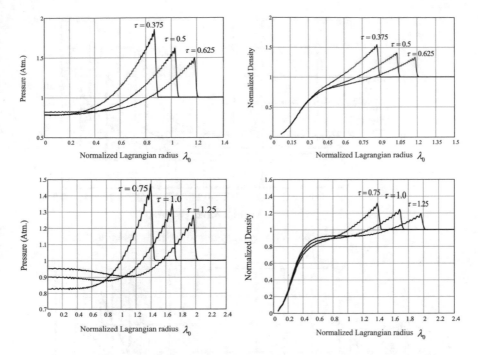

Fig. 6.6a Plots of pressure and density as a function of normalized Lagrangian radius λ_0 are shown for the point source explosion at the times indicated. For the numerical procedure the following parameters apply; $\lambda_s = 0.054$, $\gamma = 1.4$, $\kappa = 1.5$ and $\Delta\lambda = 5.4 \times 10^{-3}$, $\Delta\tau = 2.5 \times 10^{-5}$ for τ in the range, $0.375 \leq \tau \leq 0.625$ while $\Delta\lambda = 0.011$, $\Delta\tau = 5 \times 10^{-5}$ for τ in the range, $0.75 \leq \tau \leq 1.25$

By using some representative results that arise from the numerical computations, a plot is shown in Fig. 6.6d of the normalized overpressure $(p_S - p_0)/p_0$ at the shock front as a function of the normalized shock radius, λ_{0S}, where r_{0S} is the radius of the shock front and $\varepsilon = (E_0/p_0)^{1/3}$, where E_0 is the total blast energy. This gives the decay of the overpressure with distance which is shown plotted using log scales for both axes as the overpressure decays quite rapidly for large values of the pressure p_S at the front.

6.9   Shock Wave from a Sphere of High-Pressure, High-Temperature Gas

In this section we will consider the sudden expansion of a sphere of high pressure air into the surrounding atmosphere. Brode has investigated this expansion and he has presented his numerical results in graphical form in two reports [4, 5]. This sudden expansion is analogous to the *shock tube* previously discussed, except in this case, it involves the bursting of a spherical diaphragm surrounding the high-pressure sphere. We will assume that the initial density inside the sphere is at

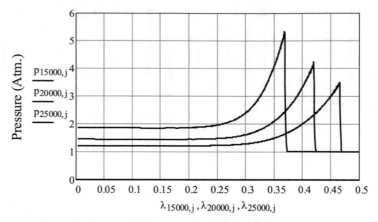

Normalized Lagrangian radius at times indicated

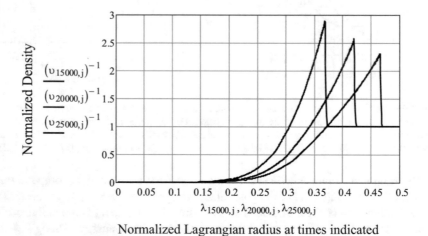

Normalized Lagrangian radius at times indicated

Fig. 6.6b Pressure and density for the point source explosion are shown plotted as a function of the normalized Lagrangian position of the fluid elements at the times indicated. For the numerical procedure the following parameters apply; $\lambda_s = 0.054$, $\Delta\lambda = 1.08 \times 10^{-3}$, $\Delta\tau = 5 \times 10^{-6}$ and $\kappa = 1.5$ (see text)

normal sea-level air density and that the internal pressure is 1000 atmospheres. This implies that the initial temperature in this *isothermal sphere* is very high, perhaps several hundreds of thousands of degrees. It is unlikely that the ideal gas equation applies at these elevated temperatures and pressures in this very hot sphere, so that the use of the equation, $p = \rho RT$, may not accurately predict the initial temperature. Temperatures of this magnitude typically occur in nuclear explosions when the fireball has expanded to several tens of meters, so that the ideal gas equation can be used as a first attempt to predict the variations in pressure, density and particle velocity for this expanding sphere and can form a basis for predicting the hydrodynamic effects of strong explosions despite the absence of a more realistic model for the equation of state for air.

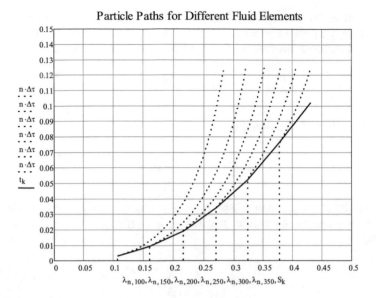

Particle Paths for Different Fluid Elements

Fig. 6.6c Some typical particle paths are shown (indicated by dash-dot lines). The solid line shows the shock path (S_k vs. t_k). For the numerical procedure the following parameters apply; $\Delta\lambda = 1.08 \times 10^{-3}$, $\Delta\tau = 5 \times 10^{-6}$, $\kappa = 1.5$ (see text)

Fig. 6.6d The normalized overpressure at the shock front versus normalized shock radius is shown

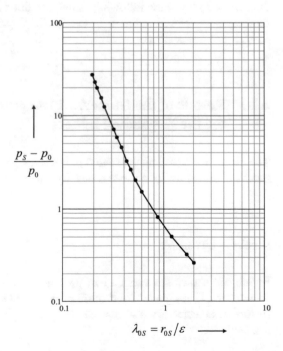

In the example to be considered here we will take the sphere to have a radius R. In order to simplify matters, let us assume $\gamma = 1.4$ for the air both inside and outside the sphere, and let us further assume that the total initial energy stored is given by;

$$E_{tot.} = \frac{pV}{\gamma - 1}$$

where $V = (4/3)\pi R^3$. Accordingly,

$$\frac{E_{tot.}}{p_0} = \frac{\widetilde{p}}{(\gamma - 1)} \frac{4}{3} \pi R^3 \tag{6.38}$$

where $\widetilde{p}$ is the initial pressure in atmospheres within the sphere and p_0 is the outside ambient air-pressure. As before, we define, $\varepsilon = (E_{Tot}/p_0)^{1/3}$, so that Eq. (6.38) becomes

$$\lambda_s = \left(\frac{\widetilde{p}}{(\gamma - 1)} \frac{4}{3} \pi \right)^{-1/3}, \tag{6.39}$$

where distance is measured in energy-reduced dimensionless units according to the relation; $\lambda_s = R/\varepsilon$. Substituting for $\widetilde{p}$ and γ we find that $\lambda_s = 0.0457$.

For the numerical procedure the sphere was divided into N spatial points, such that, $\Delta\lambda = \lambda_s/N$ and the following initial conditions were used: $\widetilde{p} = 1000$, $\widetilde{v} = 1$ for $0 \leq \lambda_0 \leq \lambda_s$ and $\widetilde{p} = 1$, $\widetilde{v} = 1$ for $\lambda_0 > \lambda_s$. In addition, the boundary condition $\widetilde{u}(0, \tau) = 0$ was applied.

6.10 Results of the Numerical Integration for the Expanding Sphere

The difference equations presented in Sect. 6.4 were numerically integrated and the results obtained for the pressure, density and particle velocity are outlined below.

6.10.1 Pressure

Figure 6.7 shows the variation in pressure as a function of normalized radius at various times (designated by τ) during the earliest stages of the expansion. The main feature as expected is the strong outward moving shock wave and an inward moving rarefaction.

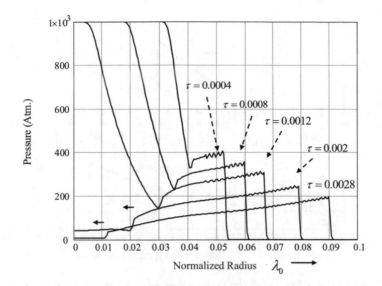

Fig. 6.7 Pressure as a function of normalized radius is shown during the earliest stages of the expansion of the isothermal sphere at the times indicated. For the numerical procedure the following parameters apply; $\Delta\lambda = 0.000457$, $\lambda_s = 0.0457$, $\kappa = 1.2$, $\Delta\tau = 2 \times 10^{-6}$ and $\gamma = 1.4$

When the rarefaction terminates at the origin one can observe the formation of an inward moving shock (indicated by the arrows) heading towards the origin. This shock collides at the origin and a reflected shock ensues which travels in the same direction as the main shock as shown in Fig. 6.8. At slightly later times (Fig. 6.9) one can follow the further progress of these shocks and one can see that the reflected shock, moving at high velocity in the high temperature environment of the sphere, catches up with the main shock. Transmitted and reflected shocks are generated when this reflected shock meets the contact surface and the smaller reflected shock heads back towards the origin as shown in Fig. 6.9.

At later times as shown in Fig. 6.10 the succession of multiple internal shocks become less pronounced, leaving the main outward moving shock as the dominant feature. In addition, the profile develops a negative phase with the pressure dropping below the pre-shock ambient air pressure and the overall profile begins to resemble the pressure profile of the point-source explosion.

Instead of plotting the quantities as a function of λ_0 which represents the initial positions of the fluid elements, one can also plot them as a function of the position of the fluid elements at the times indicated. When implemented we obtain the plots shown in Fig. 6.8a which corresponds to the data already presented in Figs. 6.7 and 6.8 for the pressure. The markers in these plots for some representative values of the pressure clearly show the position of the contact surface, and since the original isothermal sphere is divided into 100 spatial points ($\lambda_s = 0.0457$ and $\Delta\lambda = 0.000457$) the position of the contact surface as a function of the time-stepping index n is given by $\lambda_{n,\,100}$.

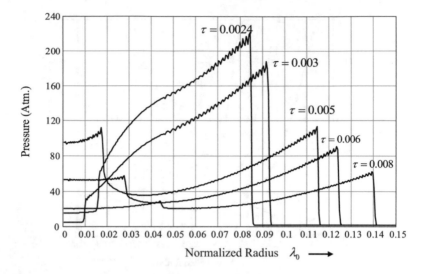

Fig. 6.8 Pressure as a function of normalized radius is shown during the early stages of the expansion of the isothermal sphere at the times indicated. For the numerical procedure the following parameters apply; $\Delta\lambda = 0.000457$, $\lambda_s = 0.0457$, $\kappa = 1.2$, $\Delta\tau = 2 \times 10^{-6}$ and $\gamma = 1.4$

At much later times as shown in Fig. 6.11 the duration of the negative phase is shown. This so-called suction-phase is a characteristic feature occurring in explosions. This phenomenon is further shown in Fig. 6.12 where, in this instance, the pressure-time history at a fixed Lagrangian position is plotted. This figure shows the arrival of the shock wave at $\lambda_0 = 1$ with the resulting compression of the air and its subsequent rapid expansion to a pressure even lower than the pre-shock ambient air pressure. One can also observe that the duration of this suction-phase is longer than the duration of the compressive phase, with the air pressure eventually returning to its pre-shock ambient value.

6.10.2 Density

The density profile at the very early stages shows the generation of a compression spike at the shock front and the subsequent production of a very low pressure within the sphere (Fig. 6.13). The generation of multiple internal shocks as in the case of the pressure profile is evident as shown in Fig. 6.14 and these eventually die away as in the case of the pressure profile. Figure 6.15 should be compared with figure 6 in the article by Brode [3] for the point-source explosion at the very late stages of the spherical expansion.

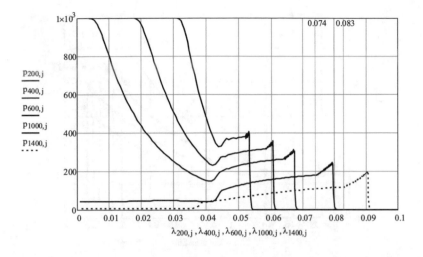

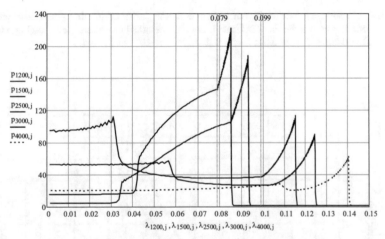

Fig. 6.8a Pressure as a function of normalized radius is shown during the early stages of the expansion of the isothermal sphere at the times indicated. For the numerical procedure the following parameters apply; $\Delta\lambda = 0.000457$, $\lambda_s = 0.0457$, $\kappa = 1.2$, $\Delta\tau = 2 \times 10^{-6}$ and $\gamma = 1.4$

6.10.3   Velocity

At the very early stages of the expansion the particle velocity profile shows the general outward movement of the air at the shock front and the subsequent reversal in the flow direction as the rapid outward expansion exhausts the air within the sphere's interior (Fig. 6.16). The formation of multiple internal shocks is shown in Fig. 6.17 as previously observed in the case of the pressure profile. At late stages as

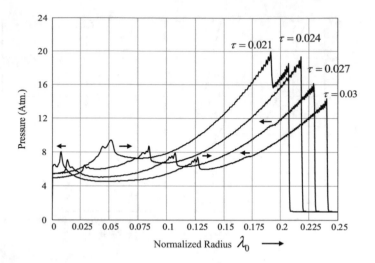

Fig. 6.9 Pressure as a function of normalized radius is shown during the continued expansion of the isothermal sphere at the times indicated. For the numerical procedure the following parameters apply; $\Delta\lambda = 0.000457$, $\lambda_s = 0.0457$, $\kappa = 1.2$, $\Delta\tau = 3 \times 10^{-6}$ and $\gamma = 1.4$

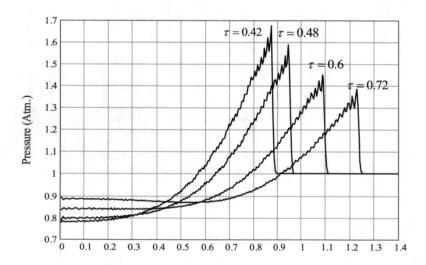

Fig. 6.10 Pressure as a function of normalized radius is shown during the late stages of the expansion of the isothermal sphere at the times indicated. For the numerical procedure the following parameters apply; $\Delta\lambda = 0.00457$, $\lambda_s = 0.0457$, $\kappa = 1.2$, $\Delta\tau = 6 \times 10^{-5}$ and $\gamma = 1.4$

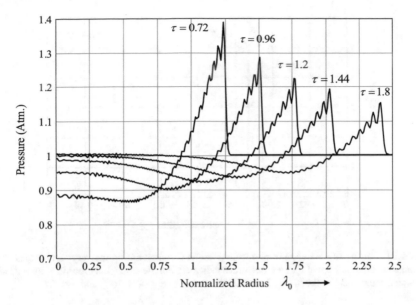

Fig. 6.11 Pressure as a function of normalized radius is shown during much later stages of the expansion of the isothermal sphere at the times indicated. For the numerical procedure the following parameters apply; $\Delta\lambda = 0.00914$, $\lambda_s = 0.0457$, $\kappa = 1.2$, $\Delta\tau = 1.2 \times 10^{-4}$ and $\gamma = 1.4$

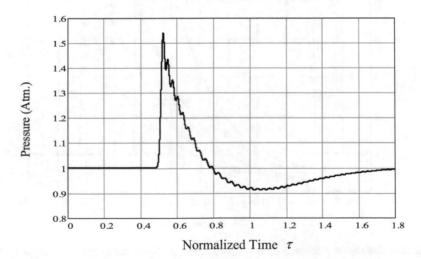

Fig. 6.12 Pressure at $\lambda_0 = 1$ as a function of normalized time for the expansion of the isothermal sphere is shown. For the numerical procedure the following parameters apply: $\Delta\lambda = 0.00914$, $\lambda_s = 0.0457$, $\kappa = 1.2$, $\Delta\tau = 1.2 \times 10^{-4}$ and $\gamma = 1.4$

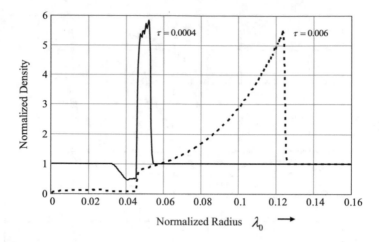

Fig. 6.13 Density as a function of normalized radius is shown during the very early stages of the expansion of the isothermal sphere at the times indicated. For the numerical procedure the following parameters apply; $\Delta\lambda = 0.000457$, $\lambda_s = 0.0457$, $\kappa = 1.2$, $\Delta\tau = 2 \times 10^{-6}$ and $\gamma = 1.4$

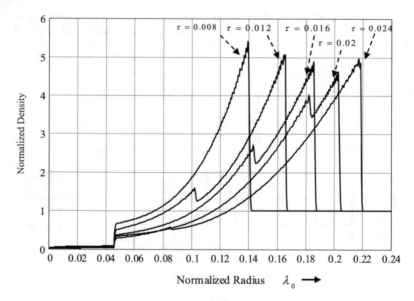

Fig. 6.14 Density as a function of normalized radius is shown during the early stages of the expansion of the isothermal sphere at the times indicated. For the numerical procedure the following parameters apply; $\Delta\lambda = 0.000457$, $\lambda_s = 0.0457$, $\kappa = 1.2$, $\Delta\tau = 2 \times 10^{-6}$ and $\gamma = 1.4$

shown in Fig. 6.18 the profile resembles the pressure profile and shows a reversal in the air flow towards the centre corresponding to the onset of the suction-phase as previously noted in the case of the pressure profile. This suction-phase is shown in Fig. 6.19 at much later times during the expansion.

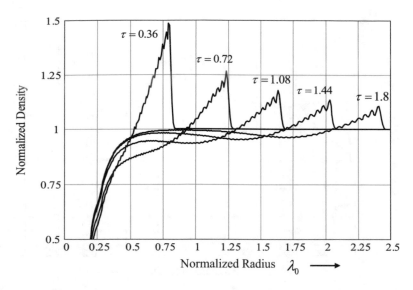

Fig. 6.15 Density as a function of normalized radius is shown during the very late stages of the expansion of the isothermal sphere at the times indicated. For the numerical procedure the following parameters apply; $\Delta\lambda = 0.00914$, $\lambda_s = 0.0457$, $\kappa = 1.2$, $\Delta\tau = 1.2 \times 10^{-4}$ and $\gamma = 1.4$

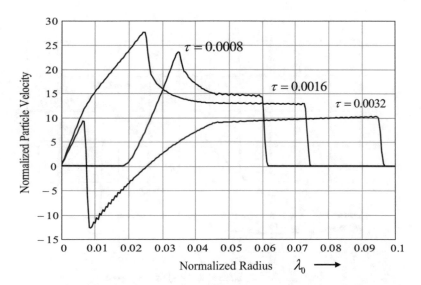

Fig. 6.16 Particle velocity as a function of normalized radius is shown during the earliest stages of the expansion of the isothermal sphere at the times indicated. For the numerical procedure the following parameters apply; $\Delta\lambda = 0.000457$, $\lambda_s = 0.0457$, $\kappa = 1.2$, $\Delta\tau = 2 \times 10^{-6}$ and $\gamma = 1.4$

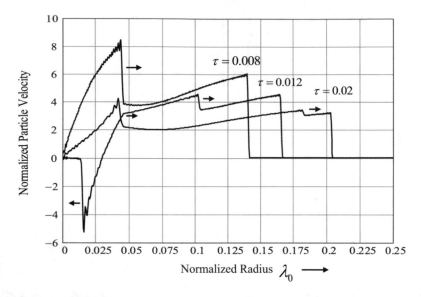

Fig. 6.17 Particle velocity as a function of normalized radius is shown during the early stages of the expansion of the isothermal sphere at the times indicated. For the numerical procedure the following parameters apply; $\Delta\lambda = 0.000457$, $\lambda_s = 0.0457$, $\kappa = 1.2$, $\Delta\tau = 2 \times 10^{-6}$ and $\gamma = 1.4$

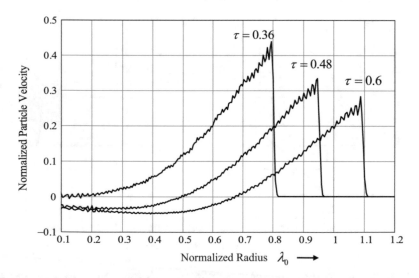

Fig. 6.18 Particle velocity as a function of normalized radius is shown during the late stages of the expansion of the isothermal sphere at the times indicated. For the numerical procedure the following parameters apply; $\Delta\lambda = 0.00457$, $\lambda_s = 0.0457$, $\kappa = 1.2$, $\Delta\tau = 6 \times 10^{-5}$ and $\gamma = 1.4$

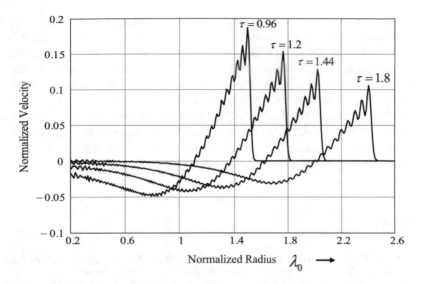

Fig. 6.19 Particle velocity as a function of normalized radius is shown during the later stages of the expansion of the isothermal sphere at the times indicated. For the numerical procedure the following parameters apply; $\Delta\lambda = 0.00914$, $\lambda_s = 0.0457$, $\kappa = 1.2$, $\Delta\tau = 1.2 \times 10^{-4}$ and $\gamma = 1.4$

6.11 A Note on Grid Size

Due to computer memory restrictions the grid size $\Delta\lambda$ and the corresponding time-step interval $\Delta\tau$ had to be increased in order to generate plots covering a larger overall time interval τ corresponding to the later stages of the expansion. This was necessary as maintaining $\Delta\lambda$ and $\Delta\tau$ at their origin values (as used for the early stages of the spherical expansion) required the use of a considerably larger number of time-steps in the numerical procedure. Attempts to implement this increased number resulted in the program terminating with the statement "not enough memory for this operation". Unfortunately, the need to increase the grid size resulted in larger oscillations and the spreading of the shock front over a similar number of grid intervals but having somewhat larger grid size.

6.12 Conclusions

In Sect. 6.10 we considered the sudden expansion of a high-pressure, high-temperature sphere into the atmosphere by numerically integrating the difference equations with artificial viscosity included. Plots of pressure, particle velocity and density were presented as a function of normalized Lagrangian radius at various time intervals during the spherical expansion. It should be noted that Chou and Huang [6] have also investigated the spherical symmetric flow from a high-pressure sphere that is suddenly released into the atmosphere. They solved this problem numerically by using the method of characteristics. Besides the main outwardly moving shock, they also established the existence of an inward travelling second shock. The method of characteristics provides accurate treatment of shock discontinuities with well defined boundaries in contrast to numerical methods using artificial viscosity where the surface of discontinuity is spread out over several mesh intervals so that there is a general lack of accuracy in determining the precise location of the shock surface. As the details of their numerical technique are outside the scope of the present text, the interested reader can obtain additional information in relation to the implementation of their numerical technique in other reports [7, 8] by these authors.

References

1. Ya. B. Zel'dovich, Yu. P. Raizer, *Physics of Shock Waves and High-Temperature Hydrodynamic Phenomena*, (Dover Publications, Inc., Mineola, New York, 2002), Chapter 1
2. J. von Neumann, *The Point Source Solution, Collected Works*, vol 6 (Pergamon Press, New York, 1976), p. 219
3. H.L. Brode, Numerical solutions of spherical blast waves. J. Appl. Phys. **26**, 766 (1955)
4. H. L. Brode, The Blast from a Sphere of High-Pressure Gas, Report No. P-582, (Rand Corporation, Santa Monica, California, January 1955)
5. H. L. Brode, The Blast Wave in Air Resulting from a High-Temperature, High-Pressure Sphere of Air, Report No. RM-1825-AEC, (Rand Corporation, Santa Monica, California, December 1956)
6. P.C. Chou, S.L. Huang, Late-stage equivalence in spherical blasts as calculated by the method of characteristics. J. Appl. Phys. **40**, 752 (1969)
7. S. L. Huang, P. C. Chou, Solution of blast waves by a constant time scheme in the method of characteristics, Report *No.* 125-9 (Drexel Institute of Technology, Philadelphia, Pennsylvania, August 1966)
8. S. L. Huang, P. C. Chou, Calculations of expanding shock waves and late-stage equivalence, Report *No.* 125-12 (Drexel Institute of Technology, Philadelphia, Pennsylvania, April 1968)

Appendix A

Further Consideration of the Piston Withdrawal Problem

In this appendix we will investigate in some detail the piston withdrawal problem presented in Sect. 2.7.2 and sketched again here in Fig. A.1; we will do so by considering a specific numerical example. We will consider a point on the head of the expansion fan at a specific time t_0 which we will assume to be at $t_0 = 1$, and we will then follow the particle-path and the negative characteristic in the x-t plane that passes through this point. We will go on to generate plots of the particle velocity, the pressure and the density based on the method of characteristics and we will compare the results with a numerical calculation using artificial viscosity that was similarly implemented in Sect. 4.8.6 but having a much smaller grid interval. Finally, we will consider this piston withdrawal problem in the case where the tube is closed at some distance from the piston [1–3].

The Expansion Fan

We will assume that the air occupying the region $x \geq 0$ at $t = 0$ is at rest with pressure, $p_0 = 1$, density, $\rho_0 = 1$ and with $\gamma = 1.4$. The ambient speed of sound in this region is given by the equation, $c_0 = \sqrt{\gamma p_0 / \rho_0}$, hence, $c_0 = \sqrt{\gamma} = 1.183$. The piston is assumed to be suddenly withdrawn at a constant speed of $u_0 = 0.9$, so that its displacement is given by the equation, $x(t) = -u_0 t$. The equation for the positive characteristic at the head of the expansion fan is given by (see Sect. 2.7.1)

$$x = c_0 t = 1.183t$$

© The Author(s), under exclusive license to Springer Nature Switzerland AG 2021
S. Prunty, *Introduction to Simple Shock Waves in Air*, Shock Wave and High
Pressure Phenomena, https://doi.org/10.1007/978-3-030-63606-7

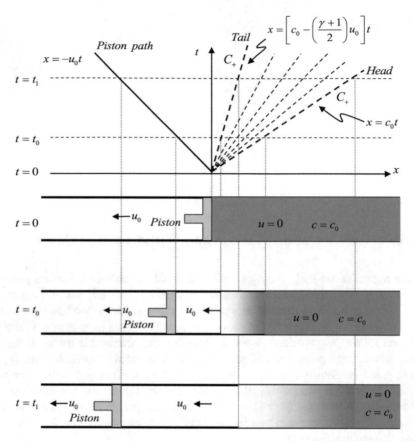

Fig. A.1 Diagram showing constant piston motion creating an expansion wave and a sketch in the x-t plane of the piston path and the expansion fan

and at $t = t_0 = 1$, $x = 1.183$ (see Fig. A.2). The negative characteristic in region $\mathfrak{R}1$ (see Fig. A.2) that passes through the point $(x, t) = (1.183, 1)$ is given by

$$x = -c_0(t - 1) + 1.183; \quad 1 \geq t \geq 0.$$

The equations for the negative characteristic according to Eq. (2.81), after passing through the point $(t_0, c_0 t_0) = (1, 1.183)$ on the positive characteristic that separates regions $\mathfrak{R}1$ and $\mathfrak{R}2$), is given by

$$x(t) = -\frac{2c_0}{\gamma - 1}t + \left(\frac{\gamma + 1}{\gamma - 1}\right)c_0 t_0 \left(\frac{t}{t_0}\right)^{\frac{3-\gamma}{1+\gamma}},$$

while the slope of the negative characteristic in $\mathfrak{R}3$ is

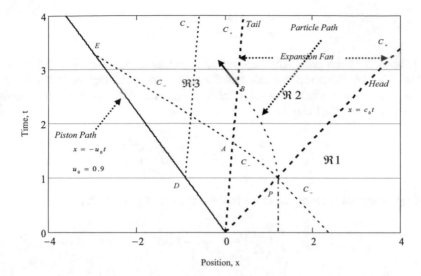

Fig. A.2 Diagram in the *x-t* plane showing the positive and negative characteristics for the piston withdrawal problem. The particle-path is also included (see text)

$$\frac{dx}{dt} = u - c,$$

where $u = -u_0$ and $c = c_0 - \left(\frac{\gamma-1}{2}\right)u_0$, hence,

$$\frac{dx}{dt} = \frac{u_0}{2}(\gamma - 3) - c_0,$$

and all negative characteristics in this region have this constant slope. The equation for the positive characteristic at the tail of the expansion fan is

$$x(t) = \left[c_0 - \left(\frac{\gamma+1}{2}\right)u_0\right]t,$$

which gives $x = 0.103$ at $t = t_0 = 1$. Let us now determine when the negative characteristic emanating from the expansion fan intersects the tail of the expansion fan at point A in Fig. A.2. Equating these values for $x(t)$ and forming the expression,

$$f(t) = \left[c_0 - \left(\frac{\gamma+1}{2}\right)u_0\right]t - \left[-\frac{2c_0}{\gamma-1}t + \left(\frac{\gamma+1}{\gamma-1}\right)c_0t_0\left(\frac{t}{t_0}\right)^{\frac{3-\gamma}{1+\gamma}}\right]$$

in order to determine the value of t that results in $f(t) = 0$; we find (after using Mathcad's *root function*, following an initial guess value for t) that $t = 1.641$ and the position of intersection, A, occurs at $x = 0.169$ (noting that $t_0 = 1$). We have already established that the negative characteristic in $\Re 3$ are straight lines with slope of

$$\frac{dx}{dt} = \left[\frac{u_0}{2}(\gamma - 3) - c_0\right]$$

and substituting the values for u_0, γ and c_0 in this latter equation, we find that $(dx/dt) = -1.903$. However, this value for the slope should correspond to the slope of the negative characteristic (P-A) at point A in Fig. A.2; differentiating this negative characteristic, we find that

$$\frac{dx}{dt} = -\frac{2c_0}{\gamma - 1} + \left(\frac{3 - \gamma}{\gamma - 1}\right)c_0\left(\frac{t}{t_0}\right)^{\frac{2(1-\gamma)}{1+\gamma}},$$

and by using this latter equation we find the slope at $t = 1.641$ as,

$$\left(\frac{dx}{dt}\right)_{t=1.641} = -1.904$$

(where $t_0 = 1$) and this is, therefore, in agreement with the slope of the negative characteristic in $\mathfrak{R}3$. By using the intersection values above, we can write the equation for the negative characteristic (A-E) in $\mathfrak{R}3$ as

$$x(t) = \left[\frac{u_0}{2}(\gamma - 3) - c_0\right](t - 1.641) + 0.169$$

and this characteristic intersects the piston's path at E in Fig. A.2 and it is easy to verify that this intersection occurs at $t = 3.281$ and, accordingly, this negative characteristic is shown plotted in the range: $1.641 \leq t \leq 3.281$.

Particle-Path

The particle velocity u within the expansion fan has the following dependence on x and t according to Eq. (2.74),

$$u = \frac{2}{\gamma + 1}\left[\frac{x}{t} - c_0\right].$$

In order to determine the path taken by particles we must solve the equation,

$$\frac{dx}{dt} = \frac{2}{\gamma + 1}\left[\frac{x}{t} - c_0\right]$$

with the initial particle position at $x = c_0 t_0$ when the particle begins to move. By carrying out a similar integration to that performed for the negative characteristic as outlined in Sect. 2.7.2, it is easy to show that the latter equation yields the following result for the particle-path,

$$x(t) = -\frac{2c_0}{\gamma - 1}t + \left(\frac{\gamma + 1}{\gamma - 1}\right)c_0 t_0 \left(\frac{t}{t_0}\right)^{\frac{2}{\gamma+1}}$$

Let us now determine the intercept where the particle-path cuts the tail of the expansion wave (that is, point B in Fig. A.2). At this point we have, $[x(t)]_{Particlepath} = [x(t)]_{Tail}$, hence, we form the function,

$$f(t) = \left[c_0 - \left(\frac{\gamma + 1}{2}\right)u_0\right]t - \left[-\frac{2c_0}{\gamma - 1}t + \left(\frac{\gamma + 1}{\gamma - 1}\right)c_0 t_0 \left(\frac{t}{t_0}\right)^{\frac{2}{\gamma+1}}\right]$$

in order to find the value of t at which the function $f(t)$ is equal to zero. It is found that $f(t) = 0$ when $t = 2.692$ (noting that $t_0 = 1$) and the slope of the particle's path can be obtained by differentiating the expression,

$$x(t) = -\frac{2c_0}{\gamma - 1}t + \left(\frac{\gamma + 1}{\gamma - 1}\right)c_0 t_0 \left(\frac{t}{t_0}\right)^{\frac{2}{\gamma+1}},$$

and when implemented, one finds that the slope is given by,

$$\frac{dx}{dt} = -\frac{2c_0}{\gamma - 1} + \frac{2c_0}{\gamma - 1}\left(\frac{t}{t_0}\right)^{\frac{-(\gamma-1)}{\gamma+1}}.$$

The slope at B corresponding to the value $t = 2.692$ (and denoted by the arrow in Fig. A.2) is found to be -0.9 (noting that $t_0 = 1$), hence, we conclude that the tail of the expansion wave moves at the same velocity as the piston, as expected.

Positive Characteristic Leaving the Piston's Surface

All positive characteristics in $\Re 3$ are straight lines with slope given by the equation

$$\frac{dx}{dt} = \left[c_0 - \left(\frac{\gamma + 1}{2}\right)u_0\right]$$

and the equation for the positive characteristic leaving the surface of the piston at $t = t_0$, according to Eq. (2.71), is

$$x(t) = -u_0 t_0 + \left[c_0 - \left(\frac{\gamma+1}{2}\right)u_0\right](t - t_0)$$

and with $t_0 = 1$ (at position D) this equation becomes

$$x(t) = -u_0 + \left[c_0 - \left(\frac{\gamma+1}{2}\right)u_0\right](t - 1)$$
$$= 0.103t - 1.003; \quad t \geq 1$$

Parameter Variations According to the Method of Characteristics

The particle velocity, the pressure and the density within the expansion fan are given by the equations (see Sect. 2.7.2);

$$u = \frac{2}{\gamma+1}\left[\frac{x}{t} - c_0\right],$$

$$\frac{p}{p_0} = \left[\left(\frac{\gamma-1}{\gamma+1}\right)\frac{x}{c_0 t} + \frac{2}{\gamma+1}\right]^{\frac{2\gamma}{\gamma-1}}$$

and

$$\frac{\rho}{\rho_0} = \left[\left(\frac{\gamma-1}{\gamma+1}\right)\frac{x}{c_0 t} + \frac{2}{\gamma+1}\right]^{\frac{2}{\gamma-1}},$$

respectively, in the range; $\left[c_0 - \left(\frac{\gamma+1}{2}\right)u_0\right]t \leq x \leq c_0 t$. In the specific case considered here, namely, at time $t = t_0 = 1$, $p_0 = 1$, $\rho_0 = 1$ and $c_0 = 1.183$, we have,

$$u(x) = \frac{2}{\gamma+1}[x - 1.183]$$

$$p(x) = \left[\left(\frac{\gamma-1}{\gamma+1}\right)\frac{x}{1.183} + \frac{2}{\gamma+1}\right]^{\frac{2\gamma}{\gamma-1}}$$

and

$$\rho(x) = \left[\left(\frac{\gamma-1}{\gamma+1}\right)\frac{x}{1.183} + \frac{2}{\gamma+1}\right]^{\frac{2}{\gamma-1}}$$

within the range; $0.103 \leq x \leq 1.183$, and these are shown plotted in Fig. A.3. With $\gamma = 1.4$, the exponents for the pressure and density are 7 and 5, respectively, which illustrates that the decrease in these quantities within the expansion fan is quite rapid.

Parameter Variations According to the Numerical Calculations

Let us now compare the above results with those generated numerically using artificial viscosity that are similar to those already presented in Sect. 4.8.6. We note, however, in this case that the piston is withdrawn at a greater velocity of 0.9 (Arb. units) in comparison to the example considered in Sect. 4.8.6, and to ensure greater accuracy, the grid increments Δx and Δt are reduced by an order of magnitude so that $\Delta x = 0.01$ and $\Delta t = 0.001$. Other parameters are $\gamma = 1.4$, $p_0 = 1$, $\rho_0 = 1$ and $\kappa = 1.2$. Figure A.4 shows the results of the numerical calculations for the particle velocity, the pressure and the density (indicated by the solid lines) at $t = 1$ (hence, 1000 time steps) and for comparison the variation of these quantities within the expansion fan, according to the method of characteristics (as shown in Fig. A.3) is included in the plots and displayed as broken lines.

Tube Closed at End

Let us now consider the slightly more complicated piston withdrawal problem where the tube is closed at some position L as illustrated in Fig. A.5 and the positive characteristics in the expansion fan reflect off the end wall. We must now deal with the network of characteristics, both positive and negative, in the region agh and designated as region $\mathfrak{R}_4$. This region has two families of curved characteristics and we will utilize the numerical method of characteristics that was briefly discussed in Sect. 2.7.4 in order to obtain a solution for this region. The solution involves the calculation of the coordinates of the network of points for the characteristics in $\mathfrak{R}_4$.

To simplify the initial numerical values, we will assume that the piston is withdrawn at a constant speed u_p of 0.75 (Arb. units) and that the speed of sound c_0 in the undisturbed air is 1.2 (Arb. units) and, as in the previous example, γ is taken as 1.4. The ambient density ρ_0 and pressure p_0 within the tube are 1 and 1.03, respectively; where the slightly higher pressure of 1.03 is in accordance with the adopted speed of sound according to the equation, $c_0 = \sqrt{\gamma p_0 / \rho_0}$.

In general, the positive and negative characteristics are

$$\left(\frac{dx}{dt}\right)_{C_+} = u + c \quad \text{and} \quad \left(\frac{dx}{dt}\right)_{C_-} = u - c,$$

respectively, and along C_+ we have the Riemann invariant

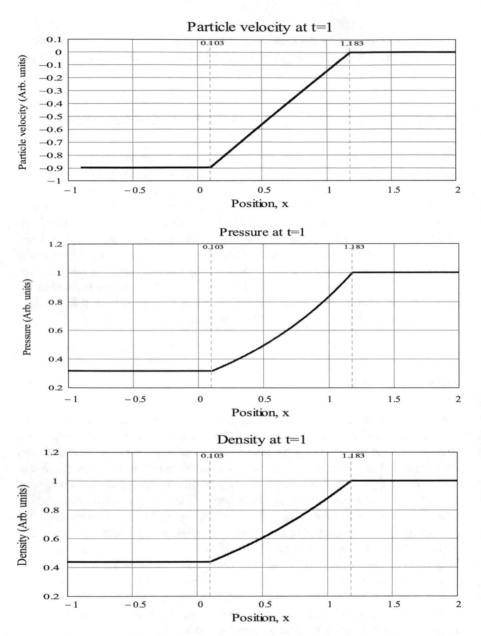

Fig. A.3 Plots of particle velocity, pressure and density for the piston withdrawal problem according to the method of characteristics. The markers denote the positions of the tail and head of the expansion fan at $t_0 = 1$ (see text)

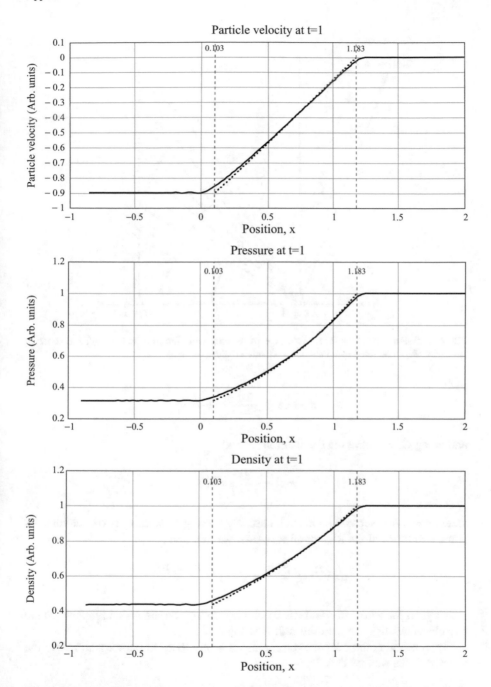

Fig. A.4 Particle velocity, pressure and density (solid lines) obtained numerically for the piston withdrawal problem. The broken lines give the variations in these parameters within the expansion fan according to the method of characteristics (see text)

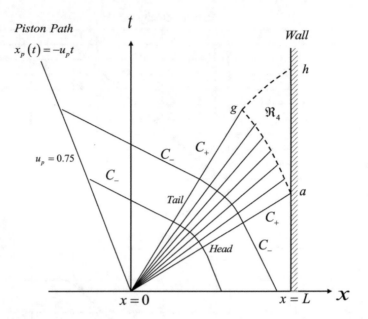

Fig. A.5 Piston withdrawal for the reflection of the expansion fan from an end wall and showing the region $\mathfrak{R}_4$ to be filled by a network of characteristics (see text)

$$R_+ = u + \frac{2c}{\gamma - 1} = u + 5c$$

and along C_- we have the Riemann invariant

$$R_- = u - \frac{2c}{\gamma - 1} = u - 5c$$

where $\gamma = 1.4$ is substituted in each case. By solving these latter two equations for u and c in terms of the Riemann invariants, we find that

$$u = \frac{R_+ + R_-}{2} \quad \text{and} \quad c = \frac{R_+ - R_-}{10},$$

which gives the values of u and c at a point in terms of the Riemann invariants on the two characteristics that pass through that point.

By substituting these latter expressions for u and c back into the equations for the characteristics we find that

$$\left(\frac{dx}{dt}\right)_{C_+} = \frac{3R_+ + 2R_-}{5} \quad \text{and} \quad \left(\frac{dx}{dt}\right)_{C_-} = \frac{2R_+ + 3R_-}{5}$$

Fig. A.6 Intersection points $(a, b, c \ldots g)$ of the positive characteristics in the expansion fan with the negative characteristic ag, and two of the points, P_1 and P_2, in the region of interest are displayed

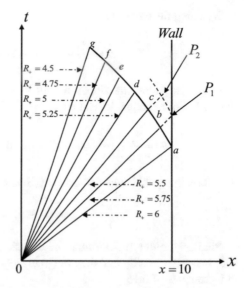

and in terms of the slopes of these characteristics as observed in a tx-diagram we write these latter equations in inverted form as

$$\left(\frac{dt}{dx}\right)_{C_+} = \frac{5}{3R_+ + 2R_-} \quad \text{and} \quad \left(\frac{dt}{dx}\right)_{C_-} = \frac{5}{2R_+ + 3R_-}.$$

The head of the expansion fan has $u = 0$ and $c = c_0 = 1.2$, hence, $R_+ = 6$ and the tail of the expansion fan has $u = -0.75$ and $c = c_0 - \left(\frac{\gamma-1}{2}\right)0.75$, hence, $R_+ = 4.5$. Let us now divide the expansion fan into seven[1] positive characteristics ranging from $R_+ = 4.5$ to $R_+ = 6$ as illustrated in Fig. A.6. The negative characteristic ag has $R_- = -6$ and we have already determined that the equation for this characteristic is given by

$$x(t) = -\frac{2c_0 t}{\gamma - 1} + \left(\frac{\gamma + 1}{\gamma - 1}\right) c_0 t_0 \left(\frac{t}{t_0}\right)^{\frac{3-\gamma}{1+\gamma}},$$

where $t_0 = L/c_0 = 10/1.2 = 8.333$, so that the coordinates (x, t) of position a as shown in Fig. A.6 is $(10, 8.333)$.

---

[1]Greater accuracy is achieved with a larger number of characteristics but at the cost of an increased number of points to be computed in region $\mathfrak{R}_4$.

By using the equations

$$u = \frac{R_+ + R_-}{2} \quad \text{and} \quad c = \frac{R_+ - R_-}{10}$$

one can determine the slopes for all seven positive characteristics as shown in Fig. A.6. For example, let us take $R_+ = 4.75$ and $R_- = -6$, hence,

$$u = \frac{R_+ + R_-}{2} = -0.625 \quad \text{and} \quad c = \frac{R_+ - R_-}{10} = 1.075,$$

therefore, this positive characteristic $0f$ has equation,

$$x = (u + c)t = 0.45t.$$

Similarly, one can determine the equations for all seven positive characteristics and it is easy to verify that they range in slopes from 0.3 at the tail to 1.2 at the head in increments of 0.15.

All that remains before we consider the network of curved characteristics in region $\mathcal{R}_4$ is to find the coordinates of the points where the expansion fan intersects the negative characteristic ag. For example, let us take the previously determined positive characteristic $0f$ whose equation is given by $x = 0.45t$ and then form the expression,

$$f(t) = 0.45t - \left[-\frac{2c_0t}{\gamma - 1} + \left(\frac{\gamma + 1}{\gamma - 1} \right) c_0 t_0 \left(\frac{t}{t_0} \right)^{\frac{3-\gamma}{1+\gamma}} \right],$$

where $t_0 = 10/1.2$ and $\gamma = 1.4$. We can determine in the usual manner the value of t that results in $f(t) = 0$; when implemented we find that $t = 11.591$ and $x = 5.216$. One can perform similar calculations for the remaining positive characteristics in the expansion fan that give the coordinates (x, t) for the intersection points a to g as shown in Table A.1.

Table A.1 Coordinates of the intersection points of the expansion fan with the negative characteristic ag (see text)

Intersection points	Coordinates (x, t)
a	(10,8.333)
b	(9.32,8.877)
c	(8.521,9.468)
d	(7.585,10.114)
e	(6.941,10.819)
f	(5.216,11.591)
g	(3.732,12.439)

Boundary Condition

The network of points that define the positive and negative characteristics in $\mathfrak{R}_4$ can now be determined. Before we proceed, however, it is necessary to recall one of our previous equations, namely, $u = (R_+ + R_-)/2$; hence, the boundary condition, $u = 0$, at the end wall gives the requirement that

$$R_- = -R_+,$$

hence, the negative characteristic reflected off the end wall carries the same magnitude as the Riemann invariant of the incident positive characteristic but bearing a negative sign, and this will be used for calculating the network of points in $\mathfrak{R}_4$.

Calculating the Coordinates of a Boundary Point

Let us now determine the coordinates of P_1 at the boundary wall as shown in Fig. A.6. The slope of the positive characteristic evaluated at b with coordinates $(9.32, 8.877)$ and that joins the points b and P_1 is

$$\left(\frac{dt}{dx}\right)_{C_+} = \frac{5}{3R_+ + 2R_-} = \frac{5}{(3 \times 5.75) + (2 \times -6)} = 0.952$$

and the slope of the positive characteristic evaluated at P_1 that joins the points b and P_1 is

$$\left(\frac{dt}{dx}\right)_{C_+} = \frac{5}{3R_+ + 2R_-} = \frac{5}{(3 \times 5.75) + (2 \times -5.75)} = 0.869$$

where the previously referred to boundary condition has been applied. The average of these slopes is 0.911, and by using the relationship,

$$\left(\frac{dt}{dx}\right)_{C_+} = \frac{\Delta t}{\Delta x}$$

we have

$$\frac{t_1 - 8.877}{10 - 9.32} = 0.911$$

for the time t_1 at P_1, hence, $t_1 = 9.496$ and the coordinates of P_1 are $(10, 9.496)$. Other boundary points are calculated in a similar manner.

Calculating the Coordinates of an Internal Point

Let us now calculate the coordinates of the internal point P_2. The slope of the negative characteristic evaluated at P_1 that joins the points P_1 and P_2 is

$$\left(\frac{dt}{dx}\right)_{C_-} = \frac{5}{2R_+ + 3R_-} = \frac{5}{(3 \times 5.75) + (3 \times -5.75)} = -0.869$$

and the slope of the negative characteristic evaluated at P_2 that joins the points P_1 and P_2 is

$$\left(\frac{dt}{dx}\right)_{C_-} = \frac{5}{2R_+ + 3R_-} = \frac{5}{(3 \times 5.5) + (3 \times -5.75)} = -0.8,$$

hence, the average slope of the negative characteristic is -0.835. The slope of the positive characteristic at c that joins the points c and P_2 is

$$\left(\frac{dt}{dx}\right)_{C_+} = \frac{5}{3R_+ + 2R_-} = \frac{5}{(3 \times 5.5) + (2 \times -6)} = 1.111$$

and the slope of the positive characteristic at P_2 that joins the points c and P_2 is

$$\left(\frac{dt}{dx}\right)_{C_+} = \frac{5}{3R_+ + 2R_-} = \frac{5}{(3 \times 5.5) + (2 \times -5.75)} = 1.0,$$

giving 1.055 as the average slope of the positive characteristic. If (x_2, t_2) is the coordinates of P_2 then the equation for the slope of the negative characteristic at P_2 can be written as

$$\frac{t_2 - 9.496}{x_2 - 10} = -0.835$$

and the equation for the slope of the positive characteristic at P_2 can be written as

$$\frac{t_2 - 9.468}{x_2 - 8.521} = 1.055.$$

By solving these simultaneous equations, we find that $x_2 = 9.189$ and $t_2 = 10.173$. This calculation for the location of P_2 is typical of that carried out for the determination of the coordinates of all internal points within $\mathfrak{R}_4$. Table A.2 gives the coordinates of all the points in $\mathfrak{R}_4$ and these are plotted in Fig. A.7, while an expanded view of this region is shown in Fig. A.8.

Table A.2 Shows the results of the numerical calculation for the coordinates of the network of points in region $\mathfrak{R}_4$

Point in region $\mathfrak{R}_4$	Coordinates (x, t) in $\mathfrak{R}_4$
P_1	(10, 9.496)
P_2	(9.189, 10.173)
P_3	(10, 10.946)
P_4	(8.225, 10.916)
P_5	(9.019, 11.8)
P_6	(10, 12.784)
P_7	(7.154, 11.682)
P_8	(7.939, 12.587)
P_9	(8.931, 13.758)
P_{10}	(10, 14.887)
P_{11}	(5.731, 12.631)
P_{12}	(6.390, 13.736)
P_{13}	(7.334, 15.092)
P_{14}	(8.393, 16.421)
P_{15}	(10, 18.213)
P_{16}	(4.139, 13.626)
P_{17}	(4.705, 14.899)
P_{18}	(5.557, 16.462)
P_{19}	(6.561, 18.016)
P_{20}	(8.117, 20.101)
P_{21}	(10, 22.325)

Comparison with the Results of the Finite Difference Calculations

Let us now compare the outcome of the calculations for the network of characteristics in $\mathfrak{R}_4$ with the results of some numerical computations that were implemented by using the finite difference equations of Sect. 4.5.3. These finite difference equations incorporates the boundary condition at the end wall with the following mathematical statement, $u_{n,\,100} \leftarrow 0$, to indicate that the velocity at all times t $(=n\Delta t)$ goes to zero at $x = 10$, noting that $x = j\Delta x$, where $\Delta x = 0.1$ and $j = 100$. In addition, the Eulerian position of a fluid element is denoted by $x_{n,\,j}$ which gives the position of a specific element of the fluid (denoted by the index j) at time t $(=n\Delta t)$ whose initial position is given by $j\Delta x$. For the numerical procedure the following parameters were adopted; $\Delta t = 0.01$, $\gamma = 1.4$, while the ambient pressure and density were taken as 1.03 and 1, respectively, to comply with the speed of sound ($c_0 = 1.2$) that was assumed for calculating the network of characteristics in $\mathfrak{R}_4$.

Suppose we now consider some of the points in region $\mathfrak{R}_4$ and let us take, for example, point P_5 with coordinates, (9.019, 11.8). The Riemann invariants passing through this point are $R_+ = 5.25$ and $R_- = -5.5$, hence, the particle velocity at this point according to the equation, $u = (R_+ + R_-)/2$, is $u = -0.125$. Since the time at this point is 11.8 and with $\Delta t = 0.01$, this gives $n = 1180$ to be used in the numerical computations.

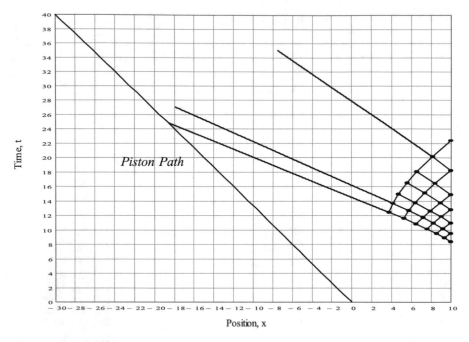

Fig. A.7 An xt-diagram showing the network of curved characteristics in region $\mathfrak{R}_4$. The piston path is also shown

As a result of the numerical computations the particle position at this time as a function of the initial position of the fluid elements (denoted by the index j) is shown in Fig. A.9 and the marker at 9.019 show the position of the particle at that time. It can be seen from this plot that this marker corresponds to the fluid element with an initial position (within the numerical approximations) at $x = 94\Delta x = 9.4$. Accordingly, the approximate particle velocity is given by $u_{1180,\ 94}$, and the numerical output gives the following result,

$$u_{1180,94} = -0.128,$$

which is in good agreement with the value of -0.125 according to the method of characteristics. The particle velocity for this fluid element is shown plotted as a function of time in Fig. A.10 and the marker at $t = 11.8$ identifies the particle velocity at this time as -0.128 as noted above.

Let us now consider another point P_{12} in $\mathfrak{R}_4$ with coordinates, $(6.390, 13.736)$. The Riemann invariants passing through this point are $R_+ = 4.75$ and $R_- = -5.5$, hence, the particle velocity according to the equation, $u = (R_+ + R_-)/2$, is $u = -0.375$. For this particular point we have, $n = 13.736/\Delta t \cong 1374$, and a numerically generated plot of $x_{1374,\ j}$ versus j is shown in Fig. A.11. The marker in

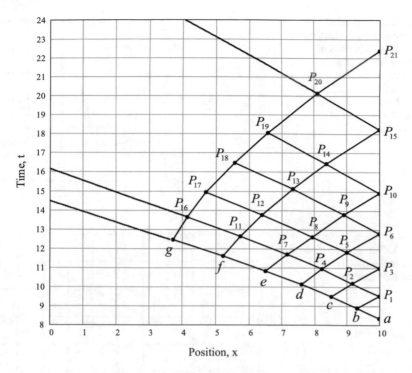

Fig. A.8 An expanded view is shown of the positive and negative characteristics in region $\mathcal{R}_4$

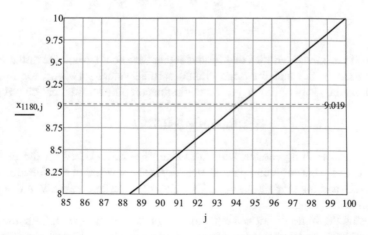

Fig. A.9 Particle position at $t = 1180\Delta t$ as a function of the initial position ($j\Delta x$) of the fluid elements

Fig. A.10 Numerical output showing the particle velocity as a function of time for the particle originally located at $x = 94\Delta x$. The markers in the plot show the particle velocity of -0.128 at $t = 11.8$ (see text)

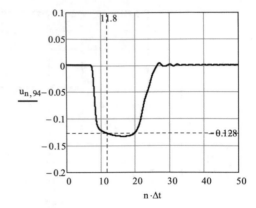

Fig. A.11 Particle position at $t = 1374\Delta t$ as a function of the initial position ($j\Delta x$) of the fluid elements

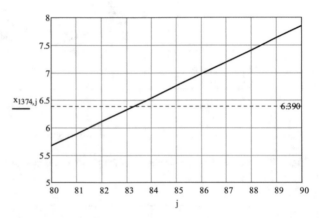

this plot at $t = 6.390$ identifies that fluid element with an approximate initial position of $x = 83\Delta x = 8.3$ and the approximate particle velocity (within the numerical approximations) is given by $u_{1374,\,83}$, while the computed numerical output gives,

$$u_{1374,83} = -0.372,$$

which is in general agreement with the value of -0.375 based on the method of characteristics. This particular particle velocity is shown plotted as a function of time in Fig. A.12 and the marker at $t = 13.736$ shows the particle velocity at this time as -0.372.

By using the finite difference equations as presented in Sect. 4.5.3 some typical particle paths were determined and they are shown in Fig. A.13 with the tube closed at $x = 10$. For comparison, the particle paths for an identical piston withdrawal problem are shown in Fig. A.14 when the tube is not closed at the end.

Fig. A.12 Numerical output showing the particle velocity as a function of time for the particle originally located at $x = 83\Delta x$. The markers in the plot show the particle velocity of -0.372 at $t = 13.736$ (see text)

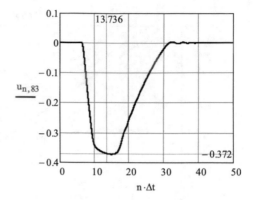

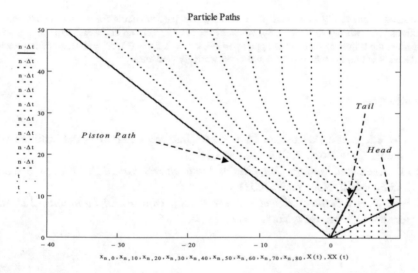

Fig. A.13 Numerically generated plots showing the particle paths for several initial particle positions ($j\Delta x$, where $j = 10, 20....80$) with the tube closed at $x = 10$. For the numerical procedure the following parameters apply: $\gamma = 1.4, \kappa = 1.2, \Delta x = 0.1, \Delta t = 0.01$, ambient pressure and density within the tube are, 1.03 and 1, respectively (see text)

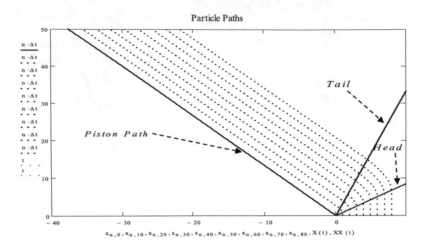

Fig. A.14 Numerically generated plots showing the particle paths for several initial particle positions ($j\Delta x$, where $j = 10, 20....80$) when the tube is NOT closed at the end. For the numerical procedure the following parameters apply: $\gamma = 1.4$, $\kappa = 1.2$, $\Delta x = 0.1$, $\Delta t = 0.01$, ambient pressure and density within the tube are, 1.03 and 1, respectively (see text)

References

[1] H.W. Liepmann, A. Roshko, *Elements of Gasdynamics* (Dover Publications Inc., Mineola, New York, 2001), Section 12-4
[2] J.D. Anderson, *Modern Compressive Flow with Historical Perspective,* 3rd edn. (McGraw-Hill, New York, 2003), Chapter 7
[3] W.F. Ames, *Numerical Methods for Partial Differential Equations*, 2nd edn. (Academic Press, New York, 1977), Section 4-5

Appendix B

Some Numerical Results for a Closed Shock Tube

In this appendix we will consider the shock tube with the same initial conditions that was discussed in Sect. 4.8.7 but in this particular case we will assume that the tube is closed at, say, $x = 10$ in the driven section as illustrated in Fig. B.1. We are also interested in determining the pressure, density and particle velocity in the tube over an extended time interval, thereby determining the evolution of these parameters over this extended period.

Since the tube is closed at both ends as shown in Fig. B.1, the boundary conditions for the numerical calculations contains the following mathematical statements; $u_{n, 0} \leftarrow 0$ and $u_{n, 100} \leftarrow 0$, where, in this case, we take $\gamma = 1.4$, $\kappa = 1.2$, $\Delta x = 0.1$ and $\Delta t = 0.005$, and a total of 160000 ($=n$) time steps is implemented for this extended period where n is the time-stepping index.

Fig. B.1 Diagram showing the initial conditions for the shock tube (see text)

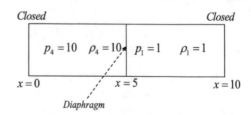

Closed   Closed

$p_4 = 10$ $\rho_4 = 10$ $p_1 = 1$ $\rho_1 = 1$

$x = 0$ $x = 5$ $x = 10$

Diaphragm

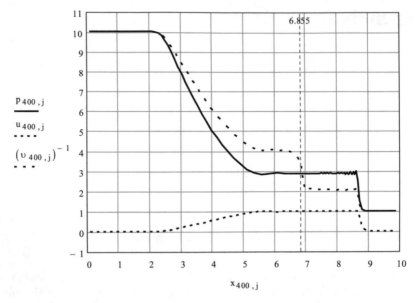

$\dfrac{p_{400,j}}{}$

$u_{400,j}$

$\left(v_{400,j}\right)^{-1}$

$x_{400,j}$

Fig. B.2 Plots of pressure, particle velocity and density as a function of particle position at $t = 400\Delta t$. For the numerical calculations the following parameters apply; $\gamma = 1.4$, $\kappa = 1.2$, $\Delta x = 0.1$ and $\Delta t = 0.005$ (see text)

The pressure, p_2, behind the advancing shock wave in terms of the diaphragm pressure ratio, p_4/p_1, is given by Eq. (4.57) and it is straightforward to show, as previously established in Sect. 4.8.7, that $p_2 = 2.848$, which is in good agreement with the numerical output as shown in Fig. B.2. Once the shock wave reflects at the end wall located at $x = 10$, the magnitude of the reflected pressure, p_3, is given by Eq. (3.47), namely,

$$\frac{p_3}{p_2} = \frac{(3\gamma - 1)p_2 - (\gamma - 1)p_1}{(\gamma - 1)p_2 + (\gamma + 1)p_1},$$

hence, we find that $p_3 = 7.012$ which is in general agreement with the numerically calculated value as shown in Fig. B.3. The variation of these parameters at much later times as a function of particle position are shown in Figs. B.4, B.5, and B.6 and the markers in all plots show the position of the contact surface at the times indicated.

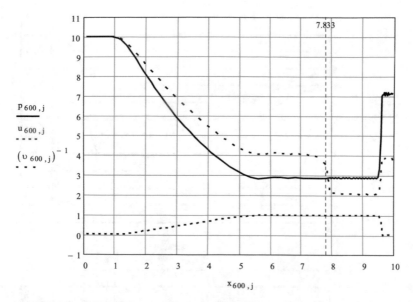

Fig. B.3 Plots of pressure, particle velocity and density as a function of particle position at $t = 600\Delta t$ after reflection from the end wall. For the numerical calculations the following parameters apply; $\gamma = 1.4$, $\kappa = 1.2$, $\Delta x = 0.1$ and $\Delta t = 0.005$ (see text)

The *contact surface* separates the driven region that has been compressed and heated by the shock wave from the driver region that has been cooled as a result of the expansion wave. No forces exist across the contact surface so that the pressure is the same on both sides and there is no change in particle velocity across it; there is, however, an abrupt change in the density and temperature that clearly distinguishes the two regions. This is clearly evident in Fig. B.6 at $t = 160000\Delta t$ where we observe that the particle velocity tends to zero while the pressure within the shock tube approaches the average of the initial values in the driver and driven section as expected. On the other hand, the density and temperature (proportional to the

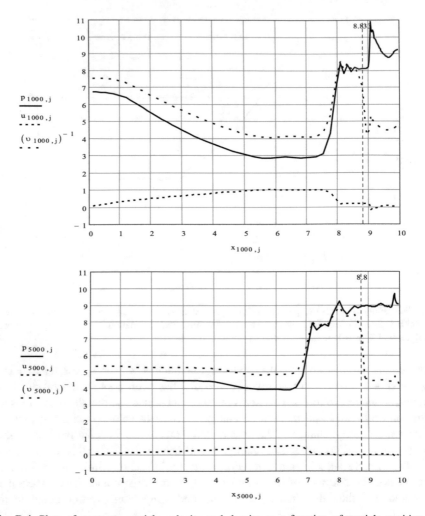

Fig. B.4 Plots of pressure, particle velocity and density as a function of particle position at $t = 1000\Delta t$ and at $t = 5000\Delta t$. For the numerical calculations the following parameters apply; $\gamma = 1.4$, $\kappa = 1.2$, $\Delta x = 0.1$ and $\Delta t = 0.005$ (see text)

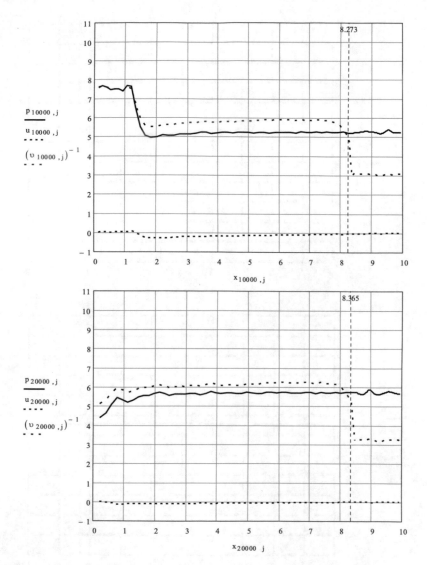

Fig. B.5 Plots of pressure, particle velocity and density as a function of particle position at $t = 10000\Delta t$ and at $t = 20000\Delta t$. For the numerical calculations the following parameters apply; $\gamma = 1.4$, $\kappa = 1.2$, $\Delta x = 0.1$ and $\Delta t = 0.005$ (see text)

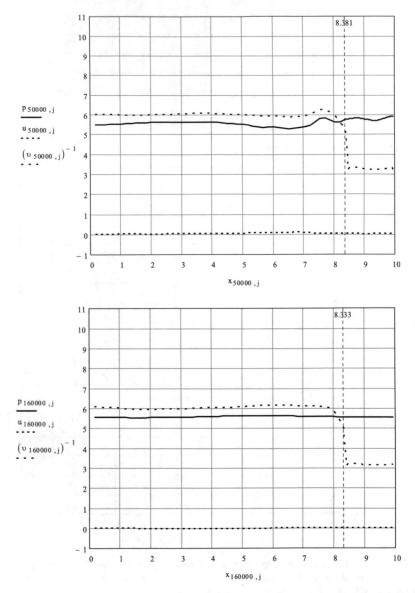

Fig. B.6 Plots of pressure, particle velocity and density as a function of particle position at $t = 50000\Delta t$ and at $t = 160000\Delta t$. For the numerical calculations the following parameters apply; $\gamma = 1.4$, $\kappa = 1.2$, $\Delta x = 0.1$ and $\Delta t = 0.005$ (see text)

product of the pressure, p, and specific volume, v) exhibit abrupt changes across the contact surface since the air is assumed to behave as an ideal fluid with zero thermal conductivity so that there is no heat transfer between the driver and driven sections to equalize the temperatures.

Index

© The Author(s), under exclusive license to Springer Nature Switzerland AG 2021
S. Prunty, *Introduction to Simple Shock Waves in Air*, Shock Wave and High
Pressure Phenomena, https://doi.org/10.1007/978-3-030-63606-7

Printed in the United States
by Baker & Taylor Publisher Services